T0190334

Tensor Calculus for Engineers and Physicists

Emil de Souza Sánchez Filho

Tensor Calculus for Engineers and Physicists

Springer

Emil de Souza Sánchez Filho
Fluminense Federal University
Rio de Janeiro, Rio de Janeiro
Brazil

ISBN 978-3-319-81056-0 ISBN 978-3-319-31520-1 (eBook)
DOI 10.1007/978-3-319-31520-1

© Springer International Publishing Switzerland 2016
Softcover reprint of the hardcover 1st edition 2016
This work is subject to copyright. All rights are reserved by the Publisher, whether the whole or part of the material is concerned, specifically the rights of translation, reprinting, reuse of illustrations, recitation, broadcasting, reproduction on microfilms or in any other physical way, and transmission or information storage and retrieval, electronic adaptation, computer software, or by similar or dissimilar methodology now known or hereafter developed.
The use of general descriptive names, registered names, trademarks, service marks, etc. in this publication does not imply, even in the absence of a specific statement, that such names are exempt from the relevant protective laws and regulations and therefore free for general use.
The publisher, the authors and the editors are safe to assume that the advice and information in this book are believed to be true and accurate at the date of publication. Neither the publisher nor the authors or the editors give a warranty, express or implied, with respect to the material contained herein or for any errors or omissions that may have been made.

Printed on acid-free paper

This Springer imprint is published by Springer Nature
The registered company is Springer International Publishing AG Switzerland

To
Sandra, Yuri, Natália and Lara

Preface

The Tensor Calculus for Engineers and Physicist provides a rigorous approach to tensor manifolds and their role in several issues of these professions. With a thorough, complete, and unified presentation, this book affords insights into several topics of tensor analysis, which covers all aspects of N-dimensional spaces.

Although no emphasis is placed on special and particular problems of Engineering or Physics, the text covers the fundamental and complete study of the aim of these fields of the science. The book makes a brief introduction to the basic concept of the tensorial formalism so as to allow the reader to make a quick and easy review of the essential topics that enable having a dominium over the subsequent themes, without needing to resort to other bibliographical sources on tensors.

This book did not have the framework of a math book, which is a work that seeks, above all else, to organize ideas and concepts in a didactic manner so as to allow the familiarity with the tensorial approach and its application of the practical cases of Physics and the areas of Engineering.

The development of the various chapters does not cling to any particular field of knowledge, and the concepts and the deductions of the equations are presented so as to permit engineers and physicists to read the text without being experts in any branch of science to which a specific topic applies.

The chapters treat the various themes in a sequential manner and the deductions are performed without omission of the intermediary steps, the subjects being treated in a didactic manner and supplemented with various examples in the form of solved exercises with the exception of Chap. 3 that broaches review topics. A few problems with answers are presented at the end of each chapter, seeking to allow the reader to improve his practice in solving exercises on the themes that were broached.

Chapter 1 is a brief introduction to the basic concepts of tensorial formalism so as to permit the reader to make a quick and easy review of the essential topics that make possible the knowledge of the subsequent themes that come later, without needing to resort to other bibliographic sources on tensors.

The concepts of covariant, absolute, and contravariant derivatives, with the detailed development of all the expressions concerning these parameters, as well as the deductions of the Christoffel symbols of the first and second kind, are the essence of Chap. 2.

Chapter 3 presents the Green, Stokes, and Gauss–Ostrogradsky theorems using a vectorial formulation.

The expansion of the concepts of the differential operators studied in Differential Calculus is performed in Chap. 4. The scalar, vectorial, and tensorial fields are defined, and the concepts and expressions for gradient, divergence, and curl are formulated. With the definition of the nabla operator, successive applications of this linear differential operator are carried out and various fundamental relations between the differential operators are deducted, defining the Laplace operator. All the formulas are deducted by means of tensorial approach.

The definition of metric spaces with several dimensions, with the introduction of Riemann curvature concept, and the Ricci tensor formulations, the scalar curvature, and the Einstein tensor are the subjects studied in detail in Chap. 5. Various particular cases of Riemann spaces are analyzed, such as the bidimensional spaces, the spaces with constant curvature, the Minkowski space, and the conformal spaces, with the definition of the Weyl tensor.

Chapter 6 broaches metric spaces provided with curvature with the introduction of the concepts of the geodesics and the geodesics and Riemann coordinate systems. The geodesics deviation and the parallelism of vectors in curved spaces are studied, with the definition of the torsion tensor concept.

The purpose of this book is to give a simple, correct, and comprehensive mathematical explanation of Tensor Calculus, and it is self-contained. Postgraduate and advanced undergraduate students and professionals will find clarity and insight into the subject of this textbook.

The preparation of a book is a hard and long work that requires the participation of other people besides the author, which are of fundamental importance in the preparation of the originals and in the tiresome task of reviewing the typing, chiefly in a text such as the one in this book. So, our sincere thanks to all those who helped in the preparation and editing of these pages.

In relation to the errors in this text which were not corrected by a more diligent review, it is stressed that they are the author's responsibility and the author apologizes for them.

Rio de Janeiro, Brazil
December 17, 2015

Emil de Souza Sánchez Filho

Historical Introduction

This brief history of Tensor Calculus broaches the development of the idea of vector and the advent of the concept of tensor in a synthetic way. The following paragraphs aim to show the history of the development of these themes in the course of time, highlighting the main stages that took place in this evolution of the mathematical knowledge. A few items of bibliographic data of the mathematicians and scientists who participated on this epic journey in a more striking manner are described.

The perception of Nature under a purely philosophical focus led Plato in 360 BC to the study of geometry. This philosopher classified the geometric figures into triangles, rectangles, and circles, and with this system, he grounded the basic concepts of geometry. Later Euclid systemized geometry in axiomatic form, starting from the fundamental concepts of points and lines.

The wise men of ancient Greece also concerned themselves with the study of the movement of bodies by means of geometric concepts. The texts of Aristotle (384–322 BC) in *Mechanics* show that he had the notion of composition of movements. In this work, Aristotle enounced in an axiomatic form that the force that moves a body is collinear with the direction of the body's movement. In a segment of *Mechanics*, he describes the velocity of two bodies in linear movement with constant proportions between each other, explaining that "*When a body moves with a certain proportion, the body needs to move in a straight line, and this is the diameter of the figure formed with the straight lines which have known proportions.*" This statement deals with the displacements of two bodies—the Greek sage acknowleding that the resultant of these displacements would be the diagonal of the rectangle (the text talks about the diameter) from the composition of the speeds.

In the Renaissance, the prominent figure of Leonardo da Vinci (1452–1519) also stood out in the field of sciences. In his writings, he reports that "*Mechanics is the paradise of mathematical science, because all the fruits of mathematics are picked here.*" Da Vinci conceived concepts on the composition of forces for maintaining the balance of the simple structures, but enunciated them in an erroneous and contradictory manner in view of the present-day knowledge.

The awakening of a new manner of facing the uniform was already blossoming in the 1600s. The ideas about the conception and study of the world were no longer conceived from the scholastic point of view, for reason more than faith had become the way to new discoveries and interpretations of the outside world. In the Netherlands, where liberal ideas were admitted and free thought could be exercised in full, the Dutch mathematician Simon Stevin (1548–1620), or Stevinus in a Latinized spelling, was the one who demonstrated in a clear manner the rule for the composition of forces, when analyzing the balance of a body located in an inclined plane and supported by weights, one hanging at the end of a lever, and the other hanging from a pulley attached to the vertical cathetus of the inclined plane. This rule is a part of the writings of Galileo Galilei (1564–1642) on the balance of bodies in a tilted plane. However, it became necessary to conceive mathematical formalism that translated these experimental verifications. The start of the concept of vector came about in an empirical mode with the formulation of the parallelogram rule, for Stevinus, in a paper published in 1586 on applied mechanics, set forth this principle of Classic Mechanics, formalizing by means of the balance of a force system the concept of a variety depending on the direction and orientation of its action, enabling in the future the theoretical preparation of the concept of vector.

The creation of the Analytical Geometry by René Du Perron Descartes (1596–1650) brought together Euclid's geometry and algebra, establishing a univocal correspondence between the points of a straight line and the real numbers. The introduction of the orthogonal coordinates system, also called Cartesian coordinates, allowed the calculation of the distance ds between two points in the Euclidean space by algebraic means, given by $ds^2 = dx^2 + dy^2 + dz^2$, where dx, dy, dz are the coordinates of the point.

The movement of the bodies was a focus of attention of the mathematicians and scientists, and a more elaborate mathematical approach was necessary when it was studied. This was taken care of by Leonhard Paul Euler (1707–1783), who conceived the concept of inertia tensor. This concept is present in his book *Theoria Motus Corporum Solidorum seu Rigidorum* (Theory of the Movement of the Solid and Rigid Bodies) published in 1760. In this paper, Euler studies the curvature lines, initiating the study of Differential Geometry. He was the most published mathematician of the all time, 860 works are known from him, and it is known that he published 560 papers during his lifetime, among books, articles, and letters.

In the early 1800s, Germany was becoming the world's largest center in mathematics. Among many of its brilliant minds, it counted Johann Karl Friedrich Gauß (1777–1855). On occupying himself with the studies of curves and surfaces, Gauß coined the term non-Euclidean geometries; in 1816, he'd already conceived concepts relative to these geometries. He prepared a theory of surfaces using curvilinear coordinates in the paper *Disquisitones Generales circa Superfícies Curvas*, published in 1827. Gauß argued that the space geometry has a physical aspect to be discovered by experimentation. These ideas went against the philosophical concepts of Immanuel Kant (1724–1804), who preconized that the conception of the space is a priori Euclidian. Gauß conceived a system of local coordinates system u, v, w located on a surface, which allowed him to calculate the distance between two points on this surface, given by the quadratic expression $ds^2 = Adu^2 + Bdv^2 + Cdw^2 + 2Edu \cdot dv + 2Fdv \cdot dw + 2Gdu \cdot dw$, where A, B, C, F, G are functions of the coordinates u, v, w.

The idea of force associated with a direction could be better developed analytically after the creation of the Analytical Geometry by Descartes. The representation of the complex numbers by means of two orthogonal axes, one axis representing the real numbers and the other axis representing the imaginary number, was developed by the Englishman John Wallis (1616–1703). This representation allowed the Frenchman Jean Robert Argand (1768–1822) to develop in 1778, in a manner independent from the Dane Gaspar Wessel (1745–1818), the mathematical operations between the complex numbers. These operations served as a framework for the Irish mathematician William Rowan Hamilton (1805–1865) to develop a more encompassing study in three dimensions, in which the complex numbers are contained in a new variety: the Quaternions.

This development came about by means of the works of Hamilton, who had the beginning of his career marked by the discovery of an error in the book *Mécanique Celeste* authored by Pierre Simon-Laplace (1749–1827), which gave him prestige in the intellectual environment. In his time, there was a great discrepancy between the mathematical production from the European continent and from Great Britain, for the golden times of Isaac Newton (1642–1727) had already passed. Hamilton studied the last advances of the continental mathematics, and between 1834 and 1835, he published the books *General Methods in Dynamics*. In 1843, he published the Quaternions Theory, printed in two volumes, the first one in 1853 and the second one in 1866, in which a theory similar to the vector theory was outlined, stressing, however, that these two theories differ in their grounds.

In the first half of the nineteenth century, the German Hermann Günther Graßmann (1809–1877), a secondary school teacher of the city of Stettin located in the region that belongs to Pomerania and that is currently a part of Poland, published the book *Die Lineale Ausdehnunsgleher ein neuer Zweig der Mathematik* (Extension Theory), in which he studies a geometry of more than three dimensions, treating *N* dimensions, and formulating a generalization of the classic geometry. To outline this theory, he used the concepts of invariants (vectors and tensors), which later enabled other scholars to develop calculus and vector analysis.

The great mathematical contribution of the nineteenth century, which definitely marked the development of Physics, is due to Georg Friedrich Bernhard Riemann (1826–1866). Riemann studied in Göttingen, where he was a pupil of Gauß, and afterward in Berlin, where he was a pupil of Peter Gustav Lejeune Dirichlet (1805–1859), and showed an exceptional capacity for mathematics when he was still young. His most striking contribution was when he submitted in December 1853 his *Habilititationsschrift* (thesis) to compete for the position of *Privatdozent* at the University of Göttingen. This thesis titled *Über die Hypothesen welche der Geometrie zu Grunde liegen* enabled a genial revolution in the structure of Physics in the beginning of the twentieth century, providing Albert Einstein (1879–1955) with the mathematical background necessary for formulating his Theory of Relativity. The exhibition of this work in a defense of thesis carried out in June 10, 1854, sought to show his capacity to teach. Gauß was a member of examination board and praised the exhibition of Riemann's new concepts. His excitement for the new formulations was expressed in words: "*... the depth of the ideas that were presented....*" This work was published 14 years later, in 1868, two years after the death of its author. Riemann generalized the geometric concepts of Gauß, conceiving a system of more general coordinates spelled as dx^i, and established a fundamental relation for the space of N dimensions, where the distance between two points ds is given by the quadratic form $ds^2 = g_{ij}dx^i dx^j$, having g_{ij} a symmetrical function, positive and defined, which characterizes the space in a unique manner. Riemann developed a non-Euclidean, elliptical geometry, different from the geometries of János Bolyai and Nikolai Ivanovich Lobachevsky. The Riemann Geometry unified these three types of geometry and generalized the concepts of curves and surfaces for hyperspaces.

The broaching of the Euclidean space in terms of generic coordinates was carried out for the first time by Gabriel Lamé (1795–1870) in his work *Leçons sur les Fonctions Inverses des Transcedentes et les Surfaces Isothermes*, published in Paris in 1857, and in another work *Leçons sur les Coordonées Curvilignes*, published in Paris in 1859.

The new experimental discoveries in the fields of electricity and magnetism made the development of an adequate mathematical language necessary to translate them in an effective way. These practical needs led the North American Josiah Willard Gibbs (1839–1903) and the Englishman Olivier Heaviside (1850–1925), in an independent manner, to reformulate the conceptions of Graßmann and Hamilton, creating the vector calculus. Heaviside had thoughts turned toward the practical cases and sought applications for the vectors and used vector calculus in electromagnetism problems in the industrial areas.

With these practical applications, the vectorial formalism became a tool to be used in problems of engineering and physics, and Edwin Bidwell Wilson, a pupil of Gibbs, developed his master's idea in the book *Vector Analysis: A Text Book for the Use for Students of Mathematics and Physics Founded* upon Lectures of Josiah Willard Gibbs, published in 1901 where he disclosed this mathematical apparatus, making it popular. This was the first book to present the modern system of vectorial analysis and became a landmark in broadcasting the concepts of calculus and vectorial analysis.

The German mathematician and prominent professor Elwin Bruno Christoffel (1829–1900) developed researches on the Invariant Theory, writing six articles about this subject. In the article *Über die Transformation der Homogenel Differentialausdrücke zweiten Grade*, published in the *Journal für Mathematik*, 70, 1869, he studied the differentiation of the symmetric tensor g_{ij} and introduced two functions formed by combinations of partial derivatives of this tensor, conceiving two differential operators called Christoffel symbols of the first and second kind, which are fundamental in Tensorial Analysis. With this, he contributed in a fundamental way to the arrival of Tensor Calculus later developed by Gregorio Ricci-Curbastro and Tullio Levi Civita. The metrics of the Riemann spaces and the Christoffel symbols are the fundaments of Tensor Calculus.

The importance of tensors in problems of Physics is due to the fact that physical phenomena are analyzed by means of models which include these varieties, which are described in terms of reference systems. However, the coordinates which are described in terms of the reference systems are not a part of the phenomena, only a tool used to represent them mathematically. As no privileged reference systems exist, it becomes necessary to establish relations which transform the coordinates from one referential system to another, so as to relate the tensors' components. These components in a coordinate system are linear and homogeneous functions of the components in another reference system.

The technological development at the end of the nineteenth century and the great advances in the theory of electromagnetism and in theoretical physics made the conception of a new mathematical tool which enabled expressing new concepts and laws imperious. The vectorial formalism did not fulfill the broad field and the variety of new knowledge that needed to be studied more and interpreted better. This tool began to be created by the Italian mathematician Gregorio Ricci-Curbastro (1853–1925), who initiated the conception of Absolute Differential Calculus in 1884. Ricci-Curbastro was a mathematical physicist par excellence. He was a pupil of the imminent Italian professors Enrico Betti (1823–1892) and Eugenio Beltrami (1835–1900). He occupied himself mainly with the Riemann geometry and the study of the quadratic differential form and was influenced by Christoffel's idea of covariant differentiation which allowed achieving great advances in geometry. He created a research group in which Tullio Levi-Civita participated and worked for 10 years (1887–1896) in the exploration of the new concepts and of an elegant and synthetic notation easily applicable to a variety of problems of mathematical analysis, geometry, and physics. In his article, *Méthodes de Calcul Differéntiel Absolu et leurs Applications*, published in 1900 in vol. 54 of the *Mathematische Annalen*, in conjunction with his pupil Levi-Civita, the applications of the differential invariants were broached, subject of the Elasticity Theory, of the Classic Mechanics and the Differential Geometry. This article is considered as the beginning of the creation of Tensor Calculus. He published the first explanation of his method in the Volume XVI of the *Bulletin des Sciences Mathématiques* (1892), applying it to problems from Differential Geometry to Mathematical Physics. The transformation law of a function system is due to Ricci-Curbastro, who published it in an article in 1887, and which is also present in another article published 1889, in which he introduces the use of upper and lower indexes, showing the differences between the contravariant and covariant transformation laws. In these papers, he exhibits the framework of Tensor Calculus.

The pupil and collaborator of Ricci-Curbastro, Tullio Levi-Civita (1873–1941) published in 1917 in the *Rediconti del Circolo Matemático di Palermo*, XLII (pp. 173–215) the article *Nozione di Parallelismo in una Varietá Qualunque e Conseguente Specificazione Geometrica della Curvatura Riemanniana*, contributing in a considerable way to the development of Tensor Calculus. In this work, he describes the parallelism in curved spaces. This study was presented in lectures addressed in two courses given at the University of Rome in the period of

1920–1921 and 1922–1923. He corresponded with Einstein, who showed great interest in the new mathematical tool. In 1925, he published the book *Lezione di Calcolo Differenziale Absoluto* which is a classic in the mathematical literature.

It was the German Albert Einstein in 1916 who called the Absolute Differential Calculus of Ricci-Curbastro and Levi-Civita Tensor Calculus, but the term tensor, such as it is understood today, had been introduced in the literature in 1908 by the physicist and crystallographer Göttingen, Waldemar Voigt (1850–1919). The development of the theoretical works of Einstein was only possible after he became aware of by means of his colleague from Zurich, Marcel Grossmann (1878–1936), head professor of descriptive geometry at the *Eidgenössische Technische Hochschule*, the article *Méthodes de Calcul Differéntiel Absolut*, which provided him the mathematical tool necessary to conceive his theory, publishing in 1916 in the *Annalen der Physik* the article *Die Grundlagen der algemeinnen Relativitatstheorie*. His contribution Tensor Calculus also came about with the conception of the summation rule incorporated to the index notation. The term *tensor* became popular mainly due to the Theory of Relativity, in which Einstein used this denomination. His researches on the gravitational field also had the help of Grossmann, Tulio Levi-Civita, and Gregorio Ricci-Curbastro, conceiving the General Relativity Theory. On the use of the Tensor Calculus in his Gravitation Theory, Einstein wrote: "*Sie bedeutet einen wahren Triumph der durch Gauss, Riemann, Christoffel, Ricci . . . begründeten Methoden des allgemeinen Differentialkalculus*."

Other notable mathematics contributed to the development of the study of tensors. The Dutch Jan Arnoldus Schouten (1873–1941), professor of the T. U. Delft, discovered independently of Levi-Civita *the parallelism* and systematized the Tensor Calculus. Schouten published in 1924 the book *Ricci-Kalkül* which became a reference work on the subject, where he innovates the tensorial notation, placing the tensor indexes in brackets to indicate that it was an antisymmetric tensor.

The Englishman Arthur Stanley Eddington (1882–1944) conceived new in Tensor Calculus and was major promoter of the Theory of Relativity to the lay public.

The German Hermann Klaus Hugo Weyl (1885–1955) published in 1913 *Die Idee der Riemannschen Fläche,* which gave a unified treatment of Riemann surfaces. He contributed to the development and disclosure of Tensor Calculus, publishing in 1918 the book *Raum-Zeit-Materie* a classic on the Theory of Relativity. Weyl was one of the greatest and most influential mathematicians of the twentieth century, with broad dominium of themes with knowledge nearing the "universalism."

The American Luther Pfahler Eisenhart (1876–1965) who contributed greatly to semi-Riemannian geometry wrote several fundamental books with tensorial approach.

The work of French mathematician Élie Joseph Cartan (1869–1951) in differential forms, one of the basic kinds of tensors used in mathematics, is principal reference in this theme. He published the famous book *Leçons sur la Géométrie des Espaces de Riemann* (first edition in 1928 and second edition in 1946).

Contents

Notations

$\mathfrak{R}$	Set of the real numbers
Z	Set of the complex numbers
$\lvert\cdots\rvert$	Determinant
$\lVert\cdots\rVert$	Modulus, absolute value
$\cdot$	Dot product, scalar product, inner product
$\times$	Cross product, vectorial product
$\otimes$	Tensorial product
$\overline{\overline{\otimes}}$	Two contractions of the tensorial product
$\delta_{ij}, \delta^{ij}, \delta_i^j$	Kronecker delta
$\delta_{ij\ldots m}$, $\delta^{ij\ldots m}$, $\delta_{i\ldots m}^{j\ldots n}$	Generalized Kronecker delta
e_{ijk}, e^{ijk}	Permutation symbol
$e_{ijk\ldots m}, e^{ijk\ldots m}$	Generalized permutation symbol
$\varepsilon_{ijk}, \varepsilon^{ijk}$	Ricci pseudotensor
$\varepsilon_{i_1 i_2 i_3 \cdots i_n}$, $\varepsilon^{i_1 i_2 i_3 \cdots i_n}$	Ricci pseudotensor for the space E_N
E_3	Euclidian space
J	Jacobian
E_N	Vectorial space or tensorial space with N dimension
$\ell n \cdots$	Natural logarithm
$\varepsilon_{ijk\ldots m}, \varepsilon^{ijk\ldots m}$	Ricci pseudotensor for the space E_N
$\dfrac{d\ldots}{dx^k}$	Differentiation with respect to variable x^k
$\phi_{,i}$	Comma notation for differentiation
$\dot{x}$	Differentiation with respect to time
$\dfrac{\partial\ldots}{\partial x^k}$	Partial differentiation with respect to variable x^k
$\partial_k \cdots$	Covariant derivative
$\dfrac{\delta\cdots}{\delta t}$	Intrinsic or absolute derivative
$\nabla\cdots$	Nabla operator

$\nabla^2 \cdots$	Laplace operator, Laplacian
$\boldsymbol{H} \cdots$	Hesse operator, Hessian
$\Box \ldots$	D'Alembert operator, D'alembertian
$div \cdots$	Divergent
$grad \cdots$	Gradient
$lap \cdots$	Laplacian
$rot \cdots$	Rotational, curl
g_{ij}, g^{ij}, g_i^j	Metric tensor
$\Gamma_{ij,k}$	Christoffel symbol of first kind
Γ_{ip}^m	Christoffel symbol of the second kind
G_{ij}, G^{ij}	Einstein tensor
G_m^k	Einstein tensor with variance (1,1)
K	Riemann curvature
R	Scalar curvature
R_{ij}	Ricci tensor of the variance (0,2)
R_j^i	Ricci tensor of the variance (1,1)
R_{ijk}^{ℓ}	Riemann–Christoffel curvature tensor, Riemann–Christoffel mixed tensor, Riemann–Christoffel tensor of the second kind, curvature tensor
R_{pijk}	Curvature tensor of variance (0, 4)
$tr \cdots$	Trace of the matrix
$W_{ijk\ell}$	Weyl curvature tensor

Greek Alphabets

Sound	Letter
Alpha	α, A
Beta	β, B
Gamma	γ, Γ
Delta	δ, Δ
Epsilon	ε, E
Zeta	ζ, Z
Eta	η, H
Theta	θ, Θ
Iota	ι, I
Kappa	κ, K
Lambda	λ, Λ
Mü	μ, M
Nü	ν, N
Ksi	ξ, Ξ
Omicron	o, O
Pi	π, Π
Rho	ρ, P

Sigma	σ, Σ
Tau	τ, T
Üpsílon	υ, Y
Phi	φ, ϕ, Φ
Khi	χ, X
Psi	ψ, Ψ
Omega	ω, Ω

Chapter 1
Review of Fundamental Topics About Tensors

1.1 Preview

This chapter presents a brief review of the fundamental concepts required for the consistent development of the later chapters. Various subjects are admitted as being previously known, which allows avoiding demonstrations that overload the text. It is assumed that the reader has full knowledge of Differential and Integral Calculus, Vectorial Calculus, Linear Algebra, and the fundamental concepts about tensors and dominium of the tensorial formalism. However, are presented succinctly the essential topics for understanding the themes that are developed in this book.

1.1.1 Index Notation and Transformation of Coordinates

On the course of the text, when dealing with the tensorial formulations, the index notation will be preferably used, and with the summation rule, for instance, $y_j = \sum_{i=1}^{3}\sum_{j=1}^{3} a_{ij}x_i = a_{ij}x_i$, where i is a free index and j is a dummy in the sense that the sum is independent of the letter used, this expression takes the forms

$$\begin{cases} y_1 = a_{11}x_1 + a_{12}x_2 + a_{13}x_3 \\ y_2 = a_{21}x_1 + a_{22}x_2 + a_{23}x_3 \\ y_3 = a_{31}x_1 + a_{32}x_2 + a_{33}x_3 \end{cases} \Rightarrow \begin{Bmatrix} y_1 \\ y_2 \\ y_3 \end{Bmatrix} = \begin{bmatrix} a_{11} & a_{12} & a_{13} \\ a_{21} & a_{22} & a_{23} \\ a_{31} & a_{32} & a_{33} \end{bmatrix} \begin{Bmatrix} x_1 \\ x_2 \\ x_3 \end{Bmatrix}$$

The transformation of the coordinates from a point in the coordinate system X^i to the coordinate system $\overline{X}^i$ given by $\overline{x}_i = a_{ij}x_j + a_{i0}$ where the terms a_{ij}, a_{i0} are constants is called affine transformation (linear). In this kind of transformation, the points of the space E_3 are transformed into points, the straight lines in straight

© Springer International Publishing Switzerland 2016

E. de Souza Sánchez Filho, *Tensor Calculus for Engineers and Physicists*,
DOI 10.1007/978-3-319-31520-1_1

lines, and the planes in planes. When $a_{i0} = 0$, this transformation is called linear and homogeneous. The term a_{i0} represents only a translation of the origin of the referential.

1.2 Space of N Dimensions

The generalization of the Euclidian space at three dimensions for a number N of dimensions is prompt, defining a space E_N. This expansion of concepts requires establishing a group of N variables $x^i, i = 1, 2, 3, \ldots N$, relative to a point $P(x^i) \in E_N$, related to a coordinate system X^i, which are called coordinates of the point in this reference system. The set of points associated in a biunivocal way to the coordinates of the reference system X^i defines the N-dimensional space E_N.

In an analogous way a subspace $E_M \subset E_N$ is defined, with $M < N$, in which the group of points $P(x^i) \in E_M$ is related biunivocally with the coordinates defined in the coordinate system X^i. To make a few specific studies easier, at times the space is divided into subspaces. The space E_N is called affine space, and if it is linked to the notion of distance between two points, then it is a metric space.

1.3 Tensors

1.3.1 Vectors

The structure of a vectorial space is defined by two algebraic operations: (a) the sum of the vectors and (b) the multiplication of vector by scalar.

The conception of vectors $\boldsymbol{u}, \boldsymbol{v}, \boldsymbol{w}$ as geometric varieties is extended to a broad range of functions, as long as the set of these functions forms a vectorial space (linear space) on a set of scalars (numbers). The functions $f, g, h, \ldots$ with continuous derivatives that fulfill certain axioms are assumed as vectors, and all the formulations and concepts developed for the geometric vectors apply to these formulations.

A vectorial space is defined by the following axioms:

1. $\boldsymbol{u} + \boldsymbol{v} = \boldsymbol{v} + \boldsymbol{u}$ or $f + g = g + f$.
2. $(\boldsymbol{u} + \boldsymbol{v}) + \boldsymbol{w} = \boldsymbol{u} + (\boldsymbol{v} + \boldsymbol{w})$ or $(f + g) + h = f + (g + h)$.
3. The null vector is such that $\boldsymbol{u} + \boldsymbol{0} = \boldsymbol{u}$ or $0 + f = f$.
4. To every vector $\boldsymbol{u}$ there is a corresponding unique vector $-\boldsymbol{u}$, such that $(-\boldsymbol{u}) + \boldsymbol{u} = \boldsymbol{0}$ or $(-f) + f = 0$.
5. $1 \cdot \|\boldsymbol{u}\| = \boldsymbol{u}$ or $1 \cdot \|f\| = \|f\|$.
6. $m(n\boldsymbol{u}) = mn(\boldsymbol{u})$ or $m(nf) = mn(f)$, where m, n are scalars.
7. $(m + n)\boldsymbol{u} = m\boldsymbol{u} + n\boldsymbol{u}$ or $(m + n)f = mf + nf$.
8. $m(\boldsymbol{u} + \boldsymbol{v}) = m\boldsymbol{u} + m\boldsymbol{v}$ or $m(f + g) = mf + mg$.

1.3.2 Kronecker Delta and Permutation Symbol

The Kronecker delta is defined by

$$\delta = \delta^{ij} = \delta^i_j = \begin{cases} 1, & i = j \\ 0, & i \neq j \end{cases} \tag{1.3.1}$$

that is symmetrical, i.e., $\delta_{ij} = \delta_{ji}, \forall i, j$. The Kronecker delta is the identity tensor. This tensor is used as a linear operator in algebraic developments, such as

$$\frac{\partial x^i}{\partial x^j}\delta_{ki} = \frac{\partial x^k}{\partial x^j} \quad \frac{\partial x^j}{\partial x^i}\delta_{ki} = \frac{\partial x^j}{\partial x^k} \quad T_{ij}\delta^{ik}u^k = T_{kj}u^k = T_j$$

The permutation symbol is defined by

$$e_{ijk} = e^{ijk} = \begin{cases} 1 & \text{is an even permutation of the indexes} \\ -1 & \text{is an odd permutation of the indexes} \\ 0 & \text{when there are repeated indexes} \end{cases} \tag{1.3.2}$$

and the generalized permutation symbol is given by

$$e_{i_1 i_2 i_3 \cdots i_n} = e^{i_1 i_2 i_3 \cdots i_n} = \begin{cases} 1 & \text{is an even permutation of the indexes} \\ -1 & \text{is an odd permutation of the indexes} \\ 0 & \text{when there are repeated indexes} \end{cases} \tag{1.3.3}$$

Figure 1.1 shows an illustration how to obtain the values of this symbol.

1.3.3 Dual (or Reciprocal) Basis

The vector $\boldsymbol{u}$ expresses itself in the Euclidean space E_3 by means of the linear combination of three linearly independent unit vectors, which form the basis of this space. For the case of oblique coordinate systems, there are two kinds of basis

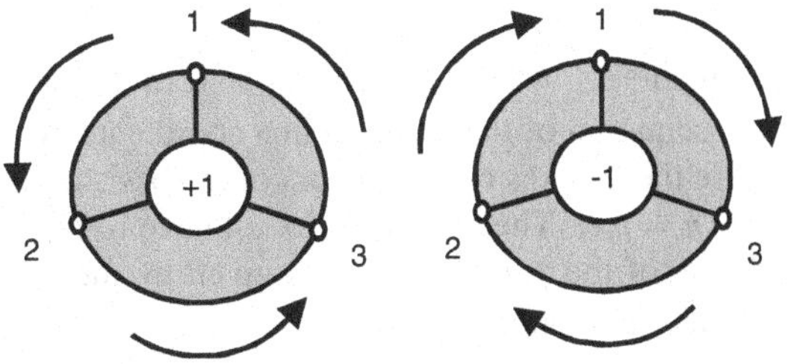

Fig. 1.1 Values of the permutation symbol

Fig. 1.2 Reciprocal basis

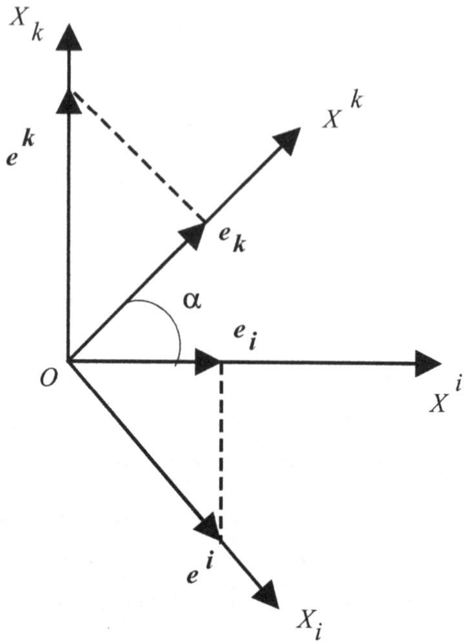

called reciprocal or dual basis. Let vector $\boldsymbol{u}$ expressed by means of their components relative to a coordinate system with orthonormal covariant basis $\boldsymbol{e}_j$:

$$\boldsymbol{u} = u_j \boldsymbol{e}_j$$

and with $\boldsymbol{e}_i \cdot \boldsymbol{e}_k = \delta_{jk}$, the dot product takes the form $\boldsymbol{u} \cdot \boldsymbol{e}_k = u_j \boldsymbol{e}_j \cdot \boldsymbol{e}_k = u_j \delta_{jk} = u_k$, which are the components' covariant of the vector $\boldsymbol{u}$. These components are the projections of this vector on the coordinate axes.

In the case of oblique coordinate system, the basis $\boldsymbol{e}_j, \boldsymbol{e}_k$ is called reciprocal basis, which fulfills the condition $\boldsymbol{e}_j \cdot \boldsymbol{e}^k = \delta_j^k$. In Fig. 1.2 the axes OX^i and OX_k are perpendicular, as are also the axes OX^k and OX_i.

This definition shows that the dot product of two reciprocal basis fulfills

$$\|\boldsymbol{e}_i\| \|\boldsymbol{e}^i\| \cos(90^{\circ} - \alpha) = 1 > 0 \Rightarrow \|\boldsymbol{e}^i\| = \frac{1}{\|\boldsymbol{e}_i\| \sin\alpha}$$

and with $\|\boldsymbol{e}_i\| = 1$ results in $\|\boldsymbol{e}^i\| > 1$, then $\boldsymbol{e}_i$ and $\boldsymbol{e}^k$ have different scales. Let the representation of the vector $\boldsymbol{u}$ in a coordinate system with covariant basis $\boldsymbol{e}_i$, $\boldsymbol{e}_j$, $\boldsymbol{e}_k$, where the indexes of the vectors of the basis indicate a cyclic permutation of i, j, k; thus, $\boldsymbol{u} = u_i \boldsymbol{e}_i$. These vectors do not have to be coplanar $\boldsymbol{e}_i \cdot \boldsymbol{e}_k \times \boldsymbol{e}_k \neq 0$; thus, the volume of the parallelepiped is given by the mixed product $\boldsymbol{e}_i \cdot \boldsymbol{e}_k \times \boldsymbol{e}_k = V$ and with the relation between the two reciprocal basis $\boldsymbol{e}_i \cdot \boldsymbol{e}^j = \delta_{ij}$ follows

$$\frac{1}{e_i} = e^i = \frac{e_j \times e_k}{V}$$

Then vector $\boldsymbol{u}$ in terms of reciprocal basis is defined by $\boldsymbol{u} = u^j \boldsymbol{e}^j$ where u^j is the components of this vector in the new basis (contravariant), having these new components expressed in terms of the original components.

Consider the representation $\boldsymbol{u}$ in terms of the two basis $\boldsymbol{u} = u_i \boldsymbol{e}_i = u^j \boldsymbol{e}^j$ and with the dot product of both sides of this expression by $\boldsymbol{e}_j$, and applying the definition of reciprocal basis $\boldsymbol{e}^j \cdot \boldsymbol{e}_i = 1$ provides

$$u^j = u_i \boldsymbol{e}_j \cdot \boldsymbol{e}_i$$

In an analogous way

$$\overline{V} = \boldsymbol{e}^i \cdot \boldsymbol{e}_j \times \boldsymbol{e}^k$$

where $\overline{V}$ is the volume of the parallelepiped defined by the mixed product of the unit vectors of the reciprocal basis. The height of the parallelepiped defined by the mixed product of the unit vectors of a base is collinear with one of the unit vectors of the reciprocal basis (Fig. 1.3).

The volume of the parallelepiped is determined by means of the mixed product of three vectors and allows assessing the relations between the same by means of the reciprocal basis in the levorotatory and dextrorotatory coordinates systems.

Consider the mixed product of the vectors of the basis of a levorotatory coordinate system

$$V = \boldsymbol{e}_i \cdot \boldsymbol{e}_j \times \boldsymbol{e}_k = \boldsymbol{e}_i \cdot (e_{123}\, e_2 e_3 \boldsymbol{e}_i)$$

which will cancel itself only if $i = 1$, whereby

$$V = e_{123}(\boldsymbol{e}_1)^2 e_2 e_3 = (\boldsymbol{e}_1)^2 e_2 e_3 \Rightarrow (\boldsymbol{e}_1)^2 = \frac{V}{e_2 e_3}$$

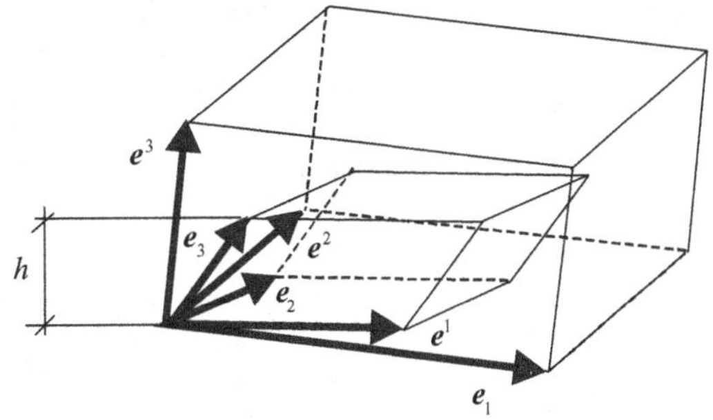

Fig. 1.3 Parallelepiped defined by means of the reciprocal basis

and for the reciprocal basis

$$\left(e^1\right)^2 = \frac{V}{\left(e_1\right)^2} = \frac{e_2 e_3}{V}$$

$$\overline{V} = e^i \cdot e^j \times e^k = e^i \cdot \left(e_{123} e^2 e^3 e^i\right) = \left(e^1\right)^2 e^2 e^3$$

$$e^1 = \frac{1}{e_1}$$

$$\overline{V} = \left(\frac{1}{e_1}\right) e^2 e^3 = \left(\frac{e_2 e_3}{V}\right) e^2 e^3 = \frac{1}{V} \Rightarrow V\overline{V} = 1 e^1 = \frac{1}{e_1}$$

For a dextrorotatory coordinate system

$$V = e_2 \cdot e_j \times e_3 = e_2 \cdot (e_{123}\, e_i e_3 e_2) e^1 = \frac{1}{e_1}$$

which cancels itself for $i = 1$, so

$$V = e_2 \cdot (-e_i e_3 e_2) = (e_2)^2 e_1 e_3 e^1 = \frac{1}{e_1}$$

$$\frac{1}{(e_2)^2} = -\frac{e_1 e_3}{V} e^1 = \frac{1}{e_1}$$

and for the case of reciprocal basis

$$\overline{V} = e^2 \cdot e^i \times e^3 = e^2\left(e_{123}\, e^i e^3 e^2\right) = -\left(e^2\right)^2 e^1 e^3$$

$$e_2 = -\frac{1}{e^2}$$

In an analogous way

$$V\overline{V} = 1$$

If e_1, e_3, e_2 are the unit vectors of an orthogonal coordinate system, then the reciprocal basis e^1, e^2, e^3 also defines this coordinate system.

1.3.3.1 Orthonormal Basis

If the basis is orthonormal

$$e_i \cdot e_j \times e_k = V = \overline{V} = 1 \qquad e^i = e_j \times e_k = e_i \qquad u^j = u_i$$

This shows that for the Cartesian vectors, it is indifferent, covariant, or contravariant, of which the basis is adopted. The vector components in terms of this basis are equal, and the orthonormal basis is defined by their unit vectors

$$e_i = \frac{u_i}{||u_i||}$$

The linear transformations $\forall m, \boldsymbol{u}, \boldsymbol{v} \in E_3$: (a) $F(m\boldsymbol{u}) = mF(\boldsymbol{u})$; (b) $F(\boldsymbol{u} \cdot \boldsymbol{v}) = F(m)$ defined in the Euclidean space E_3 are also defined in the vectorial space E_3^*, for there is an intrinsic correspondence between these two spaces. The rules of calculus in E_3^* are analogous to those of E_3, so these parameters are isomorphous.

The existence of this duality is extended to the case of a vectorial space of finite dimension E_N^*, having $E_N^* \subset \mathfrak{R}$ or $E_N^* \subset Z$, for this space is dual to the Euclidean space E_N.

1.3.3.2 Transformation Law of Vectors

The transformation of the coordinates from one point in the coordinate system X^i to the coordinate system $\overline{X}^i$ is given by $\overline{x}_i = \frac{\partial x^i}{\partial \overline{x}^j} x_j$, where $\frac{\partial x^i}{\partial \overline{x}^j} = \cos \alpha_{ij}$ are the matrix rotation elements, and its terms are the director cosines of the angles between the coordinate axes.

In this linear and homogeneous transformation, the points of the space E_3 are transformed into points expressed in terms of the new coordinates. Thus, the unit vectors of $\overline{X}^i$ and of $\overline{X}^i$ transform according to the law $\overline{\boldsymbol{e}}_i = \frac{\partial x^j}{\partial \overline{x}^i} \boldsymbol{e}_j$, where the values of $\frac{\partial x^j}{\partial \overline{x}^i} = \cos \overline{x}_i x_j$ are the components of the unit vectors $\overline{\boldsymbol{e}}_i$ in the coordinate system X^i. For the position vector, $\boldsymbol{u}$ provides $\overline{u}_i = \frac{\partial x^j}{\partial \overline{x}^i} u_j$. In the case of the inverse transformation, i.e., of $\overline{X}^i$ to X^i, provides analogously $\boldsymbol{e}_j = \frac{\partial \overline{x}^i}{\partial x^j} \overline{\boldsymbol{e}}_i$, following for the components $\frac{\partial \overline{x}^i}{\partial x^j} = \cos x_j \overline{x}_i$ of the unit vectors $\boldsymbol{e}_j$ in the coordinate system $\overline{X}^i$.

The determinant of the rotation matrix $\left| \frac{\partial x^i}{\partial \overline{x}^j} \right|$ assumes the value $+1$ in the case of the transformation taking place between coordinate systems of the same direction, which is then called proper transformation (rotation). Otherwise $\left| \frac{\partial x^i}{\partial \overline{x}^j} \right| = -1$, and the transformation is called improper transformation (reflection).

1.3.3.3 Covariant and Contravariant Vectors

The representation of the vectors in oblique coordinate systems highlights various characteristics which are more general than the Cartesian representation. In these systems the vectors are expressed by means of two kinds of components. Let the representation of vector $\boldsymbol{u}$ in the plane coordinate system of oblique axes OX^iX^j that

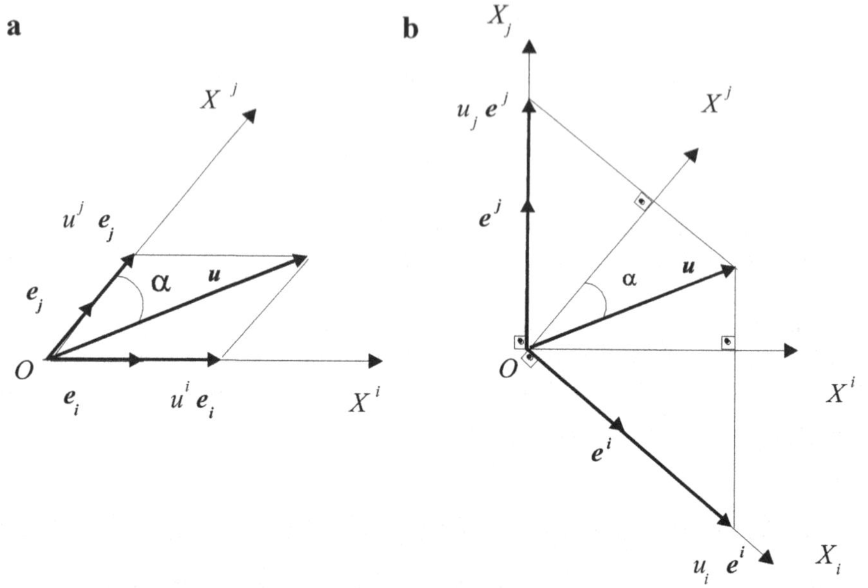

Fig. 1.4 Vector components: (**a**) contravariant, (**b**) covariant

form an angle α, with basis vectors $\boldsymbol{e}_i$, $\boldsymbol{e}_j$ (Fig. 1.4). The contravariant components are obtained by means of straight lines parallel to the axes OX^i and OX^j and graphed, respectively, as u^i, u^j (indicated with upper indexes). The covariant componentsare obtained by means of projection on the axes OX^i and OX^j given, respectively, by u_i, u_j (indicated with lower indexes).

The projection of vector $\boldsymbol{u}$ on an axis provides its component on this axis, and by means of the dot product of $\boldsymbol{u} = u^i\boldsymbol{e_i}$ and $\boldsymbol{e}^j$:

$$\boldsymbol{u} \cdot \boldsymbol{e}^j = u^i\boldsymbol{e}_i \cdot \boldsymbol{e}^j = u^i\left(\boldsymbol{e}_i \cdot \boldsymbol{e}^j\right) = u^i\delta_{ij} = u^i$$

that is the contravariant component of vector and in the same way by the covariant component

$$\boldsymbol{u} \cdot \boldsymbol{e}_j = u_i\boldsymbol{e}^i \cdot \boldsymbol{e}_j = u_i\left(\boldsymbol{e}^i \cdot \boldsymbol{e}_j\right) = u_i\delta_{ij} = u_i$$

Thus, the vector is defined by its components

$$\boldsymbol{u} = u^i\boldsymbol{e_i} = u_i\boldsymbol{e}^i$$

These components are not, in general, equal, and in the case of $\alpha = 90^{\mathrm{o}}$ (Cartesian coordinate systems), the equality of these components is verified.

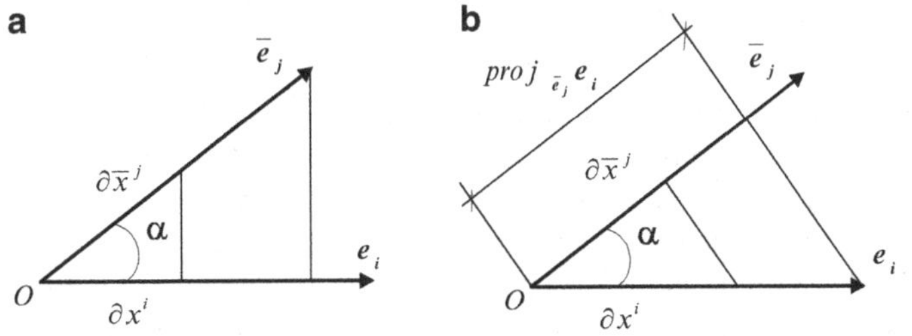

Fig. 1.5 Transformation of coordinates: (**a**) covariant, (**b**) contravariant

1.3.3.4 Transformation Law of Covariant Vectors

The transformation law of base $\boldsymbol{e}_i$ of an axis OX^i for a new axis $O\overline{X}^j$, with base $\bar{\boldsymbol{e}}_j$ (Fig. 1.5a), is given by

$$\bar{\boldsymbol{e}}_j = \left(\text{proj}_{\boldsymbol{e}_i}\|\bar{\boldsymbol{e}}_j\|\right)\boldsymbol{e}_i = (1\cos\alpha)\boldsymbol{e}_i$$

$$\cos\alpha = \frac{\partial x^i}{\partial \overline{x}^j}$$

$$\bar{\boldsymbol{e}}_j = \frac{\partial x^i}{\partial \overline{x}^j}\boldsymbol{e}_i$$

that is the transformation law of the covariant basis. For the vector $\boldsymbol{u}$ the transformation of its covariant components is given by $\bar{u}_j = \frac{\partial x^i}{\partial \overline{x}^j} u_i$, where the variables relative to the original axis in relation to which the transformation performed are found in the numerator of the equation.

1.3.3.5 Transformation Law of Contravariant Vectors

The projection of the vector $\boldsymbol{e}_i$ on the axis $O\overline{X}^j$ (Fig. 1.5b) provides

$$\bar{\boldsymbol{e}}_j = \left(\text{proj}_{\bar{\boldsymbol{e}}_j}\|\boldsymbol{e}_i\|\right)\boldsymbol{e}_i = (1\cos\alpha)\boldsymbol{e}_i$$

$$\cos\alpha = \frac{\partial \overline{x}^j}{\partial x^i}$$

$$\bar{\boldsymbol{e}}_j = \frac{\partial \overline{x}^j}{\partial x^i}\boldsymbol{e}_i$$

that is the transformation law of the contravariant basis. For the vector $\boldsymbol{u}$ follows the transformation law of its contravariant components $\overline{u}^j = \frac{\partial \overline{x}^j}{\partial x^i} u^i$, where the variables relative to the new axis, for which the transformation is carried out, are found in the numerator of the expression.

1.3.4 Multilinear Forms

The tensors of the order p are multilinear forms, which are vectorial functions, and linear in each variable considered separately. The concept of tensor is conceived by means of the following approaches: (a) the tensor is a variety that obeys a transformation law when changing the coordinate system; (b) this variety is invariant for any coordinate system; and (c) there is an equivalence between these definitions (equivalence law). A tensor of the order p is defined by a multilinear function with N^p components in the space E_N, where $R\left(1-\frac{N}{2}\right)=0$ represents its order, which is maintained invariant if a change of the coordinate system occurs, and on the rotation of the reference axes (linear and homogeneous transformation) its coordinates modify according to a certain law.

Consider the space E_N and the coordinate system X^i, $i=1,2,3,\ldots N$, defined in this space, where there are N equations that relate the coordinates of the points in E_N, given by continuously differentiable functions

$$\overline{x}^i = x^i\left(x^j\right) \quad i,j=1,2,3,\ldots N \tag{1.3.4}$$

that transform these functions to a new coordinate system $\overline{X}^i$. These transformations of coordinates require only that N functions $x^i(x^j)$ be independent. The necessary and sufficient condition for this transformation to be possible is that $J=\left|\frac{\partial x^i}{\partial \overline{x}^j}\right| \neq 0$. The inverse function has an inverse Jacobian $\overline{J}=\left|\frac{\partial \overline{x}^i}{\partial x^i}\right|$ and implies that $J\overline{J}=1$.

1.3.4.1 Transformation Law of the Second-Order Tensors

Let the position vector $u_i(x^i)$ expressed in the coordinate system X^i of base $\mathbf{e}^i$ and a new coordinate system $\overline{X}^i$, with same origin, in which the vector is expressed by $\overline{u}_i(\overline{x}^i)$. Consider the elements $\frac{\partial x^k}{\partial \overline{x}^i}$ of the rotation matrix that relates the coordinates of these two systems, then follow by means of the transformation law of covariant vectors

$$\overline{u}_i = \frac{\partial x^k}{\partial \overline{x}^i} u_k \quad i,k=1,2,3 \tag{1.3.5}$$

$$\overline{v}_j = \frac{\partial x^\ell}{\partial \overline{x}^j} v_\ell \quad j,\ell=1,2,3 \tag{1.3.6}$$

The vectors $\overline{u}_i(x^i)$ and $\overline{v}_i(x^i)$ define the transformation of the second-order tensor in terms of its covariant components

$$\overline{T}_{ij} = \overline{u}_i \overline{v}_j = \frac{\partial x^k}{\partial \overline{x}^i} \frac{\partial x^\ell}{\partial \overline{x}^j} u_k v_\ell = \frac{\partial x^k}{\partial \overline{x}^i} \frac{\partial x^\ell}{\partial \overline{x}^j} T_{k\ell} \tag{1.3.7}$$

and for the contravariant components provides an analogous manner

$$\overline{T}^{ij} = \overline{u}^i \overline{v}^j = \frac{\partial \overline{x}^i}{\partial x^k} \frac{\partial \overline{x}^j}{\partial x^\ell} u^k v^\ell = \frac{\partial \overline{x}^i}{\partial x^k} \frac{\partial \overline{x}^j}{\partial x^\ell} T^{k\ell} \tag{1.3.8}$$

In a same way, it follows for the transformation law in terms of the mixed components

$$\overline{T}^i_j = \overline{u}^i \overline{v}_j = \frac{\partial \overline{x}^i}{\partial x^k} \frac{\partial x^\ell}{\partial \overline{x}^j} u^k v_\ell = \frac{\partial \overline{x}^i}{\partial x^k} \frac{\partial x^\ell}{\partial \overline{x}^j} T^k_\ell \tag{1.3.9}$$

1.3.4.2 Transformation Law of the Third-Order Tensors

The transformations of the vectors ***u***, ***v***, ***w*** in terms of their covariant components are given by

$$\overline{u}_\ell = \frac{\partial x^i}{\partial \overline{x}^\ell} u_i \quad \overline{v}_m = \frac{\partial x^j}{\partial \overline{x}^m} v_j \quad \overline{w}_n = \frac{\partial x^k}{\partial \overline{x}^n} w_k$$

following by substitution

$$\overline{T}_{\ell mn} = \overline{u}_\ell \overline{v}_m \overline{w}_n = \frac{\partial x^i}{\partial \overline{x}^\ell} \frac{\partial x^j}{\partial \overline{x}^m} \frac{\partial x^k}{\partial \overline{x}^n} u_i v_j w_k$$

that leads to the following transformation law for the covariant components of the third-order tensors

$$\overline{T}_{\ell mn} = \frac{\partial x^i}{\partial \overline{x}^\ell} \frac{\partial x^j}{\partial \overline{x}^m} \frac{\partial x^k}{\partial \overline{x}^n} T_{ijk}$$

and for the contravariant components

$$\overline{T}^{\ell mn} = \frac{\partial \overline{x}^\ell}{\partial x^i} \frac{\partial \overline{x}^m}{\partial x^j} \frac{\partial \overline{x}^n}{\partial x^k} T_{ijk}$$

and in an analogous way, for the mixed components

$$\overline{T}_\ell^{mn} = \frac{\partial x^\ell}{\partial \overline{x}^i} \frac{\partial \overline{x}^m}{\partial x^j} \frac{\partial \overline{x}^n}{\partial x^k} T_i^{jk} \quad \overline{T}^n_{\ell m} = \frac{\partial x^\ell}{\partial \overline{x}^i} \frac{\partial x^m}{\partial \overline{x}^j} \frac{\partial \overline{x}^n}{\partial x^k} T^k_{ij}$$

$$\overline{T}^m_{\ell n} = \frac{\partial x^\ell}{\partial \overline{x}^i} \frac{\partial \overline{x}^m}{\partial x^j} \frac{\partial x^n}{\partial \overline{x}^k} T^j_{ik} \quad \overline{T}^{\ell m}_n = \frac{\partial \overline{x}^i}{\partial \overline{x}^\ell} \frac{\partial \overline{x}^m}{\partial x^j} \frac{\partial x^n}{\partial \overline{x}^k} T^{ij}_k$$

1.3.4.3 Inverse Transformation

Let the inverse transformation of the vectors $\boldsymbol{u}$ and $\boldsymbol{v}$ of the coordinate system $\overline{X}^i$ for the coordinate system X^i, given by the covariant components of the vectors

$$u_i = \frac{\partial \overline{x}^k}{\partial x^i} \overline{u}_k \quad v_j = \frac{\partial \overline{x}^\ell}{\partial x^j} \overline{v}_\ell \tag{1.3.10}$$

It follows that

$$T_{ij} = u_i v_j = \frac{\partial \overline{x}^k}{\partial x^i} \frac{\partial \overline{x}^\ell}{\partial x^j} \overline{u}_k \overline{v}_\ell = \frac{\partial \overline{x}^k}{\partial x^i} \frac{\partial \overline{x}^\ell}{\partial x^j} \overline{T}_{k\ell} \tag{1.3.11}$$

Expression (1.3.11) allows concluding that a second-order tensor can be interpreted as a transformation in the linear space E_3, which associates the vector $\boldsymbol{u}$ to the vector $\boldsymbol{v}$ by means of the tensorial product and that this linear and homogeneous transformation has an inverse transformation. The inverse transformations are defined for the contravariant and mixed components in an analogous way. Expressions (1.3.7) and (1.3.11) show that if the components of a second-order tensor are null in a coordinates system, they will cancel each other in any other coordinate system. For the definition of the transformation law of second-order tensor to be valid, it is necessary that the transitive property apply to the linear operators (Fig. 1.6).

1.3.4.4 Transitive Property

Let a second-order tensor $T_{k\ell}$ defined in the coordinate system X^i, that is expressed in the coordinate system $\overline{X}^i$ by means of the expression (1.3.7), and with the transformation of $\overline{X}^i$ for $\widetilde{X}^i$

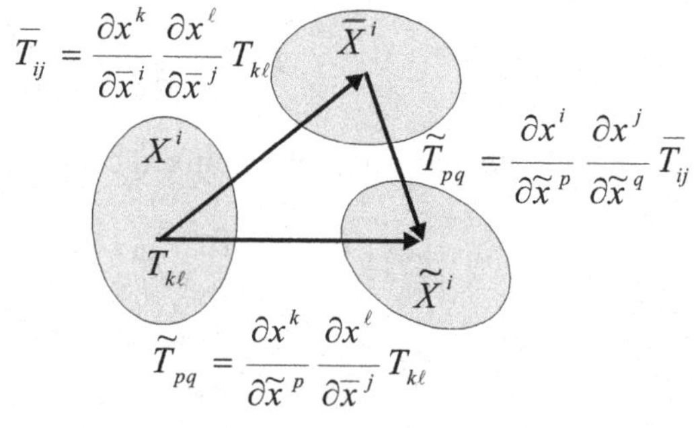

Fig. 1.6 Transitive property of the second-order tensors

$$\widetilde{T}_{pq} = \frac{\partial x^i}{\partial \widetilde{x}^p}\frac{\partial x^j}{\partial \widetilde{x}^q}\overline{T}_{ij} \tag{1.3.12}$$

However, the tensor $\widetilde{T}_{pq}$ can be expressed in terms of tensor $T_{k\ell}$, thereby avoiding the intermediary transformation, so substituting expression (1.3.7) in expression (1.3.12), it follows that

$$\widetilde{T}_{pq} = \frac{\partial x^k}{\partial \overline{x}^i}\frac{\partial x^\ell}{\partial \overline{x}^j}\frac{\partial \overline{x}^i}{\partial \widetilde{x}^p}\frac{\partial \overline{x}^j}{\partial \widetilde{x}^q}T_{k\ell} \tag{1.3.13}$$

and simplifying

$$\frac{\partial x^k}{\partial \overline{x}^i}\frac{\partial \overline{x}^i}{\partial \widetilde{x}^p} = \frac{\partial x^k}{\partial \widetilde{x}^p} \qquad \frac{\partial x^\ell}{\partial \overline{x}^j}\frac{\partial \overline{x}^j}{\partial \widetilde{x}^q} = \frac{\partial x^\ell}{\partial \overline{x}^j} \tag{1.3.14}$$

Then

$$\widetilde{T}_{pq} = \frac{\partial x^k}{\partial \widetilde{x}^p}\frac{\partial x^\ell}{\partial \overline{x}^j}T_{k\ell} \tag{1.3.15}$$

Expression (1.3.15) is the transformation law of the second-order tensor of the coordinate system X^i for the coordinate system $\widetilde{X}^i$, which proves that the transitive property applies to these tensors. This property is also valid when using the contravariant and mixed components.

The tensors studied in this book belong to metric spaces. If a variety is a tensor with respect to the linear transformations, it will be a tensor with respect to all the orthogonal linear transformations, but the inverse usually does not occur. The tensors are produced in spaces more general than the vectorial space. Table 1.1 shows the covariant, contravariant, and mixed tensors and their transformation laws for the space E_N.

1.3.4.5 Multiplication of a Tensor by a Scalar

It is the multiplication that provides a new tensor as a result, which components are the components of the original tensor multiplied by the scalar. Let the tensor T_{ijk}

Table 1.1 Kinds of tensors

Tensor	Expression	Transformation law
Covariant	$T_{ij\cdots k}$	$\overline{T}_{rs\cdots t} = \frac{\partial x^i}{\partial \overline{x}^r}\frac{\partial x^j}{\partial \overline{x}^s}\cdots\frac{\partial x^k}{\partial \overline{x}^t}T_{ij\cdots k}$
Contravariant	$T^{ij\cdots k}$	$\overline{T}^{rs\cdots t} = \frac{\partial \overline{x}^r}{\partial x^i}\frac{\partial \overline{x}^s}{\partial \overline{x}^j}\cdots\frac{\partial \overline{x}^t}{\partial x^p}T^{ij\cdots k}$
Mixed	$T^{k\ell\cdots h}_{ij\cdots f}$	$\overline{T}^{mn\cdots h}_{rs\cdots t} = \frac{\partial x^i}{\partial \overline{x}^r}\frac{\partial x^j}{\partial \overline{x}^s}\frac{\partial \overline{x}^m}{\partial x^k}\frac{\partial \overline{x}^n}{\partial x^\ell}\cdots\frac{\partial x^f}{\partial \overline{x}^t}\frac{\partial \overline{x}^h}{\partial x^h}T^{k\ell\cdots h}_{ij\cdots f}$

and the scalar m which product P_{ijk} is given by $P_{ijk} = mT_{ijk}$. For demonstrating this expression represents a tensor, all that is needed is to apply the tensor transformation law to the same.

1.3.4.6 Addition and Subtraction of Tensors

The addition of tensors of the same order and the same type is given by

$$T_{ij}^{k} = A_{ij}^{k} + B_{ij}^{k}$$

The addition of the mixed tensors given by the previous expression provides as a result a mixed tensor of the third order, which is twice covariant and once contravariant. To demonstrate this expression represents a tensor, all that is needed is to apply the tensor transformation law to the same.

The subtraction is defined in the same way as the addition, however, admitting that a tensor is multiplied by the scalar -1. As an example $T_{ij}^{k} = A_{ij}^{k} + (-1)B_{ij}^{k}$; thus, this expression provides as a result a mixed tensor of the third order, which is twice covariant and once contravariant.

To demonstrate that previous expression represents a tensor all that is needed is to carry out the analysis developed for the addition considering the negative sign.

1.3.4.7 Contraction of Tensors

The contraction of a tensor is carried out when two of its indexes are made equal, a covariant index and a contravariant index, and thus reducing the order of this tensor in two. For instance, the tensor $T_{ij}^{k\ell}$ contracted in the indexes ℓ and j results as $T_{i\ell}^{kj} = T_{ij}^{kj} = T_{i}^{k}$.

1.3.4.8 Outer Product of Tensors

The outer product is the product of two tensors that provide as a new tensor, which order is the sum of the order of these two tensors. Let, for example, the tensor $A_{ij\ldots}^{k\ldots}$ with variance index number (p, q) and the tensor $B_{\ldots rs}^{\ldots \ell m}$ with variance index number (u, v), which if multiplied provides a tensor $T_{ij\ldots rs}^{k\ldots \ell m} = A_{ij\ldots}^{k\ldots} B_{\ldots rs}^{\ldots \ell m}$ with variance index number $(p + u, q + v)$. The order of the tensor is given by the sum of these two indexes. To demonstrate that the previous expression is a tensor, all that is needed is to apply the tensor transformation law to the same.

1.3.4.9 Inner Product of Tensors

The inner product of two tensors is defined as the tensor obtained after the contracting of the outer product of these tensors. Let, for example, tensors A_{ij} and B_k^ℓ which the outer product is $P_{ijk}^\ell = A_{ij}B_k^\ell$ that provides as a result a tensor of the fourth order, which contracted in the indexes ℓ and k provide the inner product $P_{ij\ell}^\ell = A_{ij}B_\ell^\ell = P_{ij}$. This shows that the resulting tensor is of the second order. To demonstrate that this expression represents a tensor, all that is needed is to apply the tensor transformation law to the same.

1.3.4.10 Quotient Law

This law allows verifying if a group of N^p functions of the coordinates of the referential system X^i has tensorial characteristics. Its application serves to test if a variety is a tensor. The systematic for applying this law is to make the dot product of the variety that is to be tested by a vector, for the outer product of two tensors generates a tensor, and then carry out the contraction of this product and afterward, by means of applying the tensor transformation law, verify if the variety fulfills this law.

Let, for example, the contravariant tensor of the first order T^k and the variety $A(i,j,\ k)$ composed of 27 functions defined in the space E_N, for which it is desired to verify if it is tensor. The fundamental premise is that the vector T^k is independent of $A(i,j,\ k)$. If the inner product $A(i,j,\ k)T^k = B^{ij}$ originates a contravariant tensor of the second order, then $A(i,j,\ k)$ has the characteristics of a tensor. Applying the transformation law of tensors to the tensor B^{ij}

$$\overline{B}^{pq} = \frac{\partial \overline{x}^p}{\partial x^i}\frac{\partial \overline{x}^q}{\partial x^j}B^{ij} = \frac{\partial \overline{x}^p}{\partial x^i}\frac{\partial \overline{x}^q}{\partial x^j}A(i,j,k)T^k$$

and for the vector T^k, it follows that

$$T^k = \frac{\partial x^k}{\partial \overline{x}^r}\overline{T}^r$$

By substitution

$$\overline{B}^{pq} = \frac{\partial \overline{x}^p}{\partial x^i}\frac{\partial \overline{x}^q}{\partial x^j}\frac{\partial x^k}{\partial \overline{x}^r}A(i,\ j,\ k)\overline{T}^r$$

and in a new coordinate system, the tensor B^{ij} is given by

$$\overline{B}^{pq} = \overline{A}(p,\ q,\ r)\overline{T}^r$$

following by substitution

$$\left[\overline{A}(p,q,r) - \frac{\partial \overline{x}^p}{\partial x^i}\frac{\partial \overline{x}^q}{\partial x^j}\frac{\partial x^k}{\partial \overline{x}^r}A(i,j,k)\right]\overline{T}^r = 0$$

As $\overline{T}^r$ is an arbitrary vector the result is

$$\overline{A}(p,q,r) = \frac{\partial \overline{x}^p}{\partial x^i}\frac{\partial \overline{x}^q}{\partial x^j}\frac{\partial x^k}{\partial \overline{x}^r}A(i,j,k)$$

that represents the transformation law of third-order tensors. This shows that the variety $A(i,j,k)$ has tensorial characteristics.

1.4 Homogeneous Spaces and Isotropic Spaces

The isotropic space has properties which do not depend on the orientation being considered, and the components of isotropic tensors do not change on an orthogonal linear transformation. The sum of isotropic tensors results in an isotropic tensor, and the product of isotropic tensors is also an isotropic tensor.

There is no isotropic tensor of the first order. The isotropic tensor of the fourth order is given by

$$T_{ijk\ell} = \lambda\delta_{ij}\delta_{k\ell} + \mu\delta_{ik}\delta_{j\ell} + \nu\delta_{i\ell}\delta_{jk} \tag{1.4.1}$$

where λ, μ, ν are scalars. The Kronecker delta δ_{ij} is the only isotropic tensor of the second order.

The homogeneous space has properties which are independent of the position of the point. The homogeneous tensors have constant components when the coordinate system is changed.

A homogeneous tensor of the fourth order is given by

$$T_{ij\ell k} = \lambda\delta_{ij}\delta_{k\ell} + \mu\left(\delta_{ik}\delta_{j\ell} + \delta_{i\ell}\delta_{jk}\right) \tag{1.4.2}$$

where λ, μ are scalars.

1.5 Metric Tensor

The study of tensors carried out in affine spaces applies to another type, called metric space, in which the length of the curves is determined by means of a variety that defines this space, in which the basic magnitudes are the length of a curve and the vector's norm, just as the angle between vectors and the angle between two curves. The distinction between these two types of spaces is of fundamental importance in the study of tensors.

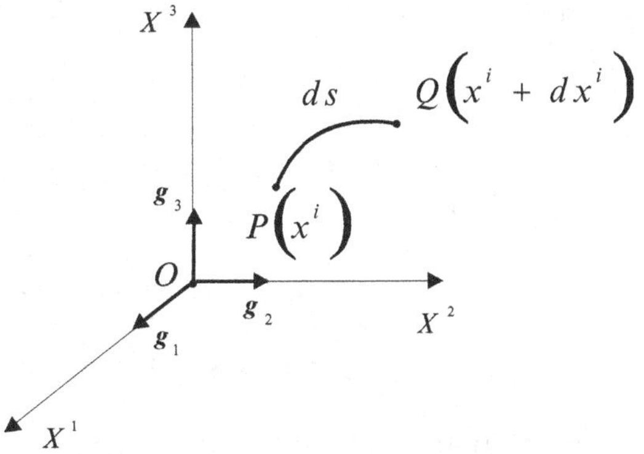

Fig. 1.7 Elementary arc of a curve

The metric space is determined by the definition of its fundamental tensor which is related with its intrinsic properties. The conception of this metric tensor, which gives an arithmetic form to the space, considers the invariance of distance between two points, the concept of distance being acquired from the space E_3. The geometry grounded in the concept of metric tensor is called Riemann geometry.

The angle between two curves is calculated by means of the dot product between vectors using the metric tensor, which awards a generalization to this tensor's formulation. Let the arc element length of a curve ds defined in the Cartesian coordinate system X^i with unit vectors $\boldsymbol{g}_1, \boldsymbol{g}_2, \boldsymbol{g}_3$ by means of its coordinates x^i (Fig. 1.7), with two neighboring points $P(x^i)$, $Q(x^i + dx^i)$, which define the position vectors $\boldsymbol{r}$ and $\boldsymbol{r} + \boldsymbol{dr}$, respectively. The coordinates of increment of the position vector $\boldsymbol{dr}$ are given by $Q - P = dx^i$; thus, $\lim\limits_{Q \to P} (Q - P) = ds$, and the dot product of this vector by himself takes the form

$$ds^2 = \boldsymbol{dr} \cdot \boldsymbol{dr} = dx^i dx^i \tag{1.5.1}$$

Consider a transformation of the coordinates $x^i = x^i(\overline{x}^i)$ for a new coordinate system $\overline{X}^i$

$$d\overline{x}^i = \frac{\partial x^i}{\partial \overline{x}^k} d\overline{x}^k \tag{1.5.2}$$

becomes

$$ds^2 = \frac{\partial x^i}{\partial \overline{x}^k} \frac{\partial x^i}{\partial \overline{x}^\ell} d\overline{x}^k d\overline{x}^\ell \tag{1.5.3}$$

Putting

$$g_{k\ell} = \frac{\partial x^i}{\partial \overline{x}^k} \frac{\partial x^i}{\partial \overline{x}^\ell} \tag{1.5.4}$$

thus the metric takes the form

$$ds^2 = g_{k\ell} d\overline{x}^k d\overline{x}^\ell \tag{1.5.5}$$

The symmetry of the variety given by expression (1.5.4) is obvious, because $g_{k\ell} = g_{\ell k}$, then

$$g_{ij} = g_{ji} = \begin{bmatrix} \boldsymbol{g}_1\boldsymbol{g}_1 & \boldsymbol{g}_1\boldsymbol{g}_2 & \boldsymbol{g}_1\boldsymbol{g}_3 \\ \boldsymbol{g}_2\boldsymbol{g}_1 & \boldsymbol{g}_2\boldsymbol{g}_2 & \boldsymbol{g}_2\boldsymbol{g}_3 \\ \boldsymbol{g}_3\boldsymbol{g}_1 & \boldsymbol{g}_3\boldsymbol{g}_2 & \boldsymbol{g}_3\boldsymbol{g}_3 \end{bmatrix} = \begin{bmatrix} g_{11} & g_{12} & g_{13} \\ g_{21} & g_{22} & g_{23} \\ g_{31} & g_{32} & g_{33} \end{bmatrix} \tag{1.5.6}$$

The analysis of expression (1.5.4) shows that $g_{k\ell}$ relates with the Jacobian $[J] = \left[\frac{\partial x^i}{\partial \overline{x}^k}\right]$ of a linear transformation by means of the following expression

$$[g_{k\ell}] = \left[\frac{\partial x^i}{\partial \overline{x}^k}\right]^{\mathrm{T}} \left[\frac{\partial x^i}{\partial \overline{x}^\ell}\right] = [J]^{\mathrm{T}}[J] \tag{1.5.7}$$

For the coordinate system X^i, the variety g_{ij} is defined by his unit vectors $\boldsymbol{g}_i, \boldsymbol{g}_j$. Consider a new coordinate system $\overline{X}^i$, with respect to which these unit vectors are expressed by

$$\overline{\boldsymbol{g}}_k = \frac{\partial x^i}{\partial \overline{x}^k}\boldsymbol{g}_i \quad \overline{\boldsymbol{g}}_\ell = \frac{\partial x^j}{\partial \overline{x}^\ell}\boldsymbol{g}_j \tag{1.5.8}$$

Thus

$$\overline{g}_{k\ell} = \left(\frac{\partial x^i}{\partial \overline{x}^k}\boldsymbol{g}_i\right)\left(\frac{\partial x^j}{\partial \overline{x}^\ell}\boldsymbol{g}_j\right) = \frac{\partial x^i}{\partial \overline{x}^k}\frac{\partial x^j}{\partial \overline{x}^\ell}(\boldsymbol{g}_i\boldsymbol{g}_j) = \frac{\partial x^i}{\partial \overline{x}^k}\frac{\partial x^j}{\partial \overline{x}^\ell}g_{ij}$$

then g_{ij} is a symmetric tensor of the second order.

The arc length is invariable when changing the coordinate system. The coefficients of $g_{k\ell}(x^i)$ are class C^2, and the N equations $x^i = x^i(\overline{x}^i)$ must satisfy the $\frac{1}{2}N(N+1)$ partial differential equations given by expression (1.5.4). However, if $g_{k\ell}(x^i)$ is specified arbitrarily, this system of $\frac{1}{2}N(N+1)$ partial differential equations, in general, has no solution. The fundamental tensor $g_{k\ell}$ related to a coordinate system X^i, in a region of the space E_N, must fulfill the following conditions:

(a) $g_{k\ell}(x^i)$ is a class C^2 function, i.e., its second-order derivatives exist and are continuous.
(b) Be symmetrical, i.e., $g_{k\ell} = g_{\ell k}$.
(c) $\det g_{k\ell} = g \neq 0$, i.e., $g_{k\ell}$ is not singular.
(d) $ds^2 = g_{k\ell} dx^k dx^\ell$ is an invariant after a change of coordinate system.

Expression (1.5.5) is put under parametric form with the coordinates $x^i = x^i(t)$ and $i = 1, 2, 3 \ldots N$, and the parameter $a \le t \le b$ provides

$$s = \int_a^b \sqrt{\left\| g_{k\ell} \frac{dx^k}{dt} \frac{dx^\ell}{dt} \right\|} dt \tag{1.5.9}$$

Admit a functional parameter $h_i = \pm 1$, so as to allow the conditions $g_{k\ell} \frac{dx^k}{dt} \frac{dx^\ell}{dt} > 0$ and $g_{k\ell} \frac{dx^k}{dt} \frac{dx^\ell}{dt} < 0$ to be be used instead of the absolute value shown in expression (1.5.9), because the use of h_i is more adequate to the algebraic manipulations; thus,

$$s = \int_a^b \sqrt{h_i g_{k\ell} \frac{dx^k}{dt} \frac{dx^\ell}{dt}} dt \tag{1.5.10}$$

The quadratic and homogeneous form $\Phi = g_{k\ell} dx^k dx^\ell$ is called metric or fundamental form of the space, being invariant after a change of coordinate system. In space E_3 with $\Phi > 0$, which provides $g > 0$, and when $\Phi = 0$, the initial and final points of the arc coincide. If $\Phi = 0$ and dx^i are not all null, the displacement between the two points is null. The possibility of Φ being undefined is admitted, for instance, in the case $\Phi = (dx^1)^2 - (dx^2)^2$, for which $dx^1 = dx^2$ results in $\Phi = 0$. This case is interpreted as having a null displacement of the point. If $dx^i \neq 0$, i.e., the displacements are not null, h_i is adopted so that $h_i \Phi > 0$.

The spaces E_N (hyperspaces) are analyzed in an analogous way to the analysis of the space E_3 by means of defining a metric, formalizing the Riemann geometry. The geometries not grounded on the concept of metric are called non-Riemann geometries.

To demonstrate that expression (1.5.10) is invariant through a change in its parametric representation, let a curve of class C^2 represented by means of the coordinates $x^i = x^i(t)$ and $a \le t \le b$. Consider a transformation for the new coordinates $\bar{x}^i = \bar{x}^i(t)$ and $\bar{a} \le \bar{t} \le \bar{b}$, where $\bar{t} = f(t)$ with $f'(t) > 0$, and in the new limits $\bar{a} = f(a)$, $\bar{b} = f(b)$. Applying the chain rule to the function $\bar{t} = f(t)$:

$$\frac{d\bar{t}}{dt} = f'(t) \Rightarrow dt = \frac{d\bar{t}}{f'(t)} \tag{1.5.11}$$

and with expression (1.5.11) in expression (1.5.10)

$$L = \int_a^b \sqrt{h_i g_{ij} \frac{dx^i}{dt} \frac{dx^j}{dt}} dt = \int_a^b \sqrt{h_i g_{ij} \frac{dx^i}{d\bar{t}} \frac{dx^j}{d\bar{t}} \left[f'(t)\right]^2} dt = \int_a^b \sqrt{h_i g_{ij} \frac{dx^i}{d\bar{t}} \frac{dx^j}{d\bar{t}}} f'(t) dt$$

$$= \int_a^b \sqrt{h_i g_{ij} \frac{dx^i}{d\bar{t}} \frac{dx^j}{d\bar{t}}} d\bar{t} = \bar{L}$$

then the value of this expression does not vary with the change of the curve's parameterization.

The metric can be written in matrix form so as to make the usual calculations easier

$$\left(\frac{ds}{dt}\right)^2 = \left\{\frac{dx^k}{dt}\right\}^{\mathrm{T}} \left[g_{ij}\right] \left\{\frac{dx^\ell}{dt}\right\} \tag{1.5.12}$$

In the space E_3, the metric is defined by

$$\frac{d\bar{t}}{dt} = f'(t) \Rightarrow dt = \frac{d\bar{t}}{f'(t)} \tag{1.5.13}$$

$$\begin{aligned} ds^2 = g_{11}dx^1dx^1 + g_{12}dx^1dx^2 + g_{13}dx^1dx^3 \\ + g_{21}dx^2dx^1 + g_{22}dx^2dx^2 + g_{23}dx^2dx^3 \\ + g_{31}dx^3dx^1 + g_{32}dx^3dx^2 + g_{33}dx^3dx^3 \end{aligned} \tag{1.5.14}$$

or

$$ds^2 = g_{ii}\left(dx^i\right)^2 + g_{kk}\left(dx^k\right)^2 + 2g_{ik}dx^i dx^k \tag{1.5.15}$$

For the particular case in which the coordinate systems are orthogonal (Fig. 1.8), the segments on the coordinate axes X^i are defined by the unit vectors $\mathbf{g}_i$ of these axes

$$ds_{(i)} = \mathbf{g}_i dx^i \tag{1.5.16}$$

which provide the metric

$$\begin{aligned} ds^2 &= \left(h_1 g_1 dx^1\right)^2 + \left(h_2 g_2 dx^2\right)^2 + \left(h_3 g_3 dx^3\right)^2 \\ &= \left(h_1 dx^1\right)^2 + \left(h_2 dx^2\right)^2 + \left(h_3 dx^3\right)^2 \end{aligned} \tag{1.5.17}$$

then the metric tensor is defined by the elements of the diagonal of the matrix

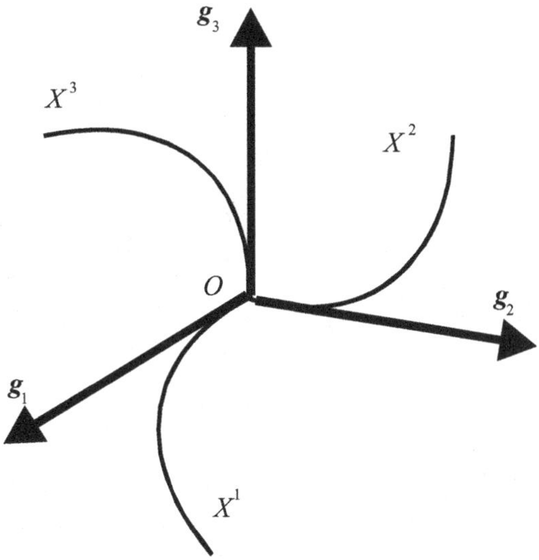

Fig. 1.8 Orthogonal coordinate systems

$$g_{ij} = \begin{bmatrix} h_1^2 & 0 & 0 \\ 0 & h_2^2 & 0 \\ 0 & 0 & h_3^2 \end{bmatrix} \tag{1.5.18}$$

where $h_1 = \sqrt{g_{11}},\ h_2 = \sqrt{g_{22}},\ h_3 = \sqrt{g_{33}}$, and $\det g_{ij} = g = g_{11}g_{22}g_{33}$.

Exercise 1.1 Let $g_{ij}x^i x^j = 0, \forall x^i, x^j$ show that $g_{k\ell} + g_{\ell k} = 0$.

Putting

$$\Phi = g_{ij}x^i x^j = 0$$

and differentiating with respect to x^k

$$\frac{\partial \Phi}{\partial x^k} = g_{ij}\frac{\partial x^i}{\partial x^k}x^j + g_{ij}x^i\frac{\partial x^j}{\partial x^k} = g_{ij}\delta^i_k x^j + g_{ij}\delta^j_k x^i = g_{kj}x^j + g_{ik}x^i = 0$$

Differentiating with respect to x^ℓ

$$\frac{\partial^2 \Phi}{\partial x^k \partial x^\ell} = g_{kj}\frac{\partial x^j}{\partial x^\ell} + g_{ik}\frac{\partial x^i}{\partial x^\ell} = g_{kj}\delta^j_\ell + g_{ik}\delta^i_\ell = g_{k\ell} + g_{\ell k} = 0 \qquad Q.E.D.$$

Exercise 1.2 Calculate the length of the curve of class C^2 given by the parametric equations $x^1 = 3 - t$, $x^2 = 6t + 3$, and $x^3 = \ell n\, t$, in the space defined by the metric tensor

$$g_{ij} = \begin{bmatrix} 12 & 4 & 0 \\ 4 & 1 & 1 \\ 0 & 1 & (x^1)^2 \end{bmatrix}$$

The metric of the space in matrix form stays

$$\left(\frac{ds}{dt}\right)^2 = \left\{\frac{dx^k}{dt}\right\}^{\mathrm{T}} \left[g_{ij}\right] \left\{\frac{dx^\ell}{dt}\right\}$$

and with the derivatives

$$\frac{dx^1}{dt} = -1 \quad \frac{dx^2}{dt} = 6 \quad \frac{dx^3}{dt} = \frac{1}{t}$$

it follows

$$\left(\frac{ds}{dt}\right)^2 = \left\{-1; \quad 6; \quad \frac{1}{t}\right\} \begin{bmatrix} 12 & 4 & 0 \\ 4 & 1 & 1 \\ 0 & 1 & (3-t)^2 \end{bmatrix} \left\{\begin{matrix} -1 \\ 6 \\ \frac{1}{t} \end{matrix}\right\} = \frac{(t+3)^2}{t^2}$$

Making $h_1 = 1$ in expression

$$s = \int_a^b \sqrt{h_i g_{k\ell} \frac{dx^k}{dt} \frac{dx^\ell}{dt}} dt \Rightarrow L = \int_1^e \left(\frac{t+3}{t}\right) dt = e + 2$$

1.5.1 Conjugated Tensor

Let the increment of the position vector expressed by means of their covariant components $\boldsymbol{dr} = r_i \boldsymbol{g}^i$ in any coordinate system, where $\boldsymbol{g}^j$ is the basis vector of this referential and with the dot product

$$\boldsymbol{dr} \cdot \boldsymbol{dr} = dx_i dx_j \left(\boldsymbol{g}^i \boldsymbol{g}^j\right) \tag{1.5.19}$$

and

$$\left(\boldsymbol{g}^i \boldsymbol{g}^j\right) = g^{ij} = g^{ji} \tag{1.5.20}$$

whose symmetry comes from the commutative property of the dot product.

This variety with properties analogous to the properties of the metric tensor is represented by nine components of a symmetrical matrix 3×3, which form a second-order contravariant tensor. It is called conjugated metric tensor; thus,

$$g^{ij} = \begin{bmatrix} g^{11} & g^{12} & g^{13} \\ g^{21} & g^{22} & g^{23} \\ g^{31} & g^{32} & g^{33} \end{bmatrix} \tag{1.5.21}$$

The definition of the conjugated of the metric tensor is given by

$$g^{ij} = \boldsymbol{g}^i \boldsymbol{g}^j \tag{1.5.22}$$

and with the relations between the reciprocal basis

$$\boldsymbol{g}^i = \frac{\boldsymbol{g}_k \times \boldsymbol{g}_\ell}{V} \quad \boldsymbol{g}^j = \frac{\boldsymbol{g}_m \times \boldsymbol{g}_n}{V} \tag{1.5.23}$$

results for the conjugated metric tensor

$$g^{ij} = \frac{1}{V^2} (\boldsymbol{g}_k \times \boldsymbol{g}_\ell)(\boldsymbol{g}_m \times \boldsymbol{g}_n) \tag{1.5.24}$$

but with the fundamental formula of the vectorial algebra

$$(\boldsymbol{g}_k \times \boldsymbol{g}_\ell) \cdot (\boldsymbol{g}_m \times \boldsymbol{g}_n) = [(\boldsymbol{g}_k \times \boldsymbol{g}_\ell) \times \boldsymbol{g}_m] \cdot \boldsymbol{g}_n \tag{1.5.25}$$

and developing the double-cross product in brackets

$$[(\boldsymbol{g}_k \times \boldsymbol{g}_\ell) \times \boldsymbol{g}_m] \cdot \boldsymbol{g}_n = (\boldsymbol{g}_k \cdot \boldsymbol{g}_m)\boldsymbol{g}_\ell - (\boldsymbol{g}_\ell \cdot \boldsymbol{g}_m)\boldsymbol{g}_k \tag{1.5.26}$$

So

$$g^{ij} = \frac{1}{V^2} [(\boldsymbol{g}_k \cdot \boldsymbol{g}_m)(\boldsymbol{g}_\ell \cdot \boldsymbol{g}_n) - (\boldsymbol{g}_\ell \cdot \boldsymbol{g}_m)(\boldsymbol{g}_k \cdot \boldsymbol{g}_n)] \tag{1.5.27}$$

The term in brackets in expression (1.5.27) is the development of the determinant

$$G^{ij} = \begin{vmatrix} \boldsymbol{g}_k \cdot \boldsymbol{g}_m & \boldsymbol{g}_k \cdot \boldsymbol{g}_n \\ \boldsymbol{g}_\ell \cdot \boldsymbol{g}_m & \boldsymbol{g}_\ell \cdot \boldsymbol{g}_n \end{vmatrix} = \begin{vmatrix} g_{km} & g_{kn} \\ g_{\ell m} & g_{\ell n} \end{vmatrix} \tag{1.5.28}$$

Then

$$g^{ij} = \frac{G_{ij}}{V^2} \tag{1.5.29}$$

Summarizing these analyses by means of the transcription of the following expressions

$$g_{ij} = \frac{G_{ij}}{\overline{V}^2} = \frac{G_{ij}}{\overline{g}} \qquad g^{ij} = \frac{G^{ij}}{V^2} = \frac{G^{ij}}{g} \tag{1.5.30}$$

Thus

$$V = \pm\sqrt{g} = \pm\sqrt{\det g^{ij}} \qquad \overline{V} = \pm\sqrt{\overline{g}} = \pm\sqrt{\det g_{ij}} \tag{1.5.31}$$

The sign $(+)$ in expressions (1.5.31) corresponds to a levorotatory coordinates, and the sign $(-)$ corresponds to a dextrorotatory coordinates. Knowing that $V\overline{V} = 1$, it follows that $g\overline{g} = 1$.

Exercise 1.3 Let $\det g_{ij}(x^n) = g(x^n)$. Calculate the derivative $\frac{\partial g}{\partial x^n}$, $n = 1, 2, \ldots$.

The matrix linked to the determinant g is a function of the variables x^n:

$$g_{ij} = g_{ij}\left(x^i\right)$$

and this determinant being a function of the matrix elements

$$g = g\left(g_{ij}\right)$$

by the chain rule

$$\frac{\partial g}{\partial x^n} = \frac{\partial g}{\partial g_{ij}} \frac{\partial g_{ij}}{\partial x^n}$$

As det g is expressed by its cofactors

$$g = g_{1k} G_{k1} = g_{11} G_{11} + g_{12} G_{21} + g_{13} G_{31} + \cdots$$

and the terms G_{k1} do not contain the terms g_{1k}, so

$$\frac{\partial g}{\partial g_{11}} = G_{11} \quad \frac{\partial g}{\partial g_{12}} = G_{21} \quad \frac{\partial g}{\partial g_{13}} = G_{31} \quad \cdots$$

Generalizing provides

$$\frac{\partial g}{\partial g_{ij}} = G_{ji}$$

By substitution

$$\frac{\partial g}{\partial x^n} = G_{ji}\frac{\partial g_{ij}}{\partial x^n}$$

Exercise 1.4 Calculate the derivative of $\det g = \begin{vmatrix} x^1x^2 & (x^1)^2 \\ (x^1)^2 & 2x^1 \end{vmatrix}$ with respect to the variable x^1.

From Exercise 1.3

$$\frac{\partial g}{\partial x^i} = G_{ji}\frac{\partial g_{ij}}{\partial x^i}$$

This expression is the sum of n determinants. Each of these determinants differs from the determinant g only in the lines and columns which are being differentiated, so

$$\frac{\partial g}{\partial x^1} = \begin{vmatrix} \frac{\partial g_{11}}{\partial x^1} & \frac{\partial g_{12}}{\partial x^1} \\ g_{21} & g_{22} \end{vmatrix} + \begin{vmatrix} g_{11} & g_{12} \\ \frac{\partial g_{21}}{\partial x^1} & \frac{\partial g_{22}}{\partial x^1} \end{vmatrix} = \begin{vmatrix} x^2 & 2x^1 \\ (x^1)^2 & 2x^1 \end{vmatrix} + \begin{vmatrix} x^1x^2 & (x^1)^2 \\ 2x^1 & 2 \end{vmatrix}$$

Exercise 1.5 Let $g = \det g_{ij}$ the determinant of the metric tensor g_{ij} and x^k an arbitrary variable. Calculate *(a)* $\frac{\partial(\ell n\, g)}{\partial g_{ij}}$ *and (b)* $\frac{\partial(\ell n\, g)}{\partial x^k}$.

(a) From Exercise 1.3

$$\frac{\partial g}{\partial g_{ij}} = G_{ji}$$

but as $g_{ij} = g_{ji}$ it follows that

$$\frac{\partial g}{\partial g_{ji}} = G_{ij}$$

Expression (1.5.30) provides

$$g^{ij} = \frac{G^{ij}}{g}$$

$$g_{ij} = \frac{G_{ij}}{\overline{g}} \Rightarrow G_{ij} = \overline{g}g_{ij}$$

By substitution

$$\frac{\partial g}{\partial g_{ji}} = \overline{g}g_{ij} = gg^{ij}$$

whereby

$$\frac{\partial(\ell n g)}{\partial g_{ij}} = \frac{1}{g}\frac{\partial g}{\partial g_{ij}} \Rightarrow \frac{\partial(\ell n\, g)}{\partial g_{ij}} = g^{ij}$$

(b) By the chain rule

$$\frac{\partial(\ell n\, g)}{\partial x^k} = \frac{\partial(\ell n\, g)}{\partial g_{ij}}\frac{\partial g_{ij}}{\partial x^k}$$

and substituting the result obtained in the previous item in this expression

$$\frac{\partial(\ell n\, g)}{\partial x^k} = g^{ij}\frac{\partial g_{ij}}{\partial x^k}$$

Exercise 1.6 Calculate the metric tensor, its conjugated tensor, and the metric for the Cartesian coor*dinate system.*

Let the Cartesian coordinates (x^1, x^2, x^3), and by the definition of the distance between two points

$$ds^2 = \left(dx^1\right)^2 + \left(dx^2\right)^2 + \left(dx^3\right)^2$$

which is the square of the metric, thus

$$ds^2 = \delta_{ij}dx^i dx^j$$

By the definition of the metric tensor and the conjugated metric tensor, then

$$g_{ij} = \delta_{ij} = \begin{bmatrix} 1 & 0 & 0 \\ 0 & 1 & 0 \\ 0 & 0 & 1 \end{bmatrix}$$

$$g^{ij} = \frac{1}{g_{ij}} = \begin{bmatrix} 1 & 0 & 0 \\ 0 & 1 & 0 \\ 0 & 0 & 1 \end{bmatrix}$$

Exercise 1.7 Calculate the metric tensor, its conjugated tensor, and the metric for the cylindrical coordinate system given by $r \equiv \overline{x}^1$, $\theta \equiv \overline{x}^2$, and $z \equiv \overline{x}^3$ where $-\infty \le r \le \infty, 0 \le \theta \le 2\pi$, and $-\infty \le z \le \infty$, which relations with the Cartesian coordinates are $x^1 = \overline{x}^1 \cos \overline{x}^2$, $x^2 = \overline{x}^1 \sin \overline{x}^2$, and $x^3 \equiv \overline{x}^3$.

With the definition of metric tensor

$$g_{ij} = \frac{\partial x^k}{\partial \bar{x}^i}\frac{\partial x^k}{\partial \bar{x}^j} = \frac{\partial x^1}{\partial \bar{x}^i}\frac{\partial x^1}{\partial \bar{x}^j} + \frac{\partial x^2}{\partial \bar{x}^i}\frac{\partial x^2}{\partial \bar{x}^j} + \frac{\partial x^3}{\partial \bar{x}^i}\frac{\partial x^3}{\partial \bar{x}^j}$$

- $i = j = 1$

$$g_{11} = \frac{\partial x^1}{\partial \bar{x}^1}\frac{\partial x^1}{\partial \bar{x}^1} + \frac{\partial x^2}{\partial \bar{x}^1}\frac{\partial x^2}{\partial \bar{x}^1} + \frac{\partial x^3}{\partial \bar{x}^1}\frac{\partial x^3}{\partial \bar{x}^1} = \left(\cos\bar{x}^2\right)^2 + \left(\sin\bar{x}^2\right)^2 + 0 = 1$$

- $i = j = 2$

$$g_{22} = \frac{\partial x^1}{\partial \bar{x}^2}\frac{\partial x^1}{\partial \bar{x}^2} + \frac{\partial x^2}{\partial \bar{x}^2}\frac{\partial x^2}{\partial \bar{x}^2} + \frac{\partial x^3}{\partial \bar{x}^2}\frac{\partial x^3}{\partial \bar{x}^2} = \left(-\bar{x}^1\sin\bar{x}^2\right)^2 + \left(\bar{x}^1\cos\bar{x}^2\right)^2 + 0 = \left(\bar{x}^1\right)^2$$

- $i = j = 3$

$$g_{33} = \frac{\partial x^1}{\partial \bar{x}^3}\frac{\partial x^1}{\partial \bar{x}^3} + \frac{\partial x^2}{\partial \bar{x}^3}\frac{\partial x^2}{\partial \bar{x}^3} + \frac{\partial x^3}{\partial \bar{x}^3}\frac{\partial x^3}{\partial \bar{x}^3} = 0 + 0 + 1 = 1$$

- $i = 1, j = 2$

$$\begin{aligned} g_{12} &= \frac{\partial x^1}{\partial \bar{x}^1}\frac{\partial x^1}{\partial \bar{x}^2} + \frac{\partial x^2}{\partial \bar{x}^1}\frac{\partial x^2}{\partial \bar{x}^2} + \frac{\partial x^3}{\partial \bar{x}^1}\frac{\partial x^3}{\partial \bar{x}^2} \\ &= \cos\bar{x}^2\left(-\bar{x}^1\sin\bar{x}^2\right)^2 + \sin\bar{x}^2\left(\bar{x}^1\cos\bar{x}^2\right)^2 + 0 = 0 \end{aligned}$$

For the other terms $g_{21} = g_{13} = g_{31} = g_{23} = g_{32} = 0$, then

$$\begin{aligned} ds^2 &= g_{11}dx^1dx^1 + g_{22}dx^2dx^2 + g_{11}dx^2dx^2 = \left(d\bar{x}^1\right)^2 + \left(\bar{x}^1\right)^2\left(d\bar{x}^2\right)^2 + \left(d\bar{x}^3\right)^2 \\ &= (dr)^2 + (rd\theta)^2 + (dz)^2 \end{aligned}$$

The metric tensor and its conjugated tensor are given, respectively, by

$$g_{ij} = \begin{bmatrix} 1 & 0 & 0 \\ 0 & r^2 & 0 \\ 0 & 0 & 1 \end{bmatrix} \quad g^{ij} = \frac{1}{g_{ij}} = \begin{bmatrix} 1 & 0 & 0 \\ 0 & \frac{1}{r^2} & 0 \\ 0 & 0 & 1 \end{bmatrix}$$

and with the base vectors

$$\mathbf{g}_i = \frac{\partial x^j}{\partial \bar{x}^i}\mathbf{e}_j$$

$$\begin{cases} i = 1 \Rightarrow \boldsymbol{g}_1 = \dfrac{\partial x^j}{\partial \overline{x}^1} \boldsymbol{e}_j \\ j = 1,2,3 \Rightarrow \boldsymbol{g}_1 = \dfrac{\partial x^1}{\partial \overline{x}^1} \boldsymbol{e}_1 + \dfrac{\partial x^2}{\partial \overline{x}^1} \boldsymbol{e}_2 \dfrac{\partial x^3}{\partial \overline{x}^1} \boldsymbol{e}_3 \\ \boldsymbol{g}_1 = \cos \overline{x}^2 \boldsymbol{e}_1 + \sin \overline{x}^2 \boldsymbol{e}_2 \end{cases}$$

$$\begin{cases} i = 2 \Rightarrow \boldsymbol{g}_2 = \dfrac{\partial x^j}{\partial \overline{x}^2} \boldsymbol{e}_j \\ j = 1,2,3 \Rightarrow \boldsymbol{g}_2 = \dfrac{\partial x^1}{\partial \overline{x}^2} \boldsymbol{e}_1 + \dfrac{\partial x^2}{\partial \overline{x}^2} \boldsymbol{e}_2 \dfrac{\partial x^3}{\partial \overline{x}^2} \boldsymbol{e}_3 \\ \boldsymbol{g}_2 = -\overline{x}^1 \sin \overline{x}^2 \boldsymbol{e}_1 + \overline{x}^1 \cos \overline{x}^2 \boldsymbol{e}_2 \end{cases}$$

$$\begin{cases} i = 3 \Rightarrow \boldsymbol{g}_3 = \dfrac{\partial x^j}{\partial \overline{x}^3} \boldsymbol{e}_j \\ j = 1,2,3 \Rightarrow \boldsymbol{g}_3 = \dfrac{\partial x^1}{\partial \overline{x}^3} \boldsymbol{e}_1 + \dfrac{\partial x^2}{\partial \overline{x}^3} \boldsymbol{e}_2 \dfrac{\partial x^3}{\partial \overline{x}^3} \boldsymbol{e}_3 \\ \boldsymbol{g}_3 = 0 + 0 + 1 \cdot \boldsymbol{e}_3 = \boldsymbol{e}_3 \end{cases}$$

By means of the dot products

$$\boldsymbol{g}_i \cdot \boldsymbol{g}_j = \delta_{ij} \quad \boldsymbol{e}_i \cdot \boldsymbol{e}_j = \delta_{ij}$$

it follows for the components of the metric tensor

$$g_{11} = \boldsymbol{g}_1 \cdot \boldsymbol{g}_1 = \left(\cos \overline{x}^2 \boldsymbol{e}_1 + \sin \overline{x}^2 \boldsymbol{e}_2\right) \cdot \left(\cos \overline{x}^2 \boldsymbol{e}_1 + \sin \overline{x}^2 \boldsymbol{e}_2\right) = 1$$
$$g_{22} = \boldsymbol{g}_2 \cdot \boldsymbol{g}_2 = \left(-\overline{x}^1 \sin \overline{x}^2 \boldsymbol{e}_1 + \overline{x}^1 \cos \overline{x}^2 \boldsymbol{e}_2\right) \cdot \left(-\overline{x}^1 \sin \overline{x}^2 \boldsymbol{e}_1 + \overline{x}^1 \cos \overline{x}^2 \boldsymbol{e}_2\right) = \left(\overline{x}^2\right)^2$$
$$g_{33} = \boldsymbol{g}_3 \cdot \boldsymbol{g}_3 = (\boldsymbol{e}_3) \cdot (\boldsymbol{e}_3) = 1$$

The other components of this tensor are null.

Exercise 1.8 Calculate the metric tensor, its conjugated tensor, and the metric for the spherical coordinate system $r \equiv \overline{x}^1, \varphi \equiv \overline{x}^2, \theta \equiv \overline{x}^3$, $-\infty \leq r \leq \infty$, and $0 \leq \varphi \leq \pi$, where $0 \leq \theta \leq 2\pi$, which relations with the Cartesian coordinates are $x^1 = \overline{x}^1 \sin \overline{x}^2 \cos \overline{x}^3$, $x^2 = \overline{x}^1 \sin \overline{x}^2 \sin \overline{x}^3$,and $x^3 \equiv x^1 \cos \overline{x}^2$.

With the definition of metric tensor

$$g_{ij} = \frac{\partial x^k}{\partial \overline{x}^i} \frac{\partial x^k}{\partial \overline{x}^j} \Rightarrow g_{ij} = \frac{\partial x^1}{\partial \overline{x}^i} \frac{\partial x^1}{\partial \overline{x}^j} + \frac{\partial x^2}{\partial \overline{x}^i} \frac{\partial x^2}{\partial \overline{x}^j} + \frac{\partial x^3}{\partial \overline{x}^i} \frac{\partial x^3}{\partial \overline{x}^j}$$

– $i = j = 1$

$$g_{11} = \frac{\partial x^1}{\partial \overline{x}^1}\frac{\partial x^1}{\partial \overline{x}^1} + \frac{\partial x^2}{\partial \overline{x}^1}\frac{\partial x^2}{\partial \overline{x}^1} + \frac{\partial x^3}{\partial \overline{x}^1}\frac{\partial x^3}{\partial \overline{x}^1}$$
$$= \left(\sin\overline{x}^2\cos\overline{x}^3\right)^2 + \left(\sin\overline{x}^2\sin\overline{x}^3\right)^2 + \left(\cos\overline{x}^2\right)^2 = 1$$

- $i = j = 2$

$$g_{22} = \frac{\partial x^1}{\partial \overline{x}^2}\frac{\partial x^1}{\partial \overline{x}^2} + \frac{\partial x^2}{\partial \overline{x}^2}\frac{\partial x^2}{\partial \overline{x}^2} + \frac{\partial x^3}{\partial \overline{x}^2}\frac{\partial x^3}{\partial \overline{x}^2}$$
$$= \left(\overline{x}^1\cos\overline{x}^2\cos\overline{x}^3\right)^2 + \left(\overline{x}^1\cos\overline{x}^2\sin\overline{x}^3\right)^2 + \left(-\overline{x}^1\sin\overline{x}^2\right)^2 = \left(\overline{x}^1\right)^2$$

- $i = j = 3$

$$g_{33} = \frac{\partial x^1}{\partial \overline{x}^3}\frac{\partial x^1}{\partial \overline{x}^3} + \frac{\partial x^2}{\partial \overline{x}^3}\frac{\partial x^2}{\partial \overline{x}^3} + \frac{\partial x^3}{\partial \overline{x}^3}\frac{\partial x^3}{\partial \overline{x}^3}$$
$$= \left(-\overline{x}^1\sin\overline{x}^2\sin\overline{x}^3\right)^2 + \left(\overline{x}^1\sin\overline{x}^2\cos\overline{x}^3\right)^2 + 0 = \left(\overline{x}^1\sin\overline{x}^2\right)^2$$

For the other terms $g_{12} = g_{21} = g_{13} = g_{31} = g_{23} = g_{32} = 0$, then

$$ds^2 = g_{11}dx^1dx^1 + g_{22}dx^2dx^2 + g_{33}dx^3dx^3 = \left(d\overline{x}^1\right)^2 + \left(\overline{x}^3\right)^2\left(d\overline{x}^2\right)^2 + \left(\overline{x}^1\sin\overline{x}^2\right)^2\left(d\overline{x}^3\right)^2$$
$$= (dr)^2 + (rd\varphi)^2 + (r\sin\theta\, d\theta)^2$$

The metric tensor and its conjugated tensor are given, respectively, by

$$g_{ij} = \begin{bmatrix} 1 & 0 & 0 \\ 0 & r^2 & 0 \\ 0 & 0 & r^2\sin^2\varphi \end{bmatrix} \qquad g^{ij} = \frac{1}{g_{ij}} = \begin{bmatrix} 1 & 0 & 0 \\ 0 & \dfrac{1}{r^2} & 0 \\ 0 & 0 & \dfrac{1}{r^2\sin^2\varphi} \end{bmatrix}$$

Exercise 1.9 Calculate the metric tensor, its conjugated tensor, and the metric for the cylindrical elliptical coordinate system $\xi \equiv \overline{x}^1, \eta \equiv \overline{x}^2$, and $z \equiv \overline{x}^3$, where $\xi \geq 0$, $0 \leq \eta \leq 2\pi$, $-\infty \leq z \leq \infty$, which relations with the Cartesian coordinates are $x^1 = \cosh\overline{x}^2\cos\overline{x}^2$, $x^2 = \sinh\overline{x}^2\sin\overline{x}^2$, $x^3 \equiv \overline{x}^3$.

With $\overline{x}^3 = $ const., the elliptical cylinder is $\overline{x}_0^1 = $ const.:

$$\left(\frac{x^1}{ch\overline{x}_0^1}\right)^2 + \left(\frac{x^2}{sh\overline{x}_0^1}\right)^2 = \left(\cos\overline{x}^2\right)^2 + \left(\sin\overline{x}^2\right)^2 = 1$$

$$dx^1 = \sinh\overline{x}^1\cos\overline{x}^2 d\overline{x}^1 \quad dx^2 = \cosh\overline{x}^1\sin\overline{x}^2 d\overline{x}^1$$
$$ds = \sqrt{(dx^1)^2 + (dx^2)^2} = \left(\cosh^2\overline{x}^1 - \cos^2\overline{x}^2\right)d\overline{x}^1$$
$$g_{11} = \left(\cosh^2\overline{x}^1 - \cos^2\overline{x}^2\right)$$

With $\overline{x}^1$ = const. the hyperbolic cylinder is $\overline{x}_0^2$ = const.:

$$\left(\frac{x^1}{\cos\overline{x}_0^2}\right)^2 - \left(\frac{x^2}{\sin\overline{x}_0^2}\right)^2 = \left(\cosh\overline{x}^1\right)^2 - \left(\sinh\overline{x}^1\right)^2 = 1$$

$$dx^1 = -\cosh\overline{x}^1 \sin\overline{x}^2 d\overline{x}^2 \quad dx^2 = \sinh\overline{x}^1 \cos\overline{x}^2 d\overline{x}^2$$

$$ds = \sqrt{(dx^1)^2 + (dx^2)^2} = \left(\cosh^2\overline{x}^1 - \cos^2\overline{x}^2\right) d\overline{x}^2$$

$$g_{22} = \left(\cosh^2\overline{x}^1 - \cos^2\overline{x}^2\right)$$

For $x^3 \equiv \overline{x}^3$ provides $dx^2 = d\overline{x}^3$, whereby $g_{33} = 1$, following

$$ds^2 = \left(\cosh^2\overline{x}^1 - \cos^2\overline{x}^2\right)^2 \left(d\overline{x}^1\right)^2 + \left(\cosh^2\overline{x}^1 - \cos^2\overline{x}^2\right)^2 \left(d\overline{x}^2\right)^2 + \left(d\overline{x}^3\right)^2$$

The metric tensor and its conjugated tensor are given, respectively, by

$$g_{ij} = \begin{bmatrix} \cosh^2\overline{x}^1 - \cos^2\overline{x}^2 & 0 & 0 \\ 0 & \cosh^2\overline{x}^1 - \cos^2\overline{x}^2 & 0 \\ 0 & 0 & 1 \end{bmatrix}$$

$$g^{ij} = \begin{bmatrix} \dfrac{1}{\cosh^2\overline{x}^1 - \cos^2\overline{x}^2} & 0 & 0 \\ 0 & \dfrac{1}{\cosh^2\overline{x}^1 - \cos^2\overline{x}^2} & 0 \\ 0 & 0 & 1 \end{bmatrix}$$

1.5.2 Dot Product in Metric Spaces

Let the vectors $\boldsymbol{u}$ and $\boldsymbol{v}$ contained in the metric space E_N defined by the fundamental tensor $g_{k\ell}$. The dot product $\boldsymbol{u} \cdot \boldsymbol{v}$ with $\boldsymbol{u} = u^i \boldsymbol{e}_i$ and $\boldsymbol{v} = v^j \boldsymbol{e}_j$ depends only on the vectors and is independent of the coordinate system in relation to which the same is specified. It is observed that only when the coordinates of the vectors are covariant and contravariant, this product is like to the dot product in Cartesian coordinates. The dot product is invariant in view of the transformation of coordinates

$$\begin{aligned} \boldsymbol{u} \cdot \boldsymbol{v} &= u^i \boldsymbol{e}_i \cdot v^j \boldsymbol{e}_j = g_{ij} u^i v^j = u_i \boldsymbol{e}^i . v_j \boldsymbol{e}^j = g^{ij} u_i v_j = u^i \boldsymbol{e}_i . v_j \boldsymbol{e}^j \\ &= g_i^j u^i v_j = u^i v_j = u_i v^j \end{aligned} \tag{1.5.32}$$

1.5.2.1 Vector Norm

The generalization of the dot product of vectors for a metric space E_N allows obtaining the norm of a vector. Let vector $\boldsymbol{v}$ with norm (modulus)

$$\|\boldsymbol{v}\| = \sqrt{\boldsymbol{v} \cdot \boldsymbol{v}} = \sqrt{v^2}$$

that is equal to the distance between the extreme points, thus, with the expression of the metrics

$$v^2 = h_i g_{k\ell} v^k v^\ell$$

results for the norm of the vector in terms of its contravariant components

$$\|\boldsymbol{v}\| = \sqrt{h_i g_{kk} v^k v^k}$$

In an analogous way for the covariant components v_k

$$\|\boldsymbol{v}\| = \sqrt{h_i g^{kk} v_k v_k}$$

and for Cartesian coordinates

$$\|\boldsymbol{v}\| = \sqrt{v^k v^k}$$

If $\boldsymbol{v}$ is a unit vector, the expressions provide

$$h_i g_{kk} v^k v^k = 1 \quad h_i g^{kk} v_k v_k = 1 \quad v^k v^k = 1$$

The properties of the vectors norm are:

(a) $\|\boldsymbol{v}\| \geq 0$, which is a trivial property, for the norm will only cancel itself if $\boldsymbol{v}$ is null.
(b) $\|m\boldsymbol{v}\| = \|m\| \|\boldsymbol{v}\|$, where m is a scalar.
(c) $\|\boldsymbol{u} + \boldsymbol{v}\| \leq \|\boldsymbol{u}\| + \|\boldsymbol{v}\|$.
(d) $\|\boldsymbol{u} \cdot \boldsymbol{v}\| \leq \|\boldsymbol{u}\| \cdot \|\boldsymbol{v}\|$, Cauchy–Schwarz inequality.

For the case of non-null vectors, the equality of the relation (d) exists only if $\boldsymbol{u} = m\boldsymbol{v}$, where m is a scalar.

Exercise 1.10 Calculate the modulus of vector $\boldsymbol{u}(1; 1; 0; 2)$ in space E_4, defined by the metric tensor

$$g_{ij} = \begin{bmatrix} -1 & 0 & 0 & 0 \\ 0 & -1 & 0 & 0 \\ 0 & 0 & -1 & 0 \\ 0 & 0 & 0 & c^2 \end{bmatrix}$$

For the line element $ds^2 = g_{ij}u^i u^j$, and developing this expression

$$ds^2 = g_{ij}u^i u^j = g_{11}u^1 u^1 + g_{22}u^2 u^2 + g_{33}u^3 u^3 + g_{44}u^4 u^4$$
$$= -1 \times 1 \times 1 - 1 \times 1 \times 1 + 0 + c^2 \times 2 \times 2 = -2 + 4c^2$$
$$ds = \sqrt{2(2c^2 - 1)}$$

1.5.2.2 Lowering of a Tensor's Indexes

By means of analysis referent to the transformation of the covariant components of the vector in their contravariant components, and vice versa, it is verified that inner product of a tensor by the metric tensor allows raising or lowering the indexes of this tensor.

For multiplying the contravariant tensor of the first order, i.e., the contravariant vector T^i by the tensor $g_{k\ell}$, results in $T^i_{k\ell} = g_{k\ell}T^i$, and for the contraction $i = \ell$, then $T^i_{ki} = g_{ki}T^i = T_k$ that is a covariant vector. The index of the original vector was lowered and its order reduced in two units.

1.5.2.3 Raising of a Tensor's Indexes

Let the covariant vector T_k, which multiplied by g^{ik}, provides as a result the tensor $g^{ik}T_k$, and changing the covariant coordinates of the vector by its contravariant coordinates

$$g^{ik}T_k = g^{ik}\left(g_{k\ell}T^\ell\right) = \delta^i_\ell T^\ell = T^i$$

that is a contravariant vector. The index of the original vector was raised. Then a covariant vector is obtained by means of the inner product of a contravariant vector, this indexes transformation process as being reciprocal. The vectors T^i and T_i are called associated vectors, and it refers to the contravariant and covariant components of the vector.

For the case of second-order tensors, an analysis is carried out that is analogous to the one developed for the vectors. Let the covariant tensor of the second-order $T_{k\ell}$ and its associated tensor $T^{ij} = g^{ik}g^{j\ell}T_{k\ell}$.

It is verified in the general case that these tensors are not conjugated tensors, for example, when $T_{ij} = mg_{ij}$, where m is a scalar, the tensor $T_{k\ell}$ will be a multiple of $g^{k\ell}$,

$$T_{k\ell} = g^{ik}g^{j\ell}T_{ij} = g^{ik}g^{j\ell}\left(m\,g_{ij}\right) = mg^{ik}\delta^\ell_i = m\,g^{\ell k} = mg^{k\ell}$$

The raising and lowering operations of the indexes of tensors are carried out adopting, firstly, a point for indicating where the position to be left empty in the index that will be raised or lowered. For example, for the tensor T_i^j, the empty position is indicated by means of the notation $T_{\bullet\, i}^{j}$, and in an equal manner $A_{\bullet\bullet p}^{rs}$ exists for the tensor A_p^{rs}.

Let the inner product of the tensor $T_{\bullet\, jk}^{i}$ by the metric tensor $g_{\ell i}$, $g_{\ell i} T_{\bullet jk}^{i} = T_{\ell jk}$ in which the upper index was lowered, and $g_{ij} T^{kj} = T_{\bullet i}^{k}$ that had an index lowered, or further, $g_{ij} g_{k\ell} T^{j\ell} = T_{ik}$, which two upper indexes were lowered.

For raising the indexes, in an analogous way to the raising of an index $g^{ij} T_{jk} = T_{\bullet k}^{i}$ or $g^{kj} T_{ij} = T_{i}^{\bullet k}$, thus $g^{ij} g^{k\ell} T_{j\ell} = T^{ik}$.

In the case in which the index is lowered and then raised, the original tensor is obtained $g_{kj} T^{ij} = T_{\bullet k}^{i}$ and next $g^{kj} T_{\bullet k}^{i} = T^{ij}$.

1.5.2.4 Tensorial Equation

If a term of a tensorial equation contains a dummy index, it can be raised or lowered, i.e., change the position without changing the value of the equation. The following example illustrates this assertion

$$A_{\bullet j}^{i} B_i = \left(g^{ki} A_{kj}\right)\left(g_{i\ell} B^{\ell}\right) = g^{ki} g_{i\ell} A_{kj} B^{\ell} = \delta_{\ell}^{k} A_{kj} B^{\ell} = A_{\ell j} B^{\ell} = A_{ij} B^{i}$$

where the index i was lowered in one tensor and raised in the tensor.

If a free index is a part of the tensorial expression, a new tensorial expression equivalent to this one can be obtained, lowering or raising this index in the members of the original expression. To illustrate this assertion, the following tensorial equation is admitted

$$T_{ijk} = A_{ij} B_k$$

which is equivalent to

$$g^{i\ell} T_{\ell jk} = g^{i\ell} A_{\ell j} B_k$$

so it results in

$$T_{\bullet jk}^{i} = A_{\bullet j}^{i} B_k$$

where the index i was raised.

Exercise 1.11 Raise and lower the indexes of vector $\boldsymbol{u}$, for the metric tensor and its conjugated tensor:

$$(a)\ u^j = \begin{Bmatrix} 3 \\ 4 \\ 5 \end{Bmatrix} \quad g_{ij} = \begin{bmatrix} 1 & 0 & 0 \\ 0 & (x^1)^2 & 0 \\ 0 & 0 & (x^1 \sin x^2)^2 \end{bmatrix}$$

$$(b)\ u_j = \begin{Bmatrix} 5 \\ 4 \\ 3 \end{Bmatrix} \quad g_{ij} = \begin{bmatrix} 1 & 0 & 0 \\ 0 & (x^1)^{-2} & 0 \\ 0 & 0 & (x^1 \sin x^2)^{-2} \end{bmatrix}$$

(a) Carrying out the following matrix multiplication provides the covariant components of the vector

$$u_i = g_{ij} u^j = \begin{bmatrix} 1 & 0 & 0 \\ 0 & (x^1)^2 & 0 \\ 0 & 0 & (x^1 \sin x^2)^2 \end{bmatrix} \begin{Bmatrix} 3 \\ 4 \\ 5 \end{Bmatrix} = \begin{Bmatrix} 3 \\ 4(x^1)^2 \\ 5(x^1 \sin x^2)^2 \end{Bmatrix}$$

(b) Carrying out the following matrix multiplication provides the contravariant components of the vector

$$u^i = g_{ij} u_j = \begin{bmatrix} 1 & 0 & 0 \\ 0 & (x^1)^{-2} & 0 \\ 0 & 0 & (x^1 \sin x^2)^{-2} \end{bmatrix} \begin{Bmatrix} 5 \\ 4 \\ 3 \end{Bmatrix} = \begin{Bmatrix} 5 \\ 4\,(x^1)^{-2} \\ 3\,(x^1 \sin x^2)^{-2} \end{Bmatrix}$$

Exercise 1.12 Given the covariant basis $\boldsymbol{g}_1 = \boldsymbol{e}_1$; $\boldsymbol{g}_2 = \boldsymbol{e}_1 + \boldsymbol{e}_2$; $\boldsymbol{g}_3 = \boldsymbol{e}_3$ and the tensor of the space $g_{ij} = \begin{bmatrix} 1 & 1 & 1 \\ 1 & 2 & 2 \\ 1 & 2 & 3 \end{bmatrix}$, calculate the vectors of the contravariant basis and the conjugated metric tensor.

The determinant of the metric tensor is given by

$$\mathrm{g} = \det \boldsymbol{g}_i = \begin{vmatrix} 1 & 1 & 1 \\ 1 & 2 & 2 \\ 1 & 2 & 3 \end{vmatrix} = 1$$

which indicates that the system is dextrorotary.

For the vectors of the contravariant basis, it follows that

$$\boldsymbol{g}^1 = \frac{\boldsymbol{g}_2 \times \boldsymbol{g}_3}{\mathrm{g}} = \begin{vmatrix} \boldsymbol{e}_1 & \boldsymbol{e}_2 & \boldsymbol{e}_3 \\ 1 & 1 & 0 \\ 1 & 1 & 1 \end{vmatrix} = \boldsymbol{e}_1 - \boldsymbol{e}_2$$

$$\boldsymbol{g}^2 = \frac{\boldsymbol{g}_3 \times \boldsymbol{g}_1}{\mathrm{g}} = \begin{vmatrix} \boldsymbol{e}_1 & \boldsymbol{e}_2 & \boldsymbol{e}_3 \\ 1 & 1 & 1 \\ 1 & 0 & 0 \end{vmatrix} = \boldsymbol{e}_2 - \boldsymbol{e}_3$$

$$\mathbf{g}^3 = \frac{\mathbf{g}_1 \times \mathbf{g}_2}{\text{g}} = \begin{vmatrix} \mathbf{e}_1 & \mathbf{e}_2 & \mathbf{e}_3 \\ 1 & 0 & 0 \\ 1 & 1 & 0 \end{vmatrix} = \mathbf{e}_3$$

then the conjugated metric tensor is given by

$$g^{ij} = \mathbf{g}^i \cdot \mathbf{g}^j = \begin{bmatrix} 2 & -1 & 0 \\ -1 & 2 & -1 \\ 0 & -1 & 1 \end{bmatrix}$$

The verification of the operation is carried out by means of the expression $g_{ij}g^{ij} = \delta^i_j$, thus

$$\begin{bmatrix} 1 & 1 & 1 \\ 1 & 2 & 2 \\ 1 & 2 & 3 \end{bmatrix} \begin{bmatrix} 2 & -1 & 0 \\ -1 & 2 & -1 \\ 0 & -1 & 1 \end{bmatrix} = \begin{bmatrix} 1 & 0 & 0 \\ 0 & 1 & 0 \\ 0 & 0 & 1 \end{bmatrix}$$

1.5.2.5 Associated Tensors

The metric tensor g_{ij} and its conjugated tensor g^{ij} relate intrinsically to each other, which allows using them for analyzing the relations between the covariant and contravariant components of the vector $\boldsymbol{u}$

$$u_i = g_{ij}u^j \quad u^i = g^{ij}u_j \tag{1.5.33}$$

These expressions generate two linear equation systems, which unknown quantities are u_1, u_2, u_3, and u^1, u^2, u^3. The solution of the system given by means of Cramer's rule and with the determinant of the metric tensor

$$g = \det g_{ij} = \begin{vmatrix} g_{11} & g_{12} & g_{13} \\ g_{21} & g_{22} & g_{23} \\ g_{31} & g_{32} & g_{33} \end{vmatrix}$$

and its cofactor

$$G^{ij} = \begin{vmatrix} g_{km} & g_{kn} \\ g_{\ell m} & g_{\ell n} \end{vmatrix}$$

thus

$$u_i = g_{ij}u^j = \frac{G_{ij}u^j}{g} \Rightarrow g_{ij} = \frac{G_{ij}}{g}$$

In an analogous way for the system given by expression 1.5.33

$$u^i = g^{ij} u_j = \frac{G^{ij} u_j}{g} \Rightarrow g^{ij} = \frac{G^{ij} u_j}{g}$$

The linear operators g_{ij}, g^{ij} allow relating the covariant and contravariant components of vector $\boldsymbol{u}$. Defining this vector by means of its covariant components and performing the dot product of this vector by the basis unit vectors $\boldsymbol{g}_i$:

$$\boldsymbol{u} \cdot \boldsymbol{g}^i = \left(u_j \boldsymbol{g}^j\right) \cdot \boldsymbol{g}^i = u_j \left(\boldsymbol{g}^j \cdot \boldsymbol{g}^i\right) = g^{ij} u_j = u^i$$

that are the contravariant components of vector $\boldsymbol{u}$.

In an analogous way for the transformation of the contravariant components in the covariant components

$$\boldsymbol{u} \cdot \boldsymbol{g}_i = \left(u^i \boldsymbol{g}_j\right) \cdot \boldsymbol{g}_i = u^i \left(\boldsymbol{g}_j \cdot \boldsymbol{g}_i\right) = g_{ij} u^j = u_i$$

These expressions relate to each other in a kind of coordinate as a function of the other, where the tensors g_{ij}, g^{ij} are the operators responsible for these transformations.

The covariant components and the contravariant components of the vector $\boldsymbol{u}$ are given, respectively, by

$$u_i = \begin{cases} u_1 = g_{11} u^1 \\ u_2 = g_{22} u^2 \\ u_3 = g_{33} u^3 \end{cases} \quad u^i = \begin{cases} u^1 = g^{11} u_1 \\ u^2 = g^{22} u_2 \\ u^3 = g^{33} u_3 \end{cases} \Rightarrow \begin{cases} g_{11} g^{11} = 1 \\ g_{22} g^{22} = 1 \\ g_{33} g^{33} = 1 \end{cases}$$

The linear operators g_{ij}, g^{ij}, g_i^j are useful in the explanation of the more general properties of tensors.

Exercise 1.13 For the vector $\boldsymbol{v} = 4\boldsymbol{g}_1 + 3\boldsymbol{g}_2$ referenced to a coordinate system, calculate their contravariant and covariant components in the referential system that have the basis vectors $\overline{\boldsymbol{g}}_1 = 3\boldsymbol{g}_1$, $\overline{\boldsymbol{g}}_2 = 6\boldsymbol{g}_1 + 8\boldsymbol{g}_2$, and $\overline{\boldsymbol{g}}_3 = \boldsymbol{g}_3$.

The metric tensor of the space is given by

$$\overline{g}_{ij} = \overline{\boldsymbol{g}}_i \cdot \overline{\boldsymbol{g}}_j = \begin{bmatrix} \overline{\boldsymbol{g}}_1\overline{\boldsymbol{g}}_1 & \overline{\boldsymbol{g}}_1\overline{\boldsymbol{g}}_2 & \overline{\boldsymbol{g}}_1\overline{\boldsymbol{g}}_3 \\ \overline{\boldsymbol{g}}_2\overline{\boldsymbol{g}}_1 & \overline{\boldsymbol{g}}_2\overline{\boldsymbol{g}}_2 & \overline{\boldsymbol{g}}_2\overline{\boldsymbol{g}}_3 \\ \overline{\boldsymbol{g}}_3\overline{\boldsymbol{g}}_1 & \overline{\boldsymbol{g}}_3\overline{\boldsymbol{g}}_2 & \overline{\boldsymbol{g}}_3\overline{\boldsymbol{g}}_3 \end{bmatrix} = \begin{bmatrix} 9 & 18 & 0 \\ 18 & 100 & 0 \\ 0 & 0 & 1 \end{bmatrix} \Rightarrow \det \overline{g}_{ij} = 576$$

For the conjugated metric tensor

$$\overline{g}_{ij} \overline{g}^{ij} = \delta_{ij} \Rightarrow \overline{g}^{ij} = \left[\overline{g}_{ij}\right]^{-1} \Rightarrow \overline{g}^{ij} = \frac{1}{576} \begin{bmatrix} 100 & -18 & 0 \\ -18 & 9 & 0 \\ 0 & 0 & 576 \end{bmatrix}$$

The vectors of the contravariant basis are given by

$$\overline{\boldsymbol{g}}^i = \overline{g}^{ij}\overline{\boldsymbol{g}}_j$$

$$\overline{\boldsymbol{g}}^i = \begin{Bmatrix} \overline{\boldsymbol{g}}^1 \\ \overline{\boldsymbol{g}}^2 \\ \overline{\boldsymbol{g}}^3 \end{Bmatrix} = \frac{1}{576}\begin{bmatrix} 100 & -18 & 0 \\ -18 & 9 & 0 \\ 0 & 0 & 576 \end{bmatrix}\begin{Bmatrix} 3\boldsymbol{g}_1 \\ 6\boldsymbol{g}_1 + 8\boldsymbol{g}_2 \\ \boldsymbol{g}_3 \end{Bmatrix} = \begin{Bmatrix} \frac{1}{3}\boldsymbol{g}_1 - \frac{1}{4}\boldsymbol{g}_2 \\ \frac{1}{8}\boldsymbol{g}_2 \\ \boldsymbol{g}_3 \end{Bmatrix}$$

With the contravariant components of $\boldsymbol{v}$

$$\overline{v}^i = \boldsymbol{v} \cdot \overline{\boldsymbol{g}}^i$$

$$\overline{v}^1 = (4\boldsymbol{g}_1 + 3\boldsymbol{g}_2) \cdot \overline{\boldsymbol{g}}^1 = \frac{7}{12} \quad \overline{v}^2 = (4\boldsymbol{g}_1 + 3\boldsymbol{g}_2) \cdot \overline{\boldsymbol{g}}^2 = \frac{3}{8} \quad \overline{v}^3 = (4\boldsymbol{g}_1 + 3\boldsymbol{g}_2) \cdot \overline{\boldsymbol{g}}^3 = 0$$

so

$$\boldsymbol{v} = \frac{7}{12}\overline{\boldsymbol{g}}^1 + \frac{3}{8}\overline{\boldsymbol{g}}^2$$

With the covariant components of $\boldsymbol{v}$

$$v_i = \boldsymbol{v} \cdot \overline{\boldsymbol{g}}_i = (4\boldsymbol{g}_1 + 3\boldsymbol{g}_2) \cdot \overline{\boldsymbol{g}}_i$$

$$v_1 = (4\boldsymbol{g}_1 + 3\boldsymbol{g}_2).\overline{\boldsymbol{g}}_1 = 12 \quad v_2 = (4\boldsymbol{g}_1 + 3\boldsymbol{g}_2).\overline{\boldsymbol{g}}_2 = 48 \quad v_3 = (4\boldsymbol{g}_1 + 3\boldsymbol{g}_2) \cdot \overline{\boldsymbol{g}}_3 = 0$$

so

$$\boldsymbol{v} = 12\overline{\boldsymbol{g}}_1 + 48\overline{\boldsymbol{g}}_2$$

Exercise 1.14 Show that in the space E_N exists $\left(g_{\ell j}g_{ik} - g_{\ell i}g_{jk}\right)g^{\ell j} = (N-1)\,g_{ik}$, where g_{ij} is the metric tensor.

Developing the given expression

$$\begin{aligned}\left(g_{\ell j}g_{ik} - g_{\ell i}g_{jk}\right)g^{\ell j} &= g_{\ell j}g_{ik}g^{\ell j} - g_{\ell i}g_{jk}g^{\ell j} = g_{\ell j}g^{\ell j}g_{ik} - g_{\ell i}g^{\ell j}g_{jk} \\ &= g_{\ell j}g^{\ell j}g_{ik} - g_{\ell i}\delta_k^\ell = g_{\ell j}g^{\ell j}g_{ik} - g_{ki} = \delta_j^j g_{ik} - g_{ki}\end{aligned}$$

as $\delta_j^j = \delta_1^1 + \delta_2^2 + \cdots + \delta_n^n = N$ for the space E_N, the result is

$$\left(g_{\ell j}g_{ik} - g_{\ell i}g_{jk}\right)g^{\ell j} = N\,g_{ik} - g_{ki}$$

but $g_{ik} = g_{ki}$; thus,

$$\left(g_{\ell j} g_{ik} - g_{\ell i} g_{jk}\right) g^{\ell j} = (N-1) g_{ik} \qquad Q.E.D.$$

Exercise 1.15 Show that in space E_N exists $g^{ij} \frac{\partial g_{ij}}{\partial x^k} + g_{ij} \frac{\partial g^{ij}}{\partial x^k} = 0$, where g_{ij} is the metric tensor.

The relation between the metric tensor and its conjugated tensor is given by

$$g_{ij} g^{ij} = \delta_j^j = \delta_1^1 + \delta_2^2 + \cdots + \delta_n^n = N$$

Differentiating this expression with respect to x^k

$$\frac{\partial \left(g_{ij} g^{ij}\right)}{\partial x^k} = \frac{\partial g_{ij}}{\partial x^k} g^{ij} + g_{ij} \frac{\partial (g^{ij})}{\partial x^k} = \frac{\partial N}{\partial x^k} = 0$$

so

$$g^{ij} \frac{\partial g_{ij}}{\partial x^k} + g_{ij} \frac{\partial g^{ij}}{\partial x^k} = 0 \qquad Q.E.D.$$

Exercise 1.16 For the symmetric tensor T_{ij}, that fulfills the condition $g_{ij} T_{\ell k} - g_{i\ell} T_{jk} + g_{jk} T_{\ell i} - g_{k\ell} T_{ij} = 0$, show that $T_{ij} = m g_{ij}$, where $m \neq 0$ is a scalar.

Multiplying the expression given by g^{ij} follows

$$g^{ij} g_{ij} T_{\ell k} - g^{ij} g_{i\ell} T_{jk} + g^{ij} g_{jk} T_{\ell i} - g^{ij} g_{k\ell} T_{ij} = \delta_i^i T_{\ell k} - \delta_j^\ell T_{jk} + \delta_k^i T_{\ell i} - g^{ij} g_{k\ell} T_{ij} = 0$$

As $\delta_j^j = \delta_1^1 + \delta_2^2 + \cdots + \delta_n^n = N$, and for $j = \ell$ and $i = k$

$$N T_{ji} = g^{ij} g_{ij} T_{ij}$$

As $ds^2 = g^{ij} T_{ij} = m_1$, where m_1 is a scalar, and with $T_{ij} = T_{ji}$ follows

$$N T_{ji} = m_1 g_{ij} \Rightarrow T_{ji} = \frac{m_1}{N} g_{ij}$$

Putting $m = \frac{m_1}{N}$

$$T_{ij} = m g_{ij} \qquad Q.E.D.$$

1.6 Angle Between Curves

The angle between two curves is defined by the angle formed by their tangent unit vectors $\boldsymbol{g}_1, \boldsymbol{g}_2$ (Fig. 1.9), by means of the dot product

$$\cos\alpha = \frac{\boldsymbol{g}_1 \cdot \boldsymbol{g}_2}{\|\boldsymbol{g}_1\|\|\boldsymbol{g}_2\|} = \boldsymbol{g}_1 \cdot \boldsymbol{g}_2$$

In differential terms this angle is calculated supposing that in the space E_3 two curves intersect in a point R, and admitting a third curve that intersects the other two at points A_1 and A_2, which distances from the point R are, $ds_{(1)}$ and $ds_{(2)}$ (Fig. 1.9). The points M, A_1, A_2 have coordinates x^i, $x^i + dx^i_{(1)}$ and $x^i + dx^i_{(2)}$, respectively.

With the cosine law

$$\cos\alpha = \lim \frac{(RA_1)^2 + (RA_2)^2 - (A_1A_2)^2}{2(RA_1)(RA_2)}$$

which in differential terms stays

$$\cos\alpha = \frac{\left(ds_{(1)}\right)^2 + \left(ds_{(2)}\right)^2 - \left(ds_{(3)}\right)^2}{2ds_{(1)}ds_{(2)}}$$

and using the basic expressions for the length of the arcs of the curves

$$\left(ds_{(1)}\right)^2 = g_{ij}dx^i_{(1)}dx^j_{(1)} \quad \left(ds_{(2)}\right)^2 = g_{ij}dx^i_{(2)}dx^j_{(2)}$$

$$\begin{aligned}\left(ds_{(3)}\right)^2 &= g_{ij}\left[\left(x^i + dx^i_{(1)}\right) - \left(x^i + dx^i_{(2)}\right)\right]^2 = g_{ij}\left(dx^i_{(1)} - dx^i_{(2)}\right)^2 \\ &= g_{ij}\left(dx^i_{(2)} - dx^i_{(1)}\right)\left(dx^j_{(2)} - dx^j_{(1)}\right)\end{aligned}$$

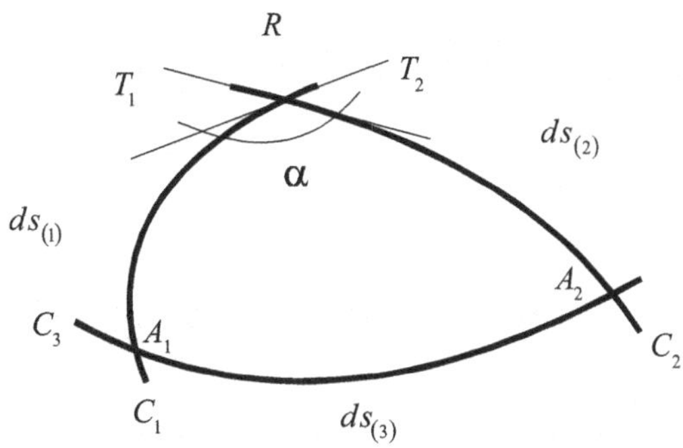

Fig. 1.9 Angle between two curves

then

$$\cos\alpha = \frac{g_{ij}\left(dx^i_{(1)}dx^j_{(2)} + dx^j_{(2)}dx^i_{(1)}\right)}{2ds_{(1)}ds_{(2)}} = \frac{g_{ij}dx^i_{(1)}dx^j_{(2)}}{ds_{(1)}ds_{(2)}}$$

Considering $u^i = \frac{dx^i_{(1)}}{ds_{(1)}}$ and $v^j = \frac{dx^j_{(2)}}{ds_{(2)}}$, which are, respectively, the contravariant unit vectors of the tangents T_1 and T_2 to the curves C_1 and C_2, respectively, provides

$$\cos\alpha = g_{ij}\left(\frac{dx^i_{(1)}}{ds_{(1)}}\right)\left(\frac{dx^j_{(2)}}{ds_{(2)}}\right)$$

If two vectors are orthogonal, then $\alpha = \frac{\pi}{2}$, so the condition of orthogonality for two directions is $g_{ij}u^iv^j = 0$. The necessary and sufficient condition so that a coordinate system is orthonormal is that $g_{ij} = 0\ \forall i \neq j$ at the points of this space. The null vector has the peculiar characteristic of being normal to itself.

Figure 1.10 illustrates the components of the differential element of arc ds with respect to the coordinate system X^i with origin at point P. The lengths of the arc elements measured with respect to the coordinate axes of the referential system are $ds_{(1)} = \sqrt{g_{11}}dx^1$, $ds_{(2)} = \sqrt{g_{22}}dx^2$, and $ds_{(3)} = \sqrt{g_{33}}dx^3$.

To prove that α is real and that $\cos\alpha \leq 1$, consider the expression

$$\cos\alpha = g_{ij}u^iv^j = g^{ij}u_iv_j = u^iv_j = u_iv^j$$

where $\boldsymbol{u}, \boldsymbol{v}$ are unit vectors. Admit that these vectors are multiplied by two non-null real numbers ℓ, m, originating $(\ell u^i + mv^i)$, as the metric of the space is positive definite, then for all the values of this pair of numbers

$$g_{ij}\left(\ell u^i + mv^i\right)\left(\ell u^j + mv^j\right) \geq 0$$

Fig. 1.10 Components of the differential arc element with respect to the X^i coordinate system

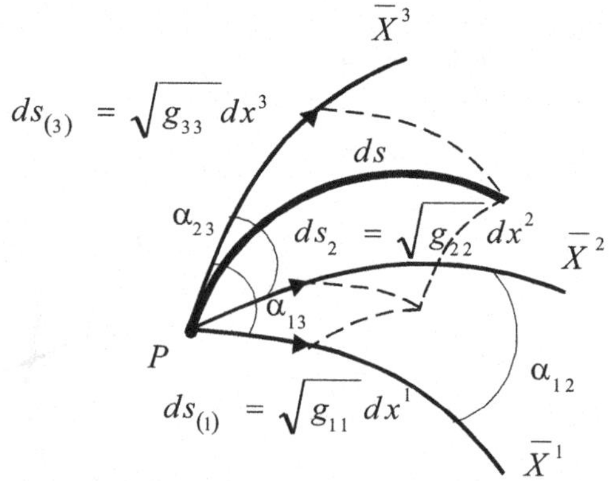

Developing this inequality

$$\ell^2 + 2\ell m \cos\alpha + m^2 \geq 0$$

for $u^i v_j = u_i v^j = \cos\alpha$, which can be written under the form

$$(\ell + m\cos\alpha)^2 + m^2(1 - \cos^2\alpha) \geq 0$$

that will be positive definite if $\cos^2\alpha \leq 1$ or $\|\cos\alpha\| \leq 1$, so α is real.

Let the modulus of a vector in terms of their contravariant components

$$v = \sqrt{\varepsilon g_{k\ell} v^k v^\ell} \tag{1.6.1}$$

and in terms of their covariant components

$$v = \sqrt{\varepsilon g^{kk} v_k v_k} \tag{1.6.2}$$

thus the angle between two curves is determined when calculating the angle between their tangent unit vector u^i, v^j, then

$$\cos\alpha = \frac{g_{ij} u^i v^j}{\sqrt{(u^i u_i)\,(v^j v_j)}} \tag{1.6.3}$$

Exercise 1.17 Let the orthogonal unit vectors u^i and v^j, calculate the norm of vector $w^i = u^i + v^i$.

The condition of orthogonality between two vectors is given by

$$g_{ij} u^i v^j = 0$$

and as u^i and v^j are unit vectors

$$g_{ij} u^i u^j = 1 \quad g_{ij} v^i v^j = 1$$

For the vector w^i

$$\begin{aligned}\|w\|^2 &= g_{ij} w^i w^j = g_{ij}(u^i + v^j)(u^j + v^i) = g_{ij} u^i u^j + g_{ij} u^i v^i + g_{ij} v^j u^j + g_{ij} v^j v^i \\ &= 1 + 0 + 0 + 1 = 2\end{aligned}$$

then

$$\|w\| = \sqrt{2}$$

Exercise 1.18 The vectors u^i and v^j are orthogonal, and each one of them has modulus ℓ, show that $\left(g_{pj}g_{ki} - g_{pk}g_{ji}\right) u^p v^j u^k v^i = \ell^4$.

The square of the modulus of the vectors is given by

$$g_{ij}u^i u^j = \ell^2 \quad g_{ij}v^i v^j = \ell^2$$

and the condition of orthogonality between these vectors is given by

$$g_{ij}u^i v^j = 0$$

Developing the given expression

$$\begin{aligned}\left(g_{pj}g_{ki} - g_{pk}g_{ji}\right) u^p v^j u^k v^i &= g_{pj}g_{ki}u^p v^j u^k v^i - g_{pk}g_{ji}u^p v^j u^k v^i \\ &= \left(g_{pj}u^p v^j\right) g_{ki}u^k v^i - \left(g_{pk}u^p u^k\right) g_{ji}v^j v^i \\ &= 0 - \ell^2 \times \ell^2\end{aligned}$$

so

$$\left(g_{pj}g_{ki} - g_{pk}g_{ji}\right) u^p v^j u^k v^i = \ell^4 \qquad Q.E.D.$$

Exercise 1.19 Given the symmetric tensor T_{ij} and the unit vectors u^i and v^j orthogonal to the vector w^k, show that $T_{ij}u^i - m_1 g_{ij}u^i + n_1 g_{ij}w^i = 0$ and $T_{ij}v^i - m_2 g_{ij}v^i + n_2 g_{ij}w^i = 0$, where $m_1 \neq m_2$ and $n_1 \neq n_2$ are scalars, then these unit vectors are orthogonal.

As u^i and v^j are unit vectors, then

$$g_{ij}u^i u^j = 1 \quad g_{ij}v^i v^j = 1$$

and the conditions of orthogonality of these unit vectors with respect to the vector w^i are

$$g_{ij}u^i w^j = 0 \quad g_{ij}v^i w^j = 0$$

Multiplying by v^j both the members of the first expression

$$T_{ij}u^i v^j - m_1 g_{ij}u^i v^j + n_1 g_{ij}w^i v^j = 0 \Rightarrow T_{ij}u^i v^j = m_1 g_{ij}u^i v^j$$

and multiplying by u^j both the members of the second expression

$$T_{ij}v^i u^j - m_2 g_{ij}v^i u^j + n_2 g_{ij}w^i u^j = 0 \Rightarrow T_{ij}v^i u^j = m_2 g_{ij}v^i u^j$$

The indexes i and j are dummies, so their position can be changed

$$T_{ij}u^i v^j = m_2 g_{ij} u^i v^j$$

thus

$$m_1 g_{ij} u^i v^j = m_2 g_{ij} u^i v^j \Rightarrow (m_1 - m_2) g_{ij} u^i v^j = 0$$

As by hypothesis $m_1 \neq m_2$, then $g_{ij}u^i v^j = 0$; this shows that the unit vectors u^i and v^j are orthogonal.

1.6.1 Symmetrical and Antisymmetrical Tensors

If the change of position of two indexes, covariant or contravariant, does not modify the tensor's components, then this is a symmetrical tensor

$$T^{pqrs}_{ijk} = T^{pqrs}_{ikj} = T^{pqrs}_{jik} = T^{pqrs}_{jki} = T^{pqrs}_{kij} = T^{pqrs}_{kji}.$$

The symmetry, a priori, does not ensure that the new variety is a tensor. Admit that $T^{pqrs}_{ijk\ell} = T^{qprs}_{ijk\ell}$, whereby by the hypothesis of this tensor's symmetry, it follows that $T^{pqrs}_{ijk\ell} - T^{qprs}_{ijk\ell} = 0$. As $T^{pqrs}_{ijk\ell}$ is a tensor, the result of the difference between the two varieties being null, and as the referential system is arbitrary, it is concluded that this result will always be null for any coordinate system, i.e., it always has the tensor null. Writing $T^{pqrs}_{ijk\ell} + 0 = T^{qprs}_{ijk\ell}$, and as the summation of tensors is a tensor, it is concluded that $T^{pqrs}_{ijk\ell}$ is a tensor.

A tensor is called antisymmetrical with respect to two of its indexes, if it changes signs on the change of position between these two indexes: $T_{ijk\ell} = -T_{\ell jki}$. The number of independent components of an antisymmetric tensor of order p in the space E_N is given by

$$n = \frac{N!}{p!(N-p)!} \tag{1.7.1}$$

Let the space E_N in which the antisymmetric pseudotensor of the third order ε^{ijk} is defined (a general definition of pseudotensors will be presented in item 1.8), and by the definition of antisymmetry, it provides six components of ε^{ijk} which are numerically equal:

$$\varepsilon^{ijk} = \varepsilon^{jki} = \varepsilon^{kij} = -\varepsilon^{ikj} = -\varepsilon^{jik} = -\varepsilon^{kji}$$

This variety has 27 components, having 21 null, for it is verified that only the six components $\varepsilon^{123} = \varepsilon^{231} = \varepsilon^{312} = -\varepsilon^{132} = -\varepsilon^{213} = -\varepsilon^{321}$ are non-null. Let, for example, a linear and homogeneous transformation be applied to the component ε^{123}:

$$\overline{\varepsilon}^{123} = \frac{\partial \overline{x}^1}{\partial x^i}\frac{\partial \overline{x}^2}{\partial x^j}\frac{\partial \overline{x}^3}{\partial x^k}\varepsilon^{123} \tag{1.7.2}$$

Developing expression (1.7.2)

$$\begin{aligned}\overline{\varepsilon}^{123} = &\left(\frac{\partial \overline{x}^1}{\partial x^1}\frac{\partial \overline{x}^2}{\partial x^2}\frac{\partial \overline{x}^3}{\partial x^3} + \frac{\partial \overline{x}^1}{\partial x^2}\frac{\partial \overline{x}^2}{\partial x^3}\frac{\partial \overline{x}^3}{\partial x^1} + \frac{\partial \overline{x}^1}{\partial x^3}\frac{\partial \overline{x}^2}{\partial x^1}\frac{\partial \overline{x}^3}{\partial x^2}\right.\\ &\left.- \frac{\partial \overline{x}^1}{\partial x^1}\frac{\partial \overline{x}^2}{\partial x^3}\frac{\partial \overline{x}^3}{\partial x^2} - \frac{\partial \overline{x}^1}{\partial x^3}\frac{\partial \overline{x}^2}{\partial x^2}\frac{\partial \overline{x}^3}{\partial x^1} - \frac{\partial \overline{x}^1}{\partial x^2}\frac{\partial \overline{x}^2}{\partial x^1}\frac{\partial \overline{x}^3}{\partial x^3}\right)\varepsilon^{123}\end{aligned} \tag{1.7.3}$$

In compact form for the component $\overline{\varepsilon}^{123}$ in the coordinate system $\overline{X}^i$

$$\overline{\varepsilon}^{123} = \left|\frac{\partial \overline{x}^k}{\partial x^\ell}\right|\varepsilon^{123} \tag{1.7.4}$$

and with

$$g_{k\ell} = \frac{\partial x^i}{\partial \overline{x}^k}\frac{\partial x^j}{\partial \overline{x}^\ell}\overline{g}_{ij}$$

It follows by means of product of determinants

$$|g_{k\ell}| = \left|\frac{\partial x^i}{\partial \overline{x}^k}\right| \cdot \left|\frac{\partial x^j}{\partial \overline{x}^\ell}\right| \cdot \left|\overline{g}_{ij}\right| \Rightarrow |g_{k\ell}| = \left|\frac{\partial x^i}{\partial \overline{x}^k}\right|^2 \left|\overline{g}_{ij}\right| \Rightarrow g = \left|\frac{\partial x^i}{\partial \overline{x}^k}\right|^2 \overline{g}$$

$$\frac{1}{\sqrt{\overline{g}}} = \left|\frac{\partial x^i}{\partial \overline{x}^k}\right| \frac{1}{\sqrt{g}} \tag{1.7.5}$$

Comparing expression (1.7.5) with the expression (1.7.4)

$$\varepsilon^{123} = \frac{1}{\sqrt{g}} \tag{1.7.6}$$

So as to generalize expression (1.7.6), this analysis is made for the other contravariant components of the pseudotensor ε^{ijk}. As this variety assumes the values 0, ± 1 as a function of the position of their indexes, it is linked to the permutation symbol e^{ijk} by means of the following relations:

$$\varepsilon^{ijk} = \begin{cases} \dfrac{+1}{\sqrt{g}}e^{ijk} & \text{is an even permutation of the indexes} \\ \dfrac{-1}{\sqrt{g}}e^{ijk} & \text{is an odd permutation of the indexes} \\ 0 & \text{when there are repeated indexes} \end{cases} \tag{1.7.7}$$

Expression (1.7.7) represents the components of the Ricci pseudotensor, also called Levi-Civita pseudotensor. The covariant components of this pseudotensor are obtained by means of the metric tensor, whereby using the approaches presented in item 1.5, it is provided for the lowering of the indexes of the pseudotensor ε^{pqr}:

$$\varepsilon_{ijk} = g_{ip}g_{jq}g_{kr}\varepsilon^{pqr} \tag{1.7.8}$$

and with the definition of the determinant of the metric tensor, and with the definition of ε^{ijk} given by the relations (1.7.7), it follows that

$$\varepsilon_{ijk} = \left|g_{ij}\right|\frac{1}{\sqrt{g}} = \sqrt{g} \tag{1.7.9}$$

In terms of the permutation symbol e_{ijk}, it is provided as the covariant coordinates of the Ricci pseudotensor

$$\varepsilon_{ijk} = \begin{cases} \sqrt{g}e_{ijk} & \text{is an even permutation of the indexes} \\ -\sqrt{g}e_{ijk} & \text{is an odd permutation of the indexes} \\ 0 & \text{when there are repeated indexes} \end{cases} \tag{1.7.10}$$

The definition of the Ricci pseudotensor presented for the space E_3 is generalized for the space E_N, in which the contravariant components and covariant of this variety are given, respectively, in terms of the permutation symbol by

$$\varepsilon^{i_1i_2i_3\cdots i_n} = \begin{cases} \dfrac{+1}{\sqrt{g}}e^{i_1i_2i_3\cdots i_n} & \text{is an even permutation of the indexes} \\ \dfrac{-1}{\sqrt{g}}e^{i_1i_2i_3\cdots i_n} & \text{is an odd permutation of the indexes} \\ 0 & \text{when there are repeated indexes} \end{cases} \tag{1.7.11}$$

$$\varepsilon_{i_1i_2i_3\cdots i_n} = \begin{cases} \sqrt{g}e_{i_1i_2i_3\cdots i_n} & \text{is an even permutation of the indexes} \\ -\sqrt{g}e_{i_1i_2i_3\cdots i_n} & \text{is an odd permutation of the indexes} \\ 0 & \text{when there are repeated indexes} \end{cases} \tag{1.7.12}$$

The conception of permutation symbol is associated to the value of a determinant, with no link to the space E_N, whereby it refers only to a symbol that seeks to simplify the calculations. With the definition of the Ricci pseudotensor in terms of this symbol, it is verified that in the relation between these two varieties exists the term $\sqrt{g}$ linked to the metric of the space. This shows the fundamental difference between the same, for the change of sign of the Ricci pseudotensor as a function of the permutations of their indexes (sign defined by the permutation symbol) indicates the orientation of the space.

With relation (1.7.10) it follows that

$$\varepsilon_{ijk}\varepsilon_{jki} = 3! = 6 \tag{1.7.13}$$

The definitions and deductions presented next seek to complement the relations between the generalized Kronecker delta and the Ricci pseudotensor in the space E_N. These expressions are called $\delta - \varepsilon$ relations.

1.6.1.1 Generalization of the Kronecker Delta

The Ricci pseudotensor represents the mixed product of three vectors $\frac{\partial x^i}{\partial x^\ell}$, $\frac{\partial x^j}{\partial x^\ell}$, $\frac{\partial x^k}{\partial x^\ell}$, where $\ell = 1, 2, 3$ indicates the components of these vectors, which comprise the lines and columns of the determinant that expresses this product, called Gram determinant, that in terms of their covariant components stays

$$\varepsilon_{ijk} = \left(\frac{\partial x^i}{\partial x^\ell} \times \frac{\partial x^j}{\partial x^\ell}\right) \cdot \frac{\partial x^k}{\partial x^\ell} = \begin{vmatrix} \frac{\partial x^i}{\partial x^1} & \frac{\partial x^j}{\partial x^1} & \frac{\partial x^k}{\partial x^1} \\ \frac{\partial x^i}{\partial x^2} & \frac{\partial x^j}{\partial x^2} & \frac{\partial x^k}{\partial x^2} \\ \frac{\partial x^i}{\partial x^3} & \frac{\partial x^j}{\partial x^3} & \frac{\partial x^k}{\partial x^3} \end{vmatrix} = \begin{vmatrix} \delta_{1i} & \delta_{1j} & \delta_{1k} \\ \delta_{2i} & \delta_{2j} & \delta_{2k} \\ \delta_{3i} & \delta_{3j} & \delta_{3k} \end{vmatrix}$$

and in terms of their contravariant components

$$\varepsilon^{pqr} = \begin{vmatrix} \frac{\partial x^p}{\partial x^1} & \frac{\partial x^q}{\partial x^1} & \frac{\partial x^r}{\partial x^1} \\ \frac{\partial x^p}{\partial x^2} & \frac{\partial x^q}{\partial x^2} & \frac{\partial x^r}{\partial x^2} \\ \frac{\partial x^p}{\partial x^3} & \frac{\partial x^q}{\partial x^3} & \frac{\partial x^r}{\partial x^3} \end{vmatrix} = \begin{vmatrix} \delta^{1p} & \delta^{1q} & \delta^{1r} \\ \delta^{2p} & \delta^{2q} & \delta^{2r} \\ \delta^{3p} & \delta^{3q} & \delta^{3r} \end{vmatrix}$$

The product of these two determinants being given by

$$\varepsilon_{ijk}\varepsilon^{pqr} = \begin{vmatrix} \delta_{1i} & \delta_{1j} & \delta_{1k} \\ \delta_{2i} & \delta_{2j} & \delta_{2k} \\ \delta_{3i} & \delta_{3j} & \delta_{3k} \end{vmatrix} \cdot \begin{vmatrix} \delta^{1p} & \delta^{1q} & \delta^{1r} \\ \delta^{2p} & \delta^{2q} & \delta^{2r} \\ \delta^{3p} & \delta^{3q} & \delta^{3r} \end{vmatrix}$$

$$\varepsilon_{ijk}\varepsilon^{pqr} = \begin{vmatrix} (\delta_{1i}\delta^{1p} + \delta_{2i}\delta^{2p} + \delta_{3i}\delta^{3p}) & (\delta_{1i}\delta^{1q} + \delta_{2i}\delta^{2q} + \delta_{3i}\delta^{3q}) & (\delta_{1i}\delta^{1r} + \delta_{2i}\delta^{2r} + \delta_{3i}\delta^{3r}) \\ (\delta_{1j}\delta^{1p} + \delta_{2j}\delta^{2p} + \delta_{3j}\delta^{3p}) & (\delta_{1j}\delta^{1q} + \delta_{2j}\delta^{2q} + \delta_{3j}\delta^{3q}) & (\delta_{1j}\delta^{1r} + \delta_{2j}\delta^{2r} + \delta_{3j}\delta^{3r}) \\ (\delta_{1k}\delta^{1p} + \delta_{2k}\delta^{2p} + \delta_{3k}\delta^{3p}) & (\delta_{1k}\delta^{1q} + \delta_{2k}\delta^{2q} + \delta_{3k}\delta^{3q}) & (\delta_{1k}\delta^{1r} + \delta_{2k}\delta^{2r} + \delta_{3k}\delta^{3r}) \end{vmatrix}$$

$$\varepsilon_{ijk}\varepsilon^{pqr} = \begin{vmatrix} \delta_{mi}\delta^{mp} & \delta_{mi}\delta^{mq} & \delta_{mi}\delta^{mr} \\ \delta_{mj}\delta^{mp} & \delta_{mj}\delta^{mq} & \delta_{mj}\delta^{mr} \\ \delta_{mk}\delta^{mp} & \delta_{mk}\delta^{mq} & \delta_{mk}\delta^{mr} \end{vmatrix}$$

$$\varepsilon_{ijk}\varepsilon^{pqr} = \begin{vmatrix} \delta_i^p & \delta_i^q & \delta_i^r \\ \delta_j^p & \delta_j^q & \delta_j^r \\ \delta_k^p & \delta_k^q & \delta_k^r \end{vmatrix} \tag{1.7.14}$$

With the expressions (1.7.10) and (1.7.7), it follows that

$$\varepsilon_{r\ell m}\varepsilon^{rst} = \delta^{rst}_{r\ell m} = \delta^{st}_{\ell m} = \begin{cases} 1 \\ 0 \\ -1 \end{cases} \tag{1.7.15}$$

The contraction of the indexes k and r of the product of two pseudotensors, given by expression (1.7.15), provides

$$\varepsilon_{ijk}\varepsilon^{pqr} = \delta^{pqk}_{ijk} = \begin{vmatrix} \delta^p_i & \delta^q_i & \delta^k_i \\ \delta^p_j & \delta^q_j & \delta^k_j \\ \delta^p_k & \delta^q_k & \delta^k_k \end{vmatrix}$$

and as $\delta^k_k = 3$ it follows that

$$\varepsilon_{ijk}\varepsilon^{pqr} = \delta^{pqk}_{ijk} = \begin{vmatrix} \delta^p_i & \delta^q_i & \delta^k_i \\ \delta^p_j & \delta^q_j & \delta^k_j \\ \delta^p_k & \delta^q_k & 3 \end{vmatrix} = \begin{vmatrix} \delta^p_i & \delta^q_i \\ \delta^p_j & \delta^q_j \end{vmatrix} = \delta^p_i\delta^q_j - \delta^q_i\delta^p_j \tag{1.7.16}$$

Analogously, and with the contraction of the indexes j and p:

$$\varepsilon_{ijk}\varepsilon^{pqr} = \delta^{jqr}_{ijk} = \begin{vmatrix} \delta^j_i & \delta^q_i & \delta^k_i \\ 3 & \delta^q_j & \delta^k_j \\ \delta^j_k & \delta^q_k & \delta^k_k \end{vmatrix} = -\begin{vmatrix} \delta^q_i & \delta^r_i \\ \delta^q_k & \delta^r_k \end{vmatrix} = \delta^r_i\delta^q_k - \delta^q_i\delta^r_k$$

The product $\varepsilon_{ijr}\varepsilon^{pqr} = \delta^{pq}_{ij}$ leads to the generalization of the Kronecker delta that has its value defined as a function of the number of permutations of their indexes. For the covariant components of this operator, $\delta_{ijpq} = \delta_{pijq}$ is provided, where the number of permutations of the indexes is even, so it is verified that this operator is symmetrical, and $\delta_{ijpq} = -\delta_{jipq} = -\delta_{ijqp}$ is antisymmetric for an odd number of permutations of the indexes. The deltas with repeated indexes are null, for example, $\delta_{11pq} = \delta_{22pq} = \delta_{ij33} = 0$. This analysis allows defining the generalized Kronecker delta in space E_N:

$$\delta^{j_1 j_2 j_3\cdots j_n}_{i_1 i_2 i_3\cdots i_n} = \begin{cases} +1 & \text{is an even permutation of } i_1 i_2 i_3\cdots, j_1 j_2 j_3\cdots \\ -1 & \text{is an odd permutation of } i_1 i_2 i_3\cdots, j_1 j_2 j_3\cdots \\ 0 & \text{when there are repeated indexes} \end{cases} \tag{1.7.17}$$

1.6.1.2 Fundamental Expressions with the Generalized Kronecker Delta

The generalized Kronecker delta in terms of the Ricci pseudotensor is given by

$$\varepsilon_{i_1 i_2 i_3 \cdots i_m}\varepsilon^{j_1 j_2 j_3 \cdots j_m} = \delta^{j_1 \cdots j_m j_{m+1} \cdots j_n}_{i_1 \cdots i_m i_{m+1} \cdots i_n} = \begin{vmatrix} \delta_{j_1 i_1} & \delta_{j_1 i_2} & \cdots & \delta_{j_1 i_n} \\ \delta_{j_2 i_1} & \delta_{j_2 i_2} & \cdots & \delta_{j_2 i_n} \\ \cdots & \cdots & \cdots & \cdots \\ \delta_{j_n i_1} & \delta_{j_n i_2} & \delta_{j_1 i_1} & \delta_{j_n i_n} \end{vmatrix} \tag{1.7.18}$$

Various fundamental expressions are obtained with the Kronecker delta δ^{pq}_{ij} that are useful in Tensor Calculus. Let, for example, the contraction of the indexes j and q of this tensor

$$\delta^{pq}_{ij} = \delta^{pj}_{ij} = \delta^p_i \delta^j_j - \delta^q_i \delta^j_j = \delta^p_i \delta^j_j - \delta^j_i \delta^p_j = 3\delta^p_i - \delta^j_i \delta^p_j = 2\delta^p_i$$

whereby

$$\delta^p_i = \frac{1}{2}\delta^{pj}_{ij} = \frac{1}{2}\left(\delta^{p1}_{i1} + \delta^{p2}_{i2} + \delta^{p3}_{i3}\right)$$

It is also verified for the contractions $j = q$ and $k = r$

$$\delta^{pqr}_{ijk} = \frac{1}{2}\delta^{pjk}_{ijk} = \delta^p_i = \frac{1}{2}\left(\delta^{p12}_{i12} + \delta^{p23}_{i23} + \delta^{p31}_{i31}\right)$$

The generalization of these expressions that involve Kronecker deltas for the space E_N is given by the following expression:

$$\delta^{p_1 p_2 p_3 \cdots p_m}_{i_1 i_2 i_3 \cdots i_m} = \frac{(N-n)!}{(N-m)!}\delta^{p_1 \cdots p_m p_{m+1} \cdots p_n}_{i_1 \cdots i_m i_{m+1} \cdots i_n} \tag{1.7.19}$$

The Kronecker delta tensor $\delta^{p_1 p_2 p_3 \cdots p_m}_{i_1 i_2 i_3 \cdots i_m}$ provided by expression (1.7.19) is of order $2(n-m)$ inferior to the order Kronecker delta tensor $\delta^{p_1 \cdots p_m p_{m+1} \cdots p_n}_{i_1 \cdots i_m i_{m+1} \cdots i_n}$, from which it was obtained by means of contractions of the indexes. In this expression for $m = 1,\ n = 3$

$$\delta^p_i = \frac{1}{(N-2)(N-1)}\delta^{pjk}_{ijk}$$

Putting $m = 1,\ n = 2$ in expression (1.7.19) results in

$$\delta^p_i = \frac{1}{(N-1)}\delta^{pj}_{ij}$$

These two examples show that δ^p_i can be obtained by two contractions of the indexes of the sixth-order tensor δ^{pqr}_{ijk} or by means of only a contraction of the indexes of the fourth-order tensor δ^{pq}_{ij}.

In expression (1.7.19) for $m = 1,\ i = p$

$$\delta_{i_1}^{i_1} = \frac{(N-n)!}{(N-1)!}\delta_{i_1 i_2 \cdots i_n}^{i_1 i_2 \cdots i_n}$$

and as $\delta_{i_1}^{i_1} = n$ it results in

$$n = \frac{(N-n)!}{(N-1)!}\delta_{i_1 i_2 \cdots i_n}^{i_1 i_2 \cdots i_n} \Rightarrow \delta_{i_1 i_2 \cdots i_n}^{i_1 i_2 \cdots i_n} = \frac{n(N-1)!}{(N-n)!}$$

$$\delta_{i_1 i_2 \cdots i_n}^{i_1 i_2 \cdots i_n} = \frac{n!}{(N-n)!} \tag{1.7.20}$$

For the inner product of the Ricci pseudotensors $\varepsilon_{i_1 i_2 \cdots i_n}$ and $\varepsilon^{i_1 i_2 \cdots i_n}$ with $N = n$ expression (1.7.20) provides

$$\varepsilon_{i_1 i_2 \cdots i_n}\varepsilon^{i_1 i_2 \cdots i_n} = \delta_{i_1 i_2 \cdots i_n}^{i_1 i_2 \cdots i_n} = n! \tag{1.7.21}$$

Expression (1.7.19) with $N = n$ provides

$$\varepsilon_{i_1 i_2 \cdots i_m i_{m+1} \cdots i_n}\varepsilon^{p_1 p_2 \cdots p_m p_{m+1} \cdots p_n} = (N-m)!\delta_{i_1 i_2 \cdots i_n}^{p_1 p_2 \cdots p_n} \tag{1.7.22}$$

Expression (1.7.22) relates in space E_N the inner product of two Ricci pseudotensors with the generalized Kronecker delta tensor.

1.6.1.3 Product of the Ricci Pseudotensor by the Generalized Kronecker Delta

The definition of the generalized Kronecker delta shows that $\delta_{q123}^{pijk} = 0$, for a dummy index will always occur when these vary. With expression (1.7.18)

$$\varepsilon_{q123}\varepsilon^{pijk} = \delta_{q123}^{pijk} = \begin{vmatrix} \delta_{pq} & \delta_{p1} & \delta_{p2} & \delta_{p3} \\ \delta_{iq} & \delta_{i1} & \delta_{i2} & \delta_{i3} \\ \delta_{jq} & \delta_{j1} & \delta_{j2} & \delta_{j3} \\ \delta_{kq} & \delta_{k1} & \delta_{k2} & \delta_{k3} \end{vmatrix} = 0$$

Developing this determinant in terms of the first column

$$\delta_{pq}\varepsilon_{ijk}\varepsilon_{123} - \delta_{iq}\varepsilon_{pjk}\varepsilon_{123} + \delta_{jq}\varepsilon_{pik}\varepsilon_{123} - \delta_{kq}\varepsilon_{pij}\varepsilon_{123} = 0$$

and with $\varepsilon_{pik} = -\varepsilon_{ipk}$

$$\varepsilon_{ijk}\delta_{pq} = \varepsilon_{pjk}\delta_{iq} + \varepsilon_{ipk}\delta_{jq} + \varepsilon_{ijp}\delta_{kq} \tag{1.7.23}$$

The symmetry of δ_{pq} allows changing the position of these indexes

$$\varepsilon_{ijk}\delta_{qp} = \varepsilon_{qjk}\delta_{ip} + \varepsilon_{iqk}\delta_{jp} + \varepsilon_{ijq}\delta_{kp} \tag{1.7.24}$$

1.6.1.4 Norm of the Antisymmetric Pseudotensor of the Second Order

A vector is represented by an oriented segment of a straight line, and its norm is given by the length of this segment. For an antisymmetric pseudotensor of the second order $\boldsymbol{A}$ associated to an axial vector $\boldsymbol{u}$ provides that its norm is linked to the area of the parallelogram which sides are the vectors that define the vectorial product $\boldsymbol{u} = \boldsymbol{v} \times \boldsymbol{w}$.

Let α the angle between the vectors $\boldsymbol{v}$ and $\boldsymbol{w}$, the square of the modulus of the cross product of these vectors is given by

$$\begin{aligned}\|\boldsymbol{u}\|^2 &= \|\boldsymbol{v} \times \boldsymbol{w}\|^2 = \|\boldsymbol{v}\|^2 \|\boldsymbol{w}\|^2 \sin^2\alpha = \|\boldsymbol{v}\|^2 \|\boldsymbol{w}\|^2 \left(1 - \cos^2\alpha\right)\\ &= \|\boldsymbol{v}\|^2 \|\boldsymbol{w}\|^2 - (\boldsymbol{v} \cdot \boldsymbol{w})^2\end{aligned}$$

thus

$$\|\boldsymbol{u}\| = \sqrt{\|\boldsymbol{v}\|^2 \|\boldsymbol{w}\|^2 - (\boldsymbol{v} \cdot \boldsymbol{w})^2} \tag{1.7.25}$$

This norm can be expressed in terms of the components of the pseudotensor $\boldsymbol{A}$. Let the components of the vectors $\boldsymbol{v}$ and $\boldsymbol{w}$ in the coordinate system X^k, so with the expression (1.7.25)

$$\begin{aligned}\|\boldsymbol{u}\|^2 &= \left(g_{i\ell}v^iv^\ell\right)\left(g_{jm}w^jw^m\right) - \left(g_{im}v^iw^m\right)\left(g_{j\ell}v^jw^\ell\right) = \left(g_{i\ell}g_{jm} - g_{im}g_{j\ell}\right)v^iv^\ell w^jw^m\\ &= \begin{vmatrix} g_{i\ell} & g_{im} \\ g_{j\ell} & g_{jm} \end{vmatrix} v^iv^\ell w^jw^m\end{aligned}$$

This determinant allows writing

$$\begin{vmatrix} g_{i\ell} & g_{im} \\ g_{j\ell} & g_{jm} \end{vmatrix} v^iv^\ell w^jw^m = \frac{1}{2}\begin{vmatrix} g_{i\ell} & g_{im} \\ g_{j\ell} & g_{jm} \end{vmatrix}\left(v^iw^j - v^jw^i\right)v^\ell w^m$$

and as

$$A^{\ell m} = \frac{1}{2}\left(v^\ell w^m - v^m w^\ell\right)$$

it follows that

$$\begin{vmatrix} g_{i\ell} & g_{im} \\ g_{j\ell} & g_{jm} \end{vmatrix} A^{ij} v^{\ell} w^{m} = \frac{1}{2} \begin{vmatrix} g_{i\ell} & g_{im} \\ g_{j\ell} & g_{jm} \end{vmatrix} A^{ij} \left(v^{\ell} w^{m} - v^{m} w^{\ell} \right)$$

then

$$\begin{vmatrix} g_{i\ell} & g_{im} \\ g_{j\ell} & g_{jm} \end{vmatrix} v^{i} w^{\ell} v^{\ell} w^{m} = \frac{1}{2} \begin{vmatrix} g_{i\ell} & g_{im} \\ g_{j\ell} & g_{jm} \end{vmatrix} A^{ij} A^{\ell m} \tag{1.7.26}$$

1.6.1.5 Generation of Tensors from the Ricci Pseudotensor

The Ricci pseudotensor generates an antisymmetric tensor from a pseudotensor (axial vector), and this pseudotensor generates an antisymmetric tensor from a pseudotensor (axial vector). This characteristic of the Ricci pseudotensor in space E_3 is generalized for the space E_N, where the known antisymmetric tensor $A_{[i_1 i_2 \cdots i_n]}$ provides

$$T^{j_1 j_2 \cdots j_{n-m}} = \frac{1}{m!} \varepsilon^{j_1 j_2 \cdots j_{n-m} i_1 i_2 \cdots i_m} A_{[i_1 i_2 \cdots i_m]} \tag{1.7.27}$$

Tensor $T^{j_1 j_2 \cdots j_{n-m}}$ is generated by the Ricci pseudotensor, which works as an operator applied to the antisymmetric tensor to produce this associated tensor.

Multiplying both the members of expression (1.7.27) by $\varepsilon_{j_1 j_2 \cdots j_{n-m} i_1 i_2 \cdots i_m}$ results in

$$\varepsilon_{j_1 j_2 \cdots j_{n-m} i_1 i_2 \cdots i_m} T^{j_1 j_2 \cdots j_{n-m}} = \frac{1}{m!} \varepsilon_{j_1 j_2 \cdots j_{n-m} i_1 i_2 \cdots i_m} \varepsilon^{j_1 j_2 \cdots j_{n-m} i_1 i_2 \cdots i_m} A_{[i_1 i_2 \cdots i_m]}$$

With expressions (1.7.19), (1.7.21), and (1.7.22), it follows that the expression for the antisymmetric tensor $A_{[i_1 i_2 \cdots i_m]}$ in terms of the Ricci pseudotensor is given by

$$A_{[i_1 i_2 \cdots i_m]} = \frac{1}{(n-m)!} \varepsilon_{j_1 j_2 \cdots j_{n-m} i_1 i_2 \cdots i_m} T^{j_1 j_2 \cdots j_{n-m}} \tag{1.7.28}$$

To illustrate the application of expression (1.7.28), let the antisymmetric tensor of the fourth order $A_{[ijk\ell]}$ with $i, j, k, \ell = 1, 2, 3, \cdots n$, to which the following five varieties are associated

$$T = \frac{1}{4!} \varepsilon^{ijk\ell} A_{[ijk\ell]} \quad T^{i} = \frac{1}{4!} \varepsilon^{ijk\ell p} A_{[jk\ell p]} \quad T^{ij} = \frac{1}{4!} \varepsilon^{ijk\ell pq} A_{[k\ell pq]}$$

$$T^{ijk} = \frac{1}{4!} \varepsilon^{ijk\ell pqr} A_{[\ell pqr]} \quad T^{ijk\ell} = \frac{1}{4!} \varepsilon^{ijk\ell pqrs} A_{[pqrs]}$$

1.7 Relative Tensors

The tensors defined in the previous items are called absolute tensors. However, other varieties with properties that are analogous to those of these tensors can be defined. The relative tensors are more general varieties, the absolute tensors being a particular case of the same.

In solving various problems that involve integration processes, the need of generalizing the concept of tensor is verified. This generalization leads to the concept of relative tensor. To exemplify the concept of the relative tensor, let a covariant tensor of the second order be defined in the space E_3, which transforms by means of the expression

$$\overline{T}_{k\ell} = \frac{\partial x^i}{\partial \overline{x}^k}\frac{\partial x^j}{\partial \overline{x}^\ell}T_{ij} \tag{1.8.1}$$

The determinants of the terms of this function are given by $\det \overline{T}_{k\ell}$, $\det\left(\frac{\partial x^i}{\partial \overline{x}^k}\right)$, $\det\left(\frac{\partial x^j}{\partial \overline{x}^\ell}\right)$, and $\det T_{ij}$. Applying the determinant product rule to the determinant terms of this expression

$$\det \overline{T}_{k\ell} = \det\left(\frac{\partial x^i}{\partial \overline{x}^k}\right)\det\left(\frac{\partial x^j}{\partial \overline{x}^\ell}\right)\det T_{ij} \tag{1.8.2}$$

the Jacobian of the inverse transformation of tensor $\overline{T}_{k\ell}$ is given by

$$J = \det\left(\frac{\partial x^m}{\partial \overline{x}^n}\right) > 0 \tag{1.8.3}$$

so

$$\det \overline{T}_{k\ell} = J^2 \det T_{ij} \tag{1.8.4}$$

Expression (1.8.4) shows that $\det \overline{T}_{k\ell}$ of a second-order tensor is not a scalar and also is not a second-order tensor of the type $\overline{T}_{pq}$. This expression is the new transformation. Assuming that $\det T_{ij} > 0$ and $\det \overline{T}_{k\ell} > 0$ provides

$$\left(\det \overline{T}_{k\ell}\right)^{\frac{1}{2}} = J\left(\det T_{ij}\right)^{\frac{1}{2}} \tag{1.8.5}$$

This shows that the definition of tensors can be expanded introducing the concept of relative tensor.

Consider the mixed tensor

$$\overline{T}^{ij..p}_{rs...v} = (J)^W \frac{\partial \overline{x}^i}{\partial x^a}\frac{\partial \overline{x}^j}{\partial x^b}\cdots\frac{\partial \overline{x}^p}{\partial x^d}\frac{\partial x^r}{\partial \overline{x}^e}\frac{\partial x^s}{\partial \overline{x}^f}\cdots\frac{\partial x^v}{\partial \overline{x}^h}T^{ab...d}_{ef...h} \tag{1.8.6}$$

that is called relative tensor of weight W or with weighing factor W. This weight is an integer number, and J is the Jacobian of the transformation. For the particular case in which $W = 0$, an absolute tensor exists.

The concept of relative tensor allows distinguishing a relative invariant of a scalar, which is an absolute invariant. To differentiate these concepts, let the relative invariant A of weight W, which transforms according to the expression

$$\overline{A} = J^W A \tag{1.8.7}$$

For the particular case in which $W = 0$, an absolute tensor exists $\overline{A} = A$ that is a scalar. For $W = 1$ provides the scalar density $\overline{A} = JA$. The definition of scalar density will be presented in detail in later paragraphs.

To illustrate the concept of relative tensor, let the metric tensor g_{ij} with $\det g_{ij} = g$. Applying a linear and homogeneous transformation to this tensor

$$\widetilde{g}_{\ell m} = \frac{\partial x^i}{\partial \widetilde{x}^\ell} \frac{\partial x^j}{\partial \widetilde{x}^m} g_{ij} \tag{1.8.8}$$

with $\det \widetilde{g}_{ij} = \widetilde{g}$, and by means of the property of the product of determinants, provides the relative scalar of weight $W = 2$

$$\widetilde{g} = J^2 g \Rightarrow \sqrt{\widetilde{g}} = J\sqrt{g} \tag{1.8.9}$$

For $J = 1$ provides $\sqrt{g}$ that is a relative tensor of unit weight, being, therefore, an invariant. With expression (1.8.9) and the condition $g\overline{g} = 1$, having $\det g^{ij} = \overline{g}$, it is verified that $\sqrt{\overline{g}}$ is a relative tensor of weight -1.

Let the Jacobian J of weight $W = 1$, which is an invariant and when changing to a new coordinate system provides for this determinant $\overline{J} = \alpha J$ being α a scalar (invariant). Raising both members of this expression to the power W

$$\overline{J}^W = \alpha^W J^W \tag{1.8.10}$$

where J^W is an invariant of weight W, thus

$$\alpha^W = \overline{J}^W J^{-W} \tag{1.8.11}$$

Consider the relative tensor T^i_{jk} of weight W that transforms by means of the expression

$$\overline{T}^i_{jk} = \alpha^W \frac{\partial x^m}{\partial \overline{x}^j} \frac{\partial x^n}{\partial \overline{x}^k} \frac{\partial \overline{x}^i}{\partial x^\ell} T^\ell_{mn} \tag{1.8.12}$$

and substituting in expression (1.8.12), the value of α^W given by expression (1.8.11) provides

$$\overline{T}^i_{jk} = \left(\overline{J}^W J^{-W}\right) \frac{\partial x^m}{\partial \overline{x}^j} \frac{\partial x^n}{\partial \overline{x}^k} \frac{\partial \overline{x}^i}{\partial x^\ell} T^\ell_{mn}$$

It follows that

$$\left(\overline{J}^{-W} \overline{T}^i_{jk}\right) = \frac{\partial x^m}{\partial \overline{x}^j} \frac{\partial x^n}{\partial \overline{x}^k} \frac{\partial \overline{x}^i}{\partial x^\ell} \left(J^{-W} T^\ell_{mn}\right) \tag{1.8.13}$$

As $\left(J^{-W} T^\ell_{mn}\right)$ is an absolute tensor, and by means of the transformation law of tensors, it is concluded that $\overline{J}^{-W} \overline{T}^i_{jk}$ is also an absolute tensor. This shows that the transformation of a relative tensor of weight W in an absolute tensor is carried out multiplying it by the invariant of unit weight raised to the power $-W$. The invariant $\sqrt{g}$ of unit weight is used to carry out this kind of transformation. This systematic allows, for instance, transforming the relative tensor T^k_{ij} of weight W into the absolute tensor A^k_{ij} by means of

$$T^k_{ij} \left(\sqrt{g}\right)^{-W} = A^k_{ij} \tag{1.8.14}$$

The operations multiplying by a scalar, addition, subtraction, contraction, outer product, and inner product are applicable to the relative tensors. These operations provide new relative tensors; as a result, the proof is analogous to the demonstrations performed for the absolute tensors.

Exercise 1.20 Show that δ_{ij} is an absolute tensor.

It is admitted firstly that δ_{ij} is a relative tensor of unit weight, being $\det \delta_{ij} = 1$, which transforms into the absolute tensor δ^*_{ij} by means of the expression

$$\delta^*_{ij} = \left(\sqrt{g}\right)^{-1} \delta_{ij}$$

As δ^*_{ij} is an isotropic tensor so δ_{ij} is also isotropic, then $\left(\sqrt{g}\right)^{-1} = 1$, which shows that δ_{ij} is an absolute tensor.

1.7.1 Multiplication by a Scalar

This operation provides as a result a relative tensor of weight W, which components are the components of the original relative tensor multiplied by the scalar.

Let, for example, the relative tensor $(J)^W T_{ij}$ and the scalar m which product P_{ij} is given by $(J)^W P_{ij} = m(J)^W T_{ij}$. To demonstrate that this expression represents a tensor, it is enough to apply the transformation law of tensors to this expression.

1.7.1.1 Addition and Subtraction

This operation is defined for relative tensors of the same order and of the same kind, such as in the case of the following mixed tensors

$$(J)^W T_{ij}^k = (J)^W A_{ij}^k + (J)^W B_{ij}^k$$

Subtraction is defined in the same way as addition, however, admitting that one of the tensors be multiplied by the scalar -1: $(J)^W T_{ij}^k = (J)^W A_{ij}^k + (-1)(J)^W B_{ij}^k$. To demonstrate that these expressions represent relative tensors, the transformation law of tensors is applied to this expression.

1.7.1.2 Outer Product

This operation is defined in the same way as the outer product of absolute tensors. Let, for example, the relative tensor $(J)^{W_1} A_{ij\ldots}^{k\ldots}$ of variance (p, q) and weight W_1, and the relative tensor $(J)^{W_2} B_{\ldots rs}^{\ldots \ell m}$ of variance (u, v) and weight W_2, which multiplied provide

$$(J)^W T_{ij\ldots rs}^{k\ldots \ell m} = \left[(J)^{W_1} A_{ij\ldots}^{k\ldots}\right]\left[(J)^{W_2} B_{\ldots rs}^{\ldots \ell m}\right]$$

that is a relative tensor of variance $(p + u, q + v)$ and weight $W = W_1 + W_2$. To demonstrate that this product is a relative tensor, the transformation law of tensors is applied to this expression.

1.7.1.3 Contraction

This operation is defined in the same way as the contraction of the absolute tensors. Let, for example, the relative tensor $(J)^W T_{k\ell m}^{ij}$, in which contracting the upper index j provides

$$(J)^W T_{j\ell m}^{ij} = (J)^W T_{\ell m}^{i}$$

that shows that the resulting relative tensor has its order reduced in two, but maintains its weight W. To demonstrate that this contraction is a relative tensor, the transformation law of tensors is applied to this expression.

1.7.1.4 Inner Product

This operation is defined in the same manner as the inner product of the absolute tensors.

Let, for example, two relative tensors $(J)^{W_1}A_{ij}$ and $(J)^{W_2}B_k^{\ell}$, whereby it follows for the outer product of these tensors

$$(J)^{W_1+W_2}P_{ijk}^{\ell} = \left[(J)^{W_1}A_{ij}\right]\left[(J)^{W_2}B_k^{\ell}\right]$$

that represents a relative tensor of the fourth order and weight $W = W_1 + W_2$, and with the contraction of the index ℓ the inner product is given by

$$(J)^{W_1+W_2}P_{ij\ell}^{\ell} = \left[(J)^{W_1}A_{ij}\right]\left[(J)^{W_2}B_{\ell}^{\ell}\right] = (J)^{W_1+W_2}P_{ij}$$

This shows that the resulting relative tensor is of the second order and weight $W = W_1 + W_2$. To demonstrate that this product is a relative tensor, the transformation law of tensors is applied to this expression.

1.7.1.5 Pseudotensor

The varieties that present a few tensorial characteristics, for example, when changing the coordinate system they follow a transformation law that differs from the transformation law of tensors by the presence of the Jacobian, are called pseudotensor (relative tensors). However, these varieties are not maintained invariant when the coordinate system is transformed.

The definitions of the antisymmetric pseudotensors ε_{ijk} and ε^{ijk} are associated, respectively, to the permutation symbols in the covariant form e_{ijk} or in the contravariant form e^{ijk}, to which correspond the values $+1$ or -1 relative to the even or odd number of permutations, respectively. The Ricci pseudotensors ε_{ijk} and ε^{ijk} are associated to the concept of space orientation.

These varieties, when changing the coordinate system, transform in the same way as the tensors, but are not invariant after these transformations. This shows that a few characteristics are similar to the tensors but vary with the change of referential, for they assume the values ± 1, so they are not tensors in the sensu *stricto* of the term.

In expression (1.7.27) it is verified that $\varepsilon^{j_1 j_2 \cdots j_{n-m} i_1 i_2 \cdots i_m}$ has weight $+1$, and the tensor $T^{j_1 j_2 \cdots j_{n-m}}$ has weight superior to the weight of the antisymmetric tensor $A_{[i_1 i_2 \cdots i_m]}$. This expression illustrates the applying of the pseudotensors.

Exercise 1.21 Show that

(a) ε_{ijk} is a covariant pseudotensor of the third order and weight -1.
(b) ε^{ijk} is a contravariant pseudotensor of the third order and weight $+1$.
(c) The absolute pseudotensors can be obtained from these pseudotensors.

(a) The definition of determinant allows writing $J\overline{\varepsilon}_{pqr}$, and as the pseudotensor ε_{ijk} assume the values $0, \pm 1$, on being applied to this variety, it provides a linear and homogeneous transformation

$$J\overline{\varepsilon}_{pqr} = \frac{\partial x^i}{\partial \overline{x}^p}\frac{\partial x^j}{\partial \overline{x}^q}\frac{\partial x^k}{\partial \overline{x}^r}\varepsilon_{ijk} \Rightarrow \overline{\varepsilon}_{pqr} = \frac{\partial x^i}{\partial \overline{x}^p}\frac{\partial x^j}{\partial \overline{x}^q}\frac{\partial x^k}{\partial \overline{x}^r}J^{-1}\varepsilon_{ijk} = \varepsilon_{pqr}$$

then ε_{ijk} is a covariant pseudotensor of the third order and weight -1.

(b) In a way that is analogous to the previous case, for defining the determinant $\overline{J}\overline{\varepsilon}^{\ell mn}$, and for the transformation law of tensors

$$\overline{J}\overline{\varepsilon}^{pqr} = \frac{\partial \overline{x}^p}{\partial x^i}\frac{\partial \overline{x}^q}{\partial x^j}\frac{\partial \overline{x}^r}{\partial x^k}\varepsilon^{ijk} \Rightarrow \overline{\varepsilon}^{pqr} = \frac{\partial \overline{x}^p}{\partial x^i}\frac{\partial \overline{x}^q}{\partial x^j}\frac{\partial \overline{x}^r}{\partial x^k}\overline{J}^{-1}\varepsilon^{ijk}$$

As $J\overline{J} = 1$ it results in

$$\overline{\varepsilon}^{pqr} = \frac{\partial \overline{x}^p}{\partial x^i}\frac{\partial \overline{x}^q}{\partial x^j}\frac{\partial \overline{x}^r}{\partial x^k}J\varepsilon^{ijk}$$

then ε^{ijk} is a contravariant pseudotensor of the third order and weight $+1$.

(c) As the pseudotensor ε_{ijk} has weight -1, it follows by the transformation law of relative tensors into absolute tensors, where the upper asterisk indicates the absolute tensor

$$\overset{*}{\varepsilon}_{ijk} = \left[\left(\sqrt{g}\right)^{-1}\right]^{-1} \quad \varepsilon_{ijk} = \sqrt{g}\varepsilon_{ijk}$$

For the relative pseudotensor ε^{ijk} the absolute pseudotensor indicated by the lower asterisk exists

$$\varepsilon_*^{ijk} = \frac{1}{\sqrt{g}}\varepsilon^{ijk}$$

Exercise 1.22 Show that g^{ij} is an absolute tensor.

Rewriting expression (1.8.9)

$$\sqrt{\overline{g}} = J\sqrt{g}$$

and with the cofactor of the matrix of tensor g^{ij} given by

$$G^{ij} = \frac{1}{2}e^{ik\ell}e^{jpq}g_{kp}g_{\ell q}$$

and in terms of Ricci's pseudotensor

$$G^{ij} = \frac{1}{2} \varepsilon^{ik\ell} \varepsilon^{jpq} g_{kp} g_{\ell q}$$

it follows that

$$g^{ij} = \frac{G^{ij}}{g} = \frac{1}{2} \varepsilon^{ik\ell} \varepsilon^{jpq} g_{kp} g_{\ell q}$$

The term to the right of this expression. is the product of two pseudotensors and the tensors, being g^{ij} the inner product of these two varieties. This expression has weight $W = 0$, then g^{ij} is an *absolute tensor*.

1.7.1.6 Scalar Capacity

Let an antisymmetric pseudotensor C^{ijk} in an affine space, for which according to expression (1.7.1) for $N = 3$ and $p = 3$, there is only one independent component. Writing C^{ijk} as a function ε^{ijk} follows

$$C^{ijk} = \varepsilon^{ijk} c$$

where c is a component of the variety and with the change of the coordinate system

$$C^{ijk} = \frac{\partial x^i}{\partial \overline{x}^p} \frac{\partial x^j}{\partial \overline{x}^q} \frac{\partial x^k}{\partial \overline{x}^r} \overline{C}^{pqr}$$

and as the antisymmetry is maintained when the reference system is changed

$$\overline{C}^{pqr} = \varepsilon^{pqr} \overline{c} \tag{1.8.15}$$

Considering the component C^{123}:

$$C^{123} = \frac{\partial x^1}{\partial \overline{x}^p} \frac{\partial x^2}{\partial \overline{x}^q} \frac{\partial x^3}{\partial \overline{x}^r} \overline{C}^{pqr} \tag{1.8.16}$$

and substituting expression (1.8.15) in expression (1.8.16)

$$c = \varepsilon^{pqr} \frac{\partial x^1}{\partial \overline{x}^p} \frac{\partial x^2}{\partial \overline{x}^q} \frac{\partial x^3}{\partial \overline{x}^r} \overline{c} \tag{1.8.17}$$

Let

$$J = \varepsilon^{pqr} \frac{\partial x^1}{\partial \overline{x}^p} \frac{\partial x^2}{\partial \overline{x}^q} \frac{\partial x^3}{\partial \overline{x}^r} \tag{1.8.18}$$

results in the following expressions

$$c = J\overline{c} \Rightarrow \overline{c} = \frac{1}{J}c = \overline{J}c \tag{1.8.19}$$

Function c is the only independent component of the antisymmetric pseudotensor C^{ijk}, which is called scalar capacity. Then a scalar capacity is a pseudotensor of weight -1. To illustrate the concept of scalar capacity, let, for example, the antisymmetric variety dV^{ijk} that defines an elementary volume in space E_3. This analysis follows the same routine presented when defining the scalar capacity. The elementary volume is obtained by means of the mixed product of three vectors that define the three reference axes in this space

$$dV^{ijk} = \begin{vmatrix} dx^1 & 0 & 0 \\ 0 & dx^2 & 0 \\ 0 & 0 & dx^3 \end{vmatrix} = dx^1 dx^2 dx^3 = dV \Rightarrow dV^{ijk} = dx^i dx^j dx^k \tag{1.8.20}$$

and with the transformation law of tensors

$$dV^{ijk} = \frac{\partial x^i}{\partial \overline{x}^p}\frac{\partial x^j}{\partial \overline{x}^q}\frac{\partial x^k}{\partial \overline{x}^r} d\overline{x}^p d\overline{x}^q d\overline{x}^r \quad dV = \frac{\partial x^1}{\partial \overline{x}^p}\frac{\partial x^2}{\partial \overline{x}^q}\frac{\partial x^3}{\partial \overline{x}^r} d\overline{x}^p d\overline{x}^q d\overline{x}^r \tag{1.8.21}$$

The antisymmetry of the pseudotensor is maintained when changing the coordinate system

$$d\overline{x}^p d\overline{x}^q d\overline{x}^r = \varepsilon^{ijk} d\overline{V} \tag{1.8.22}$$

and substituting expression (1.8.21) in expression (1.8.22)

$$dV = \varepsilon^{ijk} \frac{\partial x^1}{\partial \overline{x}^p}\frac{\partial x^2}{\partial \overline{x}^q}\frac{\partial x^3}{\partial \overline{x}^r} d\overline{V}$$

results in the following expressions

$$dV = J d\overline{V} \Rightarrow d\overline{V} = \frac{1}{J} dV \quad \therefore d\overline{x}^1 d\overline{x}^2 d\overline{x}^3 = \frac{1}{J} dx^1 dx^2 dx^3 \tag{1.8.23}$$

This shows that the elementary volume in an affine space is a pseudoscalar of weight -1. In a more restricted manner, it says that the volume is a scalar capacity. The term *capacity* comes from the association of the volume (capacity, content) to the variety being analyzed. It is concluded that the integration of expression (1.8.23), which represents a scalar field, is a pseudoscalar.

1.7.1.7 Scalar Density

Let the antisymmetric pseudotensor D_{ijk}, for which an analysis analogous to the one developed when defining the scalar capacity is carried out

$$D_{123} = \mathcal{D} = \varepsilon_{ijk} \frac{\partial \overline{x}^i}{\partial x^1} \frac{\partial \overline{x}^j}{\partial x^2} \frac{\partial \overline{x}^k}{\partial x^3} \overline{\mathcal{D}}$$

$$\overline{J} = \varepsilon_{pqr} \frac{\partial \overline{x}^1}{\partial x^p} \frac{\partial \overline{x}^2}{\partial x^q} \frac{\partial \overline{x}^3}{\partial x^r} = \varepsilon_{ijk} \frac{\partial \overline{x}^i}{\partial x^1} \frac{\partial \overline{x}^j}{\partial x^2} \frac{\partial \overline{x}^k}{\partial x^3}$$

then

$$\mathcal{D} = \overline{J}\overline{\mathcal{D}} \Rightarrow \mathcal{D} = \frac{1}{J}\overline{\mathcal{D}} \Rightarrow \overline{\mathcal{D}} = J\mathcal{D}$$

Function $\mathcal{D}$ is the unique component of the antisymmetric pseudotensor D_{ijk}, which is called scalar density. Then a scalar density is a pseudotensor of weight $+1$. To illustrate the concept of scalar density, let, for example, a body of elementary mass dm in the affine space E_3. This mass is determined by means of *density* (specific mass) $\rho(x^1, x^2, x^3)$ and the elementary volume dV, thus $dm = \rho(x^1, x^2, x^3)\, dV$.

Considering that the mass is invariable (is a scalar)

$$dm = \rho\left(x^1, x^2, x^3\right) dV = \overline{\rho}\left(x^1, x^2, x^3\right) d\overline{V} \tag{1.8.24}$$

and as dV is a scalar capacity of weight -1, substituting expression (1.8.23) in expression (1.8.24) provides

$$\rho\left(x^1, x^2, x^3\right) dV = \overline{\rho}\left(x^1, x^2, x^3\right) \frac{1}{\overline{J}} dV \Rightarrow \rho\left(x^1, x^2, x^3\right) = J\overline{\rho}\left(x^1, x^2, x^3\right) \tag{1.8.25}$$

This shows that the *density* in an affine space is a pseudoscalar of weight $+1$, this variety being called scalar density. The term *density* is not physically correct, for in truth $\rho(x^1, x^2, x^3)$ such as it is presented defines the body's specific mass.

The concepts shown for the elementary volume and for *density* in the affine space E_3 can be generalized for the space E_N. The varieties that transform by means of the expressions with structure analogous to the structures of expressions (1.8.23) and (1.8.25) are called, respectively, scalar capacity and scalar density in space E_N.

1.7.1.8 Tensorial Capacity

Let the space E_3 where product of a scalar capacity c by the tensor T_{ij}^k exists, which defines a tensorial density C_{ij}^k given by

$$C^k_{ij} = cT^k_{ij} \tag{1.8.26}$$

and for a new coordinate system

$$\overline{C}^r_{pq} = \frac{\partial x^p}{\partial \overline{x}^i}\frac{\partial x^q}{\partial \overline{x}^j}\frac{\partial \overline{x}^r}{\partial x^k} C^k_{ij} \tag{1.8.27}$$

it follows that

$$\overline{C}^r_{pq} = \frac{\partial x^p}{\partial \overline{x}^i}\frac{\partial x^q}{\partial \overline{x}^j}\frac{\partial \overline{x}^r}{\partial x^k} cT^k_{ij} = \frac{\partial x^p}{\partial \overline{x}^i}\frac{\partial x^q}{\partial \overline{x}^j}\frac{\partial \overline{x}^r}{\partial x^k} J\overline{c}T^k_{ij} \Rightarrow \overline{C}^r_{pq} = J\overline{c}\overline{T}^r_{pq} \tag{1.8.28}$$

Expression (1.8.28) shows that C^k_{ij} transforms in accordance with a law that is similar to the transformation law of scalar capacity; however, it does not represent a relative scalar but a relative tensor of weight $+1$. The generalization of the concepts of tensorial capacity for the space E_N is immediate.

1.7.1.9 Tensorial Density

Let, for example, the space E_3 where the product of a scalar density $\mathcal{D}$ by the tensor T^k_{ij} exists, which defines a tensorial density D^k_{ij} given by

$$D^k_{ij} = \mathcal{D}T^k_{ij} \tag{1.8.29}$$

and for a new coordinate system

$$\overline{D}^r_{pq} = \frac{\partial x^p}{\partial \overline{x}^i}\frac{\partial x^q}{\partial \overline{x}^j}\frac{\partial \overline{x}^r}{\partial x^k} D^k_{ij} = \frac{\partial x^p}{\partial \overline{x}^i}\frac{\partial x^q}{\partial \overline{x}^j}\frac{\partial \overline{x}^r}{\partial x^k} \mathcal{D}T^k_{ij} = \frac{\partial x^p}{\partial \overline{x}^i}\frac{\partial x^q}{\partial \overline{x}^j}\frac{\partial \overline{x}^r}{\partial x^k}\frac{1}{J}\overline{\mathcal{D}}T^k_{ij}$$

$$\overline{D}^r_{pq} = J^{-1}\overline{\mathcal{D}T}^r_{pq} \tag{1.8.30}$$

Expression (1.8.30) shows that D^k_{ij} transforms in accordance with a law that is to the scalar density transformation law. However, it does not represent a relative scalar but a relative tensor of weight -1. The generalization of the concepts of tensorial density for the space E_N is immediate.

The outer products between these varieties (pseudotensors and tensors) result in

Scalar capacity × scalar density	**= scalar**
Scalar capacity × tensor	**= tensorial capacity**
Scalar density × tensor	**= tensorial density**
Pseudotensor × pseudotensor	**= tensor**
Tensor × pseudotensor	**= pseudotensor**

1.8 Physical Components of a Tensor

In mathematics the approach to the problems, in general, is carried out by means of nondimensional parameters. In physics and engineering the parameters have magnitude and dimensions, for example, $\mathrm{N/mm^2}, \mathrm{m/s}$, etc. The analysis of a physical problem by tensorial means requires that the parameters being studied be invariant when changing the coordinate system. It happens that the axes of the coordinate systems generally do not have the same dimensions. A Cartesian coordinate system has axes that define lengths, but, for example, a spherical coordinate system has two axes that express nondimensional coordinates, the same occurring in the cylindrical coordinate system with one of their axes. Therefore, the components of a tensor have dimensions, and when the coordinate system is changed, these components vary in magnitude and dimension.

To express the transformation of tensors in a consistent way (in magnitude and dimension), and that these varieties can be added after a change of the coordinate systems, the same must be expressed in terms of their physical components.

1.8.1 Physical Components of a Vector

The concept of geometric vector is associated to the idea of displacement, its transformation law being $dx^k = \frac{\partial x^k}{\partial \overline{x}^i} d\overline{x}^i$ where the coefficients $\frac{\partial x^k}{\partial \overline{x}^i}$ are constants. With respect to a Cartesian coordinates, the term $dx^j \boldsymbol{g}^j$ represents a displacement in terms of the unit vectors of the coordinate axes

$$dx^j \boldsymbol{g}^j = dx^1 \boldsymbol{i} + dx^2 \boldsymbol{j} + dx^3 \boldsymbol{k} \tag{1.9.1}$$

However, this term in a curvilinear coordinates does not represent a displacement, so $\boldsymbol{g}^k$ will not be a unit vector in this coordinate system. This shows that the vector must be written in terms of components that express a displacement, called the physical components of the vector.

Consider the vector $\boldsymbol{u}$ with physical components u_j^*, which can be written in terms of these components and of their contravariant components

$$\boldsymbol{u} = \overset{*}{u_k} \boldsymbol{e}_k = u^k \boldsymbol{g}_k \tag{1.9.2}$$

Comparing expressions (1.9.1) and (1.9.2)

$$dx^j \boldsymbol{g}^j = dx^{*j} \boldsymbol{e}_j \tag{1.9.3}$$

where dx^{*j} are the physical components which by analogy correspond to the displacement dx^k, and with the unit vectors of base $\boldsymbol{g}_k, \boldsymbol{e}_k$, the components u^k, u_k^*

are obtained in terms of the unit vector $\boldsymbol{g_k}$. Let $\boldsymbol{g_i} \cdot \boldsymbol{g}_j = g_{ij}$ then $\|\boldsymbol{g}_j\| = \sqrt{g_{(jj)}}$, where the indexes shown in parenthesis do not indicate a summation in j. As the unit vector $\boldsymbol{e}_j$ is collinear with $\boldsymbol{g_i}$, thus

$$\boldsymbol{g}_j = \sqrt{g_{(jj)}}\boldsymbol{e}_j \tag{1.9.4}$$

and with expression (1.9.4) in expression (1.9.3)

$$dx^{*j} = \sqrt{g_{(jj)}}dx^j \tag{1.9.5}$$

In an analogous way, by means of expression (1.9.2)

$$u_k^* = \sqrt{g_{(kk)}}u^k \tag{1.9.6}$$

The physical components u_k^* have the characteristics of displacement, so they can be added vectorially (parallelogram rule), denoting the contravariant physical components of the vector. These components are not unique. Let another variety of components $\tilde{u}_k$ that represents the projection of vector $\boldsymbol{u}$ on the direction of the unit vector $\boldsymbol{e_k}$. Consider $\tilde{\boldsymbol{e}}_k$ the reciprocal unit vector of $\boldsymbol{e_k}$, whereby, for this reciprocal basis,

$$\widetilde{u}_k = \boldsymbol{u} \cdot \boldsymbol{e}_k \tag{1.9.7}$$

$$\boldsymbol{u} = u_k\boldsymbol{g}^k = \widetilde{u}_k\widetilde{\boldsymbol{e}}_k \tag{1.9.8}$$

but $\tilde{\boldsymbol{e}}_k$ is collinear with $\boldsymbol{g}^k$ thus

$$\boldsymbol{e}_k = \frac{1}{\widetilde{\boldsymbol{e}}_k}$$

$$\boldsymbol{g}_k = \sqrt{g_{(kk)}}\boldsymbol{e}_k = \frac{\sqrt{g_{(kk)}}}{\widetilde{\boldsymbol{e}}_k} \Rightarrow \widetilde{\boldsymbol{e}}_k = \frac{\sqrt{g_{(kk)}}}{\boldsymbol{g}_k}$$

$$\widetilde{\boldsymbol{e}}_k = \sqrt{g_{(kk)}}\boldsymbol{g}^k \tag{1.9.9}$$

where the indexes shown in parenthesis do not indicate summation in k, and with expression (1.9.9) in expression (1.9.7)

$$u_k = \sqrt{g_{(kk)}}\widetilde{u}_k \tag{1.9.10}$$

The physical components $\tilde{u}_k$ are the covariant components of vector $\boldsymbol{u}$. Putting $u_k = g_{(kk)}u^k$ and with the expressions (1.9.6) and (1.9.10) then in an orthogonal coordinate system $u_k^* = \widetilde{u}_k$. This shows that the distinction between the covariant and contravariant basis disappears when the coordinate system is orthogonal.

Fig. 1.11 Physical components of the vector $\boldsymbol{u}$ in the curvilinear coordinate system X^i

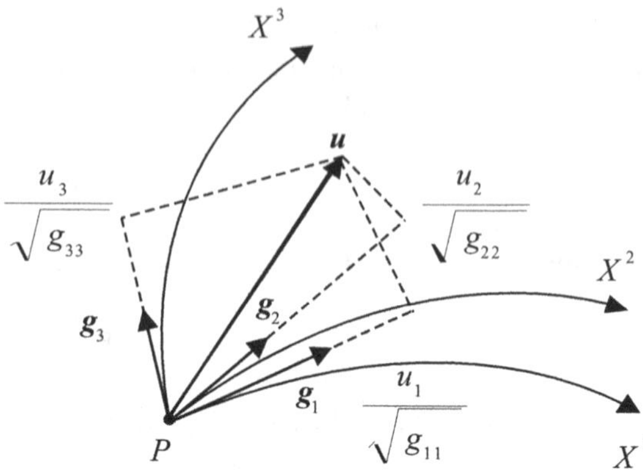

Figure 1.11 shows the physical components of vector $\boldsymbol{u}$ in the curvilinear coordinate system X^i. The components $\frac{u_k}{\sqrt{g_{(kk)}}}$ (expression (1.9.10)) represent the lengths of the projections which are orthogonal to the coordinate axes of the referential system. The components $\sqrt{g_{(kk)}}u^k$ (expression (1.9.6)) represent the lengths the of the sides of the parallelepiped, which diagonal is the vector $\boldsymbol{u}$.

Exercise 1.23 Calculate the contravariant, covariant, and physical components of the velocity vector of a point $\bar{v}^i = \frac{d\bar{x}^i}{dt}$, in terms of the cylindrical coordinates of the point $x^i(r, \theta, z)$.

Cartesian to cylindrical	Cylindrical to Cartesian
$\bar{x}^1 = x^1 \cos x^2 = r\cos\theta$	$x^1 = \sqrt{(\bar{x}^1)^2 + (\bar{x}^2)^2} = r$
$\bar{x}^2 = x^2 \sin x^2 = r\sin\theta$	$x^2 = \operatorname{arctg}\frac{\bar{x}^2}{\bar{x}^1} = \theta$
$\bar{x}^3 = x^3 = z$	$x^3 = \bar{x}^3 = z$

The Cartesian coordinates of the vector are $\bar{v}^i = \frac{d\bar{x}^i}{dt}$, and for the cylindrical coordinates

$$v^i = \frac{\partial x^i}{\partial \bar{x}^j}\bar{v}^j \Rightarrow v^i = \frac{\partial x^i}{\partial \bar{x}^j}\frac{d\bar{x}^j}{dt} = \frac{dx^i}{dt}$$

This shows that the contravariant components of *vector* $\boldsymbol{v}$ are derivatives with respect to the time of the position vector defined by the coordinates $\bar{x}^i$, then

$$\{v^i\} = \left\{\frac{dx^1}{dt}, \frac{dx^2}{dt}, \frac{dx^3}{dt}\right\} = \left\{\frac{dr}{dt}, \frac{d\theta}{dt}, \frac{dz}{dt}\right\}$$

For the covariant components in terms of the cylindrical coordinates

$$v_i = \frac{\partial \overline{x}^i}{\partial x^i}\overline{v}^j = g_{ij}v^j$$

$$g_{ij} = \begin{bmatrix} 1 & 0 & 0 \\ 0 & r^2 & 0 \\ 0 & 0 & 1 \end{bmatrix}$$

Developing the expression of v_i

$$v_1 = g_{11}v^1 + g_{12}v^2 + g_{13}v^3 = v^1 = \frac{dx^1}{dt} = \frac{dr}{dt}$$

$$v_2 = g_{12}v^1 + g_{22}v^2 + g_{23}v^3 = \left(x^1\right)^2\frac{dx^2}{dt} = r^2\frac{d\theta}{dt}$$

$$v_3 = g_{31}v^1 + g_{23}v^2 + g_{33}v^3 = \frac{dx^3}{dt} = \frac{dz}{dt}$$

whereby

$$\{v^i\} = \left\{\frac{dx^1}{dt}, \left(x^1\right)^2\frac{dx^2}{dt}, \frac{dx^3}{dt}\right\} = \left\{\frac{dr}{dt}, r^2\frac{d\theta}{dt}, \frac{dz}{dt}\right\}$$

The vector norm is given by

$$\|\boldsymbol{v}\| = \sqrt{\left(\frac{dx^1}{dt}\right)^2 + \left[(x^1)^2\frac{dx^2}{dt}\right]^2 + \left(\frac{dx^3}{dt}\right)^2}$$

whereby for its physical components

$$\{v\} = \left\{\frac{dx^1}{dt}, x^1\frac{dx^2}{dt}, \frac{dx^3}{dt}\right\} = \left\{\frac{dr}{dt}, r\frac{d\theta}{dt}, \frac{dz}{dt}\right\}$$

Exercise 1.24 Let the vector $\boldsymbol{u} = 3\boldsymbol{g}_1 + \boldsymbol{g}_2 + 2\boldsymbol{g}_3$, having $\boldsymbol{g}_1 = 2\boldsymbol{e}_1, \boldsymbol{g}_2 = 2\boldsymbol{e}_1 + \boldsymbol{e}_2$, and $\boldsymbol{g}_3 = 2\boldsymbol{e}_1 + \boldsymbol{e}_2 + 3\boldsymbol{e}_3$, where $\boldsymbol{e}_1, \boldsymbol{e}_2, \boldsymbol{e}_3$ are orthonormal vectors, calculate their contravariant physical components.

From the covariant basis

$$\boldsymbol{g}_1 \cdot \boldsymbol{g}_1 = 2\boldsymbol{e}_1.2\boldsymbol{e}_1 = 4 \Rightarrow \sqrt{g_{11}} = 2$$

$$\boldsymbol{g}_2 \cdot \boldsymbol{g}_2 = 2\boldsymbol{e}_1.2\boldsymbol{e}_1 + \boldsymbol{e}_2 \cdot \boldsymbol{e}_2 = 4 + 1 = 5 \Rightarrow \sqrt{g_{22}} = \sqrt{5}$$

$$\boldsymbol{g}_3 \cdot \boldsymbol{g}_3 = 2\boldsymbol{e}_1.2\boldsymbol{e}_1 + \boldsymbol{e}_2 \cdot \boldsymbol{e}_2 + 3\boldsymbol{e}_3 \cdot 3\boldsymbol{e}_3 = 4 + 1 + 9 = 14 \Rightarrow \sqrt{g_{33}} = \sqrt{14}$$

follows

$$u^{*1} = u^1\sqrt{g_{11}} = 3 \times 2 = 6$$
$$u^{*2} = u^2\sqrt{g_{22}} = 2 \times \sqrt{5} = 2\sqrt{5}$$
$$u^{*3} = u^3\sqrt{g_{33}} = 1 \times \sqrt{14} = \sqrt{14}$$

1.8.1.1 Physical Components of the Second-Order Tensor

The contravariant physical components of the vectors $\boldsymbol{u}$ and $\boldsymbol{v}$ are given by expression (1.9.10)

$$u_i = \sqrt{g_{(ii)}}\,\widetilde{u}_i \Rightarrow \widetilde{u}_i = \frac{u_i}{\sqrt{g_{(ii)}}}$$
$$v_j = \sqrt{g_{(jj)}}\,\widetilde{v}_j \Rightarrow \widetilde{v}_j = \frac{v_j}{\sqrt{g_{(jj)}}}$$

For the second-order tensor

$$\widetilde{T}_{ij} = \widetilde{u}_i\widetilde{v}_j$$
$$\widetilde{T}_{ij} = \frac{u_i}{\sqrt{g_{(ii)}}}\frac{v_j}{\sqrt{g_{(jj)}}} = \frac{1}{\sqrt{g_{(ii)}}\sqrt{g_{(jj)}}}\left(u_iv_j\right) = \frac{T_{ij}}{\sqrt{g_{(ii)}g_{(jj)}}}$$
$$T_{ij} = \sqrt{g_{(ii)}g_{(jj)}}\,\widetilde{T}_{ij} \tag{1.9.11}$$

In a related manner, for the contravariant physical components

$$^{*}T^{ij} = \sqrt{g_{(ii)}g_{(jj)}}\,\widetilde{T}^{ij} \tag{1.9.12}$$

The obtaining of the physical components of tensors of a higher order follows the analogous way to that of the second-order tensors.

1.9 Tests of the Tensorial Characteristics of a Variety

The transformation law of the tensors and the quotient law allow establishing a group of functions N^p of the coordinates of the referential system X^i which are the components of a tensor. The tensorial nature of the functions that fulfill these requirements is highlighted by the invariance that this variety has when there is a change of the coordinate system. However, the evaluation if a variety has tensorial characteristics by means of the quotient law is not wholly complete, as it will be shown next applying to the group of N^2 components of a variety T_{pq}, for which it is desired to search if it has the characteristics of a tensor. Multiplying T_{pq} by an arbitrary vector v^p and admitting by hypothesis that $T_{pq}v^p v^q = m$, where m is a scalar, it provides for a new coordinate system $\overline{T}_{ij}\overline{v}^i\overline{v}^j = \overline{m}$, and as m is an invariant, then $m = \overline{m}$, by means of the transformation law of vectors

$$T_{pq}v^p v^q = \frac{\partial \overline{x}^i}{\partial x^p}\frac{\partial \overline{x}^j}{\partial x^q}\overline{T}_{ij}v^p v^q$$

Then

$$\left(T_{pq} - \frac{\partial \overline{x}^i}{\partial x^p}\frac{\partial \overline{x}^j}{\partial x^q}\overline{T}_{ij}\right)v^p v^q = 0 \tag{1.10.1}$$

The summation rule is applied varying the indexes p and q, so the product $v^p v^q$ is not, in general, null. Consider the vectors v^i with unit components $(1, 0, 0\cdots 0)$, $(0, 1, 0\cdots 0)$, and $(0, 0, 0\cdots 1)$, the term in parenthesis of expression (1.10.1) stays

$$\left(T_{11} - \frac{\partial \overline{x}^i}{\partial x^1}\frac{\partial \overline{x}^j}{\partial x^1}\overline{T}_{ij}\right)v^1 v^1 = 0$$

and as $v^1 v^1 \neq 0$

$$T_{11} - \frac{\partial \overline{x}^i}{\partial x^1}\frac{\partial \overline{x}^j}{\partial x^1}\overline{T}_{ij} = 0 \tag{1.10.2}$$

In an analogous way it results in

$$T_{22} - \frac{\partial \overline{x}^i}{\partial x^2}\frac{\partial \overline{x}^j}{\partial x^2}\overline{T}_{ij} = 0 \tag{1.10.3}$$

and so successively for the other values assumed for the indexes. This shows that for $p = q$ the terms in parenthesis from expression (1.10.1) cancel each other. However, for $p \neq q$ the complementary analysis of this expression behavior becomes necessary.

Let vector v^i with components $(v^1, v^2, 0, \cdots 0)$, whereby from expression (1.10.1) for $p, q = 1, 2$, it follows that

$$\begin{aligned}&\left(T_{11} - \frac{\partial \overline{x}^i}{\partial x^1}\frac{\partial \overline{x}^j}{\partial x^1}\overline{T}_{ij}\right)v^1 v^1 + \left(T_{12} - \frac{\partial \overline{x}^i}{\partial x^1}\frac{\partial \overline{x}^j}{\partial x^2}\overline{T}_{ij}\right)v^1 v^2 \\ &\quad + \left(T_{21} - \frac{\partial \overline{x}^i}{\partial x^2}\frac{\partial \overline{x}^j}{\partial x^1}\overline{T}_{ij}\right)v^2 v^1 + \left(T_{22} - \frac{\partial \overline{x}^i}{\partial x^2}\frac{\partial \overline{x}^j}{\partial x^2}\overline{T}_{ij}\right)v^2 v^2 = 0\end{aligned} \tag{1.10.4}$$

Expressions (1.10.2) and (1.10.3) simplify expression (1.10.4), for the coefficients of the terms $v^p v^q$ are null for $p = q$. For $p \neq q$ with $\overline{T}_{ij} = \overline{T}_{ji}$

$$\frac{\partial \overline{x}^i}{\partial x^1}\frac{\partial \overline{x}^j}{\partial x^2}\overline{T}_{ij} = \frac{\partial \overline{x}^i}{\partial x^2}\frac{\partial \overline{x}^j}{\partial x^1}\overline{T}_{ij}$$

and with the hypothesis of symmetry results in

$$\frac{\partial \overline{x}^i}{\partial x^1}\frac{\partial \overline{x}^j}{\partial x^2}\overline{T}_{ij} = \frac{\partial \overline{x}^i}{\partial x^2}\frac{\partial \overline{x}^j}{\partial x^1}\overline{T}_{ji}$$

Expression (1.10.4) is rewritten as

$$\left[(T_{12}+T_{21}) - \left(\overline{T}_{ij}+\overline{T}_{ji}\right)\frac{\partial \overline{x}^i}{\partial x^1}\frac{\partial \overline{x}^j}{\partial x^2}\right]v^1 v^2 = 0$$

and as the components v^1 and v^2 are arbitrary, for $v^1 = v^2 = 1$

$$T_{12}+T_{21} = \left(\overline{T}_{ij}+\overline{T}_{ji}\right)\frac{\partial \overline{x}^i}{\partial x^1}\frac{\partial \overline{x}^j}{\partial x^2} \tag{1.10.5}$$

Generalizing expression (1.10.5) for the variation of the indexes $p, q = 1, 2, 3, \ldots$, it results in

$$T_{pq}+T_{qp} = \left(\overline{T}_{ij}+\overline{T}_{ji}\right)\frac{\partial \overline{x}^i}{\partial x^p}\frac{\partial \overline{x}^j}{\partial x^q} \tag{1.10.6}$$

Expression (1.10.6) is the transformation law of second-order tensors, for the term $\left(T_{pq}+T_{qp}\right)$ represents the symmetric part of tensor $2T_{pq}$. However, the antisymmetric part of this tensor is not contained in this analysis, whereby it cannot be concluded that this portion has tensorial characteristics. It is concluded that only the symmetric part of the N^2 components of variety T_{pq} is a tensor, for when applying the quotient law to this portion it transforms according to the transformation law of second-order tensors. This is the reason why the quotient law must be applied with caution, so as to avoid evaluation errors when checking the tensorial characteristics of a variety.

The transformation law of tensors and the consideration of invariance of the variety when having a linear transformation form the criterion that is most appropriate to evaluate if the N^p components of this variety have tensorial characteristics.

Problems

1.1 Use the index notation to write:

(a) $\begin{Bmatrix} \dfrac{d\overline{x}^1}{dt} \\ \dfrac{d\overline{x}^2}{dt} \end{Bmatrix} = \begin{bmatrix} a_{11} & a_{12} \\ a_{21} & a_{22} \end{bmatrix}\begin{Bmatrix} x_1 \\ x_2 \end{Bmatrix}$; (b) $\Phi = x_1^2 + x_2^2 + 2x_1x_2$

Answer: (a) $\overline{x}_{,t} = a_{ij}x_j$; (b) $\Phi = x_ix_j$.

1.2 Let a_{ij} constant $\forall i, j$, calculate $\frac{\partial\left(a_{ij}x_i x_j\right)}{\partial x_k}$) where $a_{ij} = a_{ji}$.
Answer: $\frac{\partial(2a_{ik}x_i)}{\partial x_\ell} = 2a_{ik}$

1.3 If $a_{ijk}x^i x^j x^k = 0\ \forall x^1, x^2, \cdots, x^n$ and a_{ijk} are constant values, show that $a_{kji} + a_{jki} + a_{ikj} + a_{ijk} + a_{kij} + a_{jik} = 0$.

1.4 Calculate for $i, j = 1, 2, 3$: (a) $\delta_{ij}A_i$, (b) $\delta_{ij}A_{ij}$, (c) δ_{ii}, (d) $\delta_{ij}\delta_{ji}$, (e) $\delta_{ij}\delta_{jk}\delta_{k\ell}$, (f) $C = a_{ijk}a_{ijk}$
Answer: (a) $\delta_{ij}A_i = A_j$, (b) $\delta_{ij}A_i = A_{ii} = A_{jj}$, (c) $\delta_i{}^i = 3$, (d) $\delta_{ij}\delta^{ijij}{}_{ji}{}^{jiji} = 3$, (e) $\delta_{ij}\delta^{ijij}{}_{jk}\delta^{jkjk}{}_{k\ell}{=}^{k\ell k\ell}\delta_{i\ell}$, and (f) 64.

1.5 Calculate the Jacobian of the linear transformations between the coordinate systems (a) $x^1 = \overline{x}^1$; $x^2 = \overline{x}^1\overline{x}^2$; $x^3 = \overline{x}^1\overline{x}^2\overline{x}^3$; (b) $x^1 = \overline{x}^1\cos\overline{x}^2\sin\overline{x}^3$; $x^2 = \overline{x}^1\sin\overline{x}^2\sin\overline{x}^3$; $x^3 = \overline{x}^1\cos\overline{x}^3$.
Answer: (a) $J = \left(\overline{x}^1\right)^2\overline{x}^2$; (b) $J = -\left(\overline{x}^1\right)^2\sin\overline{x}^3$.

1.6 Given the tensor $T_{k\ell} = \begin{bmatrix} 1 & 0 & 0 \\ 0 & 2 & 1 \\ 0 & 1 & 3 \end{bmatrix}$ in the coordinate system X^i, calculate the components of this tensor in the coordinate system $\overline{X}^i$, with the relations between the coordinates of these systems given by $x^1 = \overline{x}^1 + \overline{x}^3$, $x^2 = \overline{x}^1 + \overline{x}^2$, $x = \overline{x}^3$.

Answer: $\overline{T}_{ij} = \begin{bmatrix} 3 & 2 & 2 \\ 2 & 2 & 1 \\ 2 & 1 & 4 \end{bmatrix}$

1.7 Given the tensor $T^{ij} = \begin{bmatrix} 1 & 1 & 5 \\ 1 & 2 & -1 \\ 5 & -1 & 3 \end{bmatrix}$ in the coordinate system X^i, calculate the components of this tensor in the coordinate system $\overline{X}^i$, with the relations between the coordinates of these systems given by $x^1 = \overline{x}^1 + 2\overline{x}^2$, $x^2 = 3\overline{x}^3$, $x = \overline{x}^3$.

Answer: $\overline{T}^{ij} = \begin{bmatrix} 25 & 8 & 2 \\ 8 & 4 & 10 \\ 2 & 10 & 3 \end{bmatrix}$

1.8 Show that (a) $\mathrm{tr}(\boldsymbol{T}) = \mathrm{tr}(\boldsymbol{S})$, where $\boldsymbol{T}$ and $\boldsymbol{S}$ are, respectively, a symmetric and an antisymmetric tensor, both of the second order; (b) $T_{ijk\ell} = 0$, being $T_{ijk\ell}$ one symmetric tensor in the indexes i, j and antisymmetric in the indexes j, ℓ.

1.9 Decompose the second-order tensor in two tensors, one symmetric and another antisymmetric

$$T_{ij} = \begin{bmatrix} -1 & 2 & 0 \\ 3 & 0 & -2 \\ 1 & 0 & 1 \end{bmatrix}$$

Answer: $\begin{bmatrix} -1 & 2.5 & 0.5 \\ 2.5 & 0 & -1 \\ 0.5 & -1 & 1 \end{bmatrix} \begin{bmatrix} 0 & -0.5 & -0.5 \\ 0.5 & 0 & -1 \\ 0.5 & 1 & 0 \end{bmatrix}$.

1.10 Consider the tensor T_{ij} that satisfies the tensorial equation $mT_{ij} + nT_{ji} = 0$, where $m > 0$ and $n > 0$ are scalars. Prove that if T_{ij} is a symmetric tensor, then $m = -n$, and $m = n$ if this is an antisymmetric tensor.

1.11 Let the Cartesian coordinate system with basis vectors $\boldsymbol{e}_1, \boldsymbol{e}_2, \boldsymbol{e}_3$. Calculate the metric tensor of the space with basis vectors $\boldsymbol{g}_1 = \boldsymbol{e}_1,\ \boldsymbol{g}_2 = \boldsymbol{e}_1 + \boldsymbol{e}_2$, and $\boldsymbol{g}_3 = \boldsymbol{e}_1 + \boldsymbol{e}_2 + \boldsymbol{e}_3$.

Answer: $g_{ij} = \begin{bmatrix} 1 & 1 & 1 \\ 1 & 2 & 2 \\ 1 & 2 & 3 \end{bmatrix}$

1.12 Let the basis vectors $\boldsymbol{e}_1, \boldsymbol{e}_2$ of the coordinate system X^i with metric tensor g_{ij} and the basis vectors $\widetilde{\boldsymbol{e}}_1 = 3\boldsymbol{e}_1 + \boldsymbol{e}_2$ and $\widetilde{\boldsymbol{e}}_2 = -\boldsymbol{e}_1 + 2\boldsymbol{e}_2$ of the coordinate system $\widetilde{X}^i$. Calculate the covariant components of the metric tensor $\widetilde{g}_{ij}$ in terms of the components of g_{ij}.

Answer: $\widetilde{g}_{11} = 9g_{11} + 6g_{12} + g_{22}$; $\widetilde{g}_{12} = \widetilde{g}_{21} = -3g_{11} + 5g_{12} + 2g_{22}$; $\widetilde{g}_{22} = g_{11} - 4g_{12} + 4g_{22}$.

1.13 Calculate the contravariant components of the vector $\boldsymbol{u} = \boldsymbol{g}_1 + 2\boldsymbol{g}_2 + \boldsymbol{g}_3$, where the covariant base vectors are $\boldsymbol{g}_1 = \boldsymbol{e}_1$, $\boldsymbol{g}_2 = \boldsymbol{e}_1 + \boldsymbol{e}_2$, $\boldsymbol{g}_3 = \boldsymbol{e}_3$, being $\boldsymbol{e}_1, \boldsymbol{e}_2, \boldsymbol{e}_3$ base vectors.

Answer: $\boldsymbol{u} = 4\boldsymbol{g}^1 + 7\boldsymbol{g}^2 + 8\boldsymbol{g}^3$.

1.14 Let the contravariant base vectors $\boldsymbol{g}_1 = \boldsymbol{e}_1$, $\boldsymbol{g}_2 = \boldsymbol{e}_1 + \boldsymbol{e}_2$, and $\boldsymbol{g}_3 = \boldsymbol{e}_1 + \boldsymbol{e}_2 + \boldsymbol{e}_3$, where $\boldsymbol{e}_1, \boldsymbol{e}_2, \boldsymbol{e}_3$ are the vectors of the one orthonormal base. Calculate the:

(a) Vectors $\boldsymbol{g}^1, \boldsymbol{g}^2, \boldsymbol{g}^3$ of the contravariant base

(b) Metric tensor and the conjugated tensor.

Answer: (a) $\begin{cases} \boldsymbol{g}^1 = \boldsymbol{e}_1 - \boldsymbol{e}_2 \\ \boldsymbol{g}^2 = \boldsymbol{e}_2 - \boldsymbol{e}_3 \\ \boldsymbol{g}^3 = \boldsymbol{e}_3 \end{cases}$; (b) $g_{ij} = \begin{bmatrix} 1 & 1 & 1 \\ 1 & 2 & 2 \\ 1 & 2 & 3 \end{bmatrix}$ $g^{ij} = \begin{bmatrix} 2 & -1 & 0 \\ -1 & 2 & 2 \\ 0 & -1 & 1 \end{bmatrix}$

1.15 Consider the coordinate system $\overline{x}^1 = x^1 \cos x^2, \overline{x}^2 = x^1 \sin x^2, \overline{x}^3 = x^3$. Calculate the arc length along the parametric curve $\overline{x}^1 = a \cos t$, $\overline{x}^2 = a \sin t$, $\overline{x}^3 = b\, t$ in the interval $0 \le t \le c$, being a, b, c positive constants.

Answer: $L = c\sqrt{a^2 + b^2}$

1.16 Calculate the angle between the vectors

(a) $\boldsymbol{u}(2; -3; 1)$, $\boldsymbol{v}(3; -1; -2)$; (b) $\boldsymbol{u}(2; 1; -5)$, $\boldsymbol{v}(5; 0; 2)$.

Answer: (a) 60°; (b) 90°

1.17 With $i, j, k = 1, 2, 3$ calculate the following expressions:

(a) $u_i v_j \delta_{ji} - v_k u_i \delta_{ki}$; (b) $\delta_{ij}\delta_{ji}$; (c) $e_{ijk} u_i u_j u_k$

Answer: (a) zero, (b) 3, (c) zero

1.18 Show that the followings expressions are invariants (a) $T_{ij}u_iv_j$; (b) T_{ii}; (c) $\det T_{ij}$.

1.19 Let the vector (a) u_i show that if $A^{ij}u_iu_j = B^{ij}u_iu_j$, then $A^{ij} + A^{ji} = B^{ij} + B^{ij}$; (b) u_i and if $A^{ij}u_iu_j$ is invariant, show that $\left(A^{ij} + A^{ji}\right)$ is a tensor.

1.20 Let T_{pqrs} an absolute tensor, show that if $T_{ijk\ell} + T_{ij\ell k} = 0$ in the coordinate system X^i, then $\overline{T}_{ijk\ell} + \overline{T}_{ij\ell k} = 0$ in another coordinate system $\overline{X}^i$.

1.21 Let the vector $\boldsymbol{u} = \boldsymbol{g}_1 + 2\boldsymbol{g}_2 + \boldsymbol{g}_3$, having $\boldsymbol{g}_1 = \boldsymbol{e}_1$, $\boldsymbol{g}_2 = \boldsymbol{e}_1 + \boldsymbol{e}_2$, and $\boldsymbol{g}_3 = \boldsymbol{e}_1 + \boldsymbol{e}_2 + \boldsymbol{e}_3$, where $\boldsymbol{e}_1, \boldsymbol{e}_2, \boldsymbol{e}_3$ are orthonormal vectors, calculate their contravariant physical components.
Answer: $u^{*1} = 1$, $u^{*2} = 2\sqrt{2}$, $u^{*3} = \sqrt{3}$

1.22 Calculate the value of the permutation symbol e_{321546}.
Answer: $e_{321546} = 1$.

1.23 Show that

$$\text{(a) } e_{ijk}e_{jki} = 6;\ \text{(b) } e_{ijk}u_ju_k = 0;\ \text{(c) } e_{r\ell m}e^{rst} = \delta^{rst}_{r\ell m} = \begin{cases} 1 \\ 0 \\ -1 \end{cases} \quad r, \ell, m, s, t = 1,$$

2, 3;
(d) $e_{r\ell m}e^{rst} = \delta^s_\ell\delta^t_m - \delta^t_\ell\delta^s_m$; (e) $e^{ij}e_{ij} = 2!$; (f) $e^{ijk\ell}e_{ijk\ell} = 4!$; (g) $e^{ijk\ell\ldots}e_{ijk\ell\ldots} = N!$ where N is the index number.

1.24 For $i, j, k, \ell = 1, 2, 3$ show that

$$\text{(a) } \delta^{ij}_{k\ell} = \begin{vmatrix} \delta^i_k & \delta^i_\ell \\ \delta^j_k & \delta^j_\ell \end{vmatrix} = \delta^i_k\delta^j_\ell - \delta^i_\ell\delta^j_k,\ \text{(b) } \delta^{ijk}_{rst} = \begin{vmatrix} \delta^i_r & \delta^i_s & \delta^i_t \\ \delta^j_r & \delta^j_s & \delta^j_t \\ \delta^k_r & \delta^k_s & \delta^k_t \end{vmatrix} = \delta^{ij}_{k\ell},\ \text{(c) } \delta^{ijk}_{ijk} = 6.$$

1.25 Calculate the determinant by means of the expansion of the permutation symbol

$$\begin{vmatrix} 1 & 1 & 0 & 1 \\ -1 & 0 & 1 & 1 \\ 0 & 1 & -1 & 1 \\ 1 & -1 & 1 & 0 \end{vmatrix} = e_{ijk\ell}a_{1i}a_{2j}a_{3k}a_{4\ell}$$

Answer: -3.

1.26 Show that (a) $\varepsilon_{ijk} = \left(\boldsymbol{e}_j \times \boldsymbol{e}_k\right) \cdot \boldsymbol{e}_i$ where $\boldsymbol{e}_i, \boldsymbol{e}_j, \boldsymbol{e}_k$ are unit vectors of one coordinate system; (b) if $\boldsymbol{u}_i$ and $\boldsymbol{u}^i$ are associate tensors and $u^i = \varepsilon^{ijk}u_jv_k$, then $u_i = \varepsilon_{ijk}u^jv^k$; (c) $\varepsilon^{ijk}u_iv_jw_k = \varepsilon_{ijk}u^iv^jw^k$.

1.27 Simplify the expression $F = \varepsilon_{ijk}\varepsilon_{pqr}A_{ip}A_{jq}A_{kr}$.

1.28 Verify if the expression $\varepsilon_{mnp}\varepsilon_{mij} + \varepsilon_{mnj}\varepsilon_{mpi} = \varepsilon_{mni}\varepsilon_{mpj}$ is correct or false. Justify the answer.

1.29 Write the tensor components $T^{ij} = \frac{1}{2}\varepsilon^{ijk}A_{k\ell}$ with $i, j, k, \ell = 1, 2, 3, 4$, and show that if $ijk\ell$ is an even permutation for the pair of 1234, then $T^{ij} = A_{k\ell}$.

Chapter 2
Covariant, Absolute, and Contravariant Derivatives

2.1 Initial Notes

The curve represented by a function $\phi(x^i)$ in a closed interval is continuous if this function is continuous in this interval. If the curve is parameterized, i.e., $\phi[x^i(t)]$ being $t \in [a, b]$, then it will be continuous if $x^i(t)$ are continuous functions in this interval, and it will be smooth if it has continuous and non-null derivatives for a value of $t \in [a, b]$. The smooth curves do not intersect, i.e., the conditions $x^i(a) = x^i(b)$ will only be satisfied if $a = b$. This condition defines a curve that can be divided into differential elements, forming curve arcs. For the case in which the initial and final points coincide, expressed by condition $a = b$, the curve is closed. The various differential elements obtained on the curve allow calculating its line integral.

The curves can be smooth by part, i.e., they are composed of a finite number of smooth parts (arc elements), connected in their initial and final point. This kind of curve can intersect in one or more points, and if their extreme points coincide, it is called a closed curve.

The differentiation condition of a function is associated to the concept of neighborhood and limit. The neighborhood of a point $P(x^i)$ is defined admitting that the very small radius ε, with which a sphere is traced, is centered on it. The interior of this sphere is this point's neighborhood of radius ε. This definition is valid in the plane, changing the sphere for a circle, and is complemented admitting a set of points, which is called an open set. The points interior to the cube shown in Fig. 2.1 form an open set, for in each point $P(x^i)$ a sphere of radius ε can be drawn in its interior, which will be contained in the cube's interior. If the cube's edges are included, the result is a closed set.

© Springer International Publishing Switzerland 2016

E. de Souza Sánchez Filho, *Tensor Calculus for Engineers and Physicists* ,
DOI 10.1007/978-3-319-31520-1_2

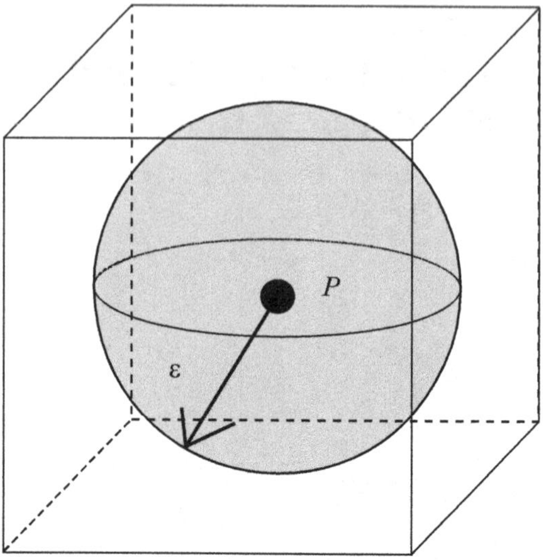

Fig. 2.1 Neighborhood of a point

2.2 Cartesian Tensor Derivative

The tensors derivative study begins with the research of what happens when a scalar function is differentiated. The behavior of this kind of function leads to the study of more general cases, such as those of the vectorial and tensorial functions.

Consider a scalar function $\phi(x^i)$ defined in a coordinate system X^i, which derivative with respect to the variable x^i is given by

$$\frac{\partial\phi(x^i)}{\partial x^i} = \phi_{,i} = G_i$$

and in another coordinate system $\overline{X}^i$ has as derivative

$$\overline{G}_i = \frac{\partial\overline{\phi}(x^i)}{\partial\overline{x}^i} = \frac{\partial\phi(x^i)}{\partial\overline{x}^i} = \frac{\partial\phi(x^i)}{\partial x_i}\frac{\partial x^i}{\partial\overline{x}^i} \Rightarrow \overline{G}_i = \frac{\partial x^i}{\partial\overline{x}^i}G_i$$

whereby

$$\overline{\phi}_{,i} = \frac{\partial\phi(x^i)}{\partial x^i}\phi_{,i}$$

that is the transformation law of vectors, so $\frac{\partial\phi(x^i)}{\partial x^i}$ is a vector and defines the gradient of the scalar function. In vectorial notation, it is graphed as $\boldsymbol{u} = \operatorname{grad}\phi(x^i)$.

Differentiating the scalar function again with respect to the variable x^j results in

$$\phi_{,ij} = \left(\frac{\partial x^k}{\partial \overline{x}^j}\frac{\partial x^m}{\partial \overline{x}^i}\right)\frac{\partial^2 \phi}{\partial \overline{x}^m \partial \overline{x}^k}$$

and for $i = j$

$$\phi_{,ii} = \left(\frac{\partial x_k}{\partial \overline{x}^i}\frac{\partial x_m}{\partial \overline{x}^i}\right)\frac{\partial^2 \phi}{\partial \overline{x}^m \partial \overline{x}^k}$$

$$\delta_{km} = \frac{\partial x^k}{\partial \overline{x}^i}\frac{\partial x^m}{\partial \overline{x}^i}$$

$$\phi_{,ii} = \delta_{km}\frac{\partial^2 \phi}{\partial \overline{x}^m \partial \overline{x}^k} = \frac{\partial^2 \phi}{\partial \overline{x}^k \partial \overline{x}^k}$$

then $\phi(x^i)_{,ii}$ is a scalar.

2.2.1 Vectors

In the study of vectors, there are two distinct manners to carry out a derivative: the derivative of a vector and the derivative of a point. In this text, a few conceptual considerations are made before calculating the derivative of a vector.

Consider the vectorial space E_N in which the scalar variable $t \in (a; b)$ is defined, where for each value of this variable limited in the open interval there is a vector $\boldsymbol{u}$ (t) embedded in this space which has a metric tensor. Let the vector $\boldsymbol{u}(t)$ defined by a continuous function of the variable t, and that admits continuous derivatives, then by means of an elementary increment Δt this vector will have an elementary increment

$$\Delta \boldsymbol{u}(t) = \boldsymbol{u}(t + \Delta t) - \boldsymbol{u}(t)$$

and when $\Delta t \to 0$ a vector $\boldsymbol{v}$ will exist such that

$$\lim\left(\frac{\Delta \boldsymbol{u}(t)}{\Delta t} - \boldsymbol{v}\right) \to 0$$

then the vector $\boldsymbol{v}$ is derived from the vector $\boldsymbol{u}(t)$ by means of the variation of the parameter t, whereby

$$\boldsymbol{v} = \frac{d\boldsymbol{u}}{dt}$$

The rules of Differential Calculus are applicable to this kind of differentiation. The other method of calculating a vector by means of a derivative is associated to the concept of punctual space E_N, with respect to which a variable $t \in (a; b)$ is associated in a univocal way to an arbitrary point $P(t)$. Taking a fixed point O contained in E_N as reference origin, the vector $\boldsymbol{r}(t) = \boldsymbol{OP}(t)$ is determined.

Let the vector $\boldsymbol{r}(t)$ defined by a continuous and derivable function, then the result is the vector $\frac{dr(t)}{dt}$ that does not depend on the origin O, but only on the point $P(t)$. This statement can be demonstrated admitting a new arbitrary point O^* for which the result is the vector

$$\boldsymbol{OP} = \boldsymbol{OO}^* + \boldsymbol{O}^*\boldsymbol{P}$$

Fixing the vector $\boldsymbol{OO}^*$ and calculating the derivative of these vectors with respect to the variable t, the result is

$$\frac{d(\boldsymbol{OP})}{dt} = \frac{d(\boldsymbol{OO}^*)}{dt} + \frac{d(\boldsymbol{O}^*\boldsymbol{P})}{dt} \Rightarrow \frac{d(\boldsymbol{OP})}{dt} = \frac{d(\boldsymbol{O}^*\boldsymbol{P})}{dt}$$

This equality proves the independence of the vector $\frac{dr(t)}{dt}$ with respect to the arbitrated origin. This vector is derived from the point $P(t)$, and in the case of the points defining a smooth curve C contained in the space E_N, continuous and differentiable, and will be tangent to the curve in each point for which this derivative was calculated.

In the general case of the vector being a function of various scalar variables $\boldsymbol{r}(x^i),\ i = 1, 2, \ldots, N$, the result is by means of the differentiation rules of Differential Calculus

$$\frac{\partial^2 \boldsymbol{r}(t)}{\partial x^i \partial x^k} = \frac{\partial^2 \boldsymbol{r}(t)}{\partial x^k \partial x^i}$$

For a vectorial function of scalar variables, the vector derivative is given by

$$d\boldsymbol{r}(t) = \frac{\partial \boldsymbol{r}(t)}{\partial x^i} dx^i$$

If the curve C is a function of the variables x^i, and these depend on the variable $t \in (a; b)$, i.e., $x^i = x^i(t)$, the curve is parameterized, then the hypothesis of Differential Calculus is applicable, and the total differential of the vector $\boldsymbol{r}[x^i(t)]$ is

$$\frac{d\boldsymbol{r}(t)}{dt} = \frac{\partial \boldsymbol{r}(t)}{\partial x^i} \frac{dx^i}{dt}$$

The rules applicable to the differentiation of vectors, and to the vectors obtained by means of differentiation of a point are the same. The concept of vector calculated by differentiation of one point can be extended to the study of tensors, where the points to be analyzed are contained in the tensorial space E_N.

Exercise 2.1 Show that derivative of the vector $\boldsymbol{r}$ with constant direction maintains its direction invariable.

Let

$$\boldsymbol{r}(t) = \psi(t)\boldsymbol{u}$$

where $\psi(t)$ is a parameterized scalar function and $\boldsymbol{u}$ is a constant vector, thus

$$\frac{d\boldsymbol{r}(t)}{dt} = \psi'(t)\boldsymbol{u} + \psi(t)\frac{d\boldsymbol{u}}{dt} = \psi'(t)\boldsymbol{u}$$

then $\boldsymbol{r}$ and $\frac{d\boldsymbol{r}(t)}{dt}$ have the same direction.

2.2.2 Cartesian Tensor of the Second Order

The tensorial functions defined by Cartesian tensors are often found in applications of physics and the areas of engineering. Let, for instance, the derivative of this kind of tensorial function that is a particular case of the derivative of tensors expressed in curvilinear coordinate systems. For the analysis of this derivative a Cartesian tensor of the second order T_{ij} is admitted, which components are functions of the coordinates x^i. The tensor with this characteristic is a function of the point considered in the space E_3. The transformation law of the tensors of the second order is given by

$$\overline{T}_{ij} = \frac{\partial x^p}{\partial \overline{x}^i}\frac{\partial x^q}{\partial \overline{x}^j}T_{pq}$$

For the Cartesian coordinate systems the coefficients $\frac{\partial x^p}{\partial \overline{x}^i}$ and $\frac{\partial x^q}{\partial \overline{x}^j}$ are constants, for they represent the variation rates for the linear transformations, so they do not depend on the point's coordinates. The derivative of this expression is given by

$$\frac{\partial \overline{T}_{ij}}{\partial \overline{x}^\ell} = \frac{\partial x^p}{\partial \overline{x}^i}\frac{\partial x^q}{\partial \overline{x}^j}\left(\frac{\partial T_{pq}}{\partial \overline{x}^m}\frac{\partial x^m}{\partial \overline{x}^\ell}\right) \Rightarrow \frac{\partial \overline{T}_{ij}}{\partial \overline{x}^\ell} = \frac{\partial x^p}{\partial \overline{x}^i}\frac{\partial x^q}{\partial \overline{x}^j}\frac{\partial x^m}{\partial \overline{x}^\ell}\frac{\partial T_{pq}}{\partial \overline{x}^m}$$

that is the transformation law of tensors of the third order, concluding that the derivative of tensor T_{ij} increased the order of this tensor in one unit. This conclusion is general and applicable to any Cartesian tensor.

This kind of derivative is not valid for the more general tensors. The concept of tensors derivative must, therefore, be generalized for tensors which components are given in curvilinear coordinate systems.

Exercise 2.2 Show that if T_{ij} is a Cartesian tensor of the second order, then $\frac{\partial^2 T_{ij}}{\partial x_k \partial x_m}$ will be a tensor of the fourth order.

The tensor is a function of the coordinates $T_{ij}(x_1, x_2, x_3)$, whereby the result is

$$\frac{\partial x_i}{\partial \overline{x}_j} = \delta_{ij} \Rightarrow \frac{\partial^2 x_i}{\partial \overline{x}_j \partial \overline{x}_k} = 0$$

and by transformation law

$$\overline{T}_{pq} = \frac{\partial x_i}{\partial \overline{x}_p} \frac{\partial x_j}{\partial \overline{x}_q} T_{ij}$$

then

$$\begin{aligned} \frac{\partial^2 \overline{T}_{pq}}{\partial \overline{x}_r \partial \overline{x}_s} &= \frac{\partial^2}{\partial \overline{x}_r \partial \overline{x}_s} \left(\frac{\partial x_i}{\partial \overline{x}_p} \frac{\partial x_j}{\partial \overline{x}_q} T_{ij} \right) = \frac{\partial x_k}{\partial \overline{x}_r} \frac{\partial}{\partial x_k} \left[\frac{\partial x_m}{\partial \overline{x}_s} \frac{\partial}{\partial x_m} \left(\frac{\partial x_i}{\partial \overline{x}_p} \frac{\partial x_j}{\partial \overline{x}_q} T_{ij} \right) \right] \\ &= \frac{\partial x_k}{\partial \overline{x}_r} \frac{\partial x_m}{\partial \overline{x}_s} \frac{\partial x_i}{\partial \overline{x}_p} \frac{\partial x_j}{\partial \overline{x}_q} \frac{\partial}{\partial x_k} \left(\frac{\partial T_{ij}}{\partial x_m} \right) \end{aligned}$$

In a mnemonic manner

$$\overline{T}_{pq,rs} = \frac{\partial x_k}{\partial \overline{x}_r} \frac{\partial x_m}{\partial \overline{x}_s} \frac{\partial x_i}{\partial \overline{x}_p} \frac{\partial x_j}{\partial \overline{x}_q} T_{ij,mk}$$

that is the transformation law of tensors of the fourth order as $T_{ij,mk} = T_{ij,km}$ the result is that $T_{ij,km} = \frac{\partial^2 T_{ij}}{\partial x_k \partial x_m}$ is a tensor of the fourth order.

2.3 Derivatives of the Basis Vectors

Consider the contravariant vector $u^k = u^k(x^i)$ defined in terms of the parametric curve $x^i = x^i(t)$, expressed with respect to the Cartesian coordinate system X^i. By means of the transformation law of vectors, the result for the curvilinear coordinate system $\overline{X}^i$ is

$$\overline{u}^\ell = \frac{\partial \overline{x}^i}{\partial x^k} u^k$$

and with the techniques of differentiation with respect to the parameter t results in

$$\frac{d\overline{u}^\ell}{dt} = \frac{\partial \overline{x}^i}{\partial x^k} \frac{du^k}{dt} + \frac{\partial^2 \overline{x}^i}{\partial x^k \partial x^\ell} \frac{dx^\ell}{dt} u^k \tag{2.3.1}$$

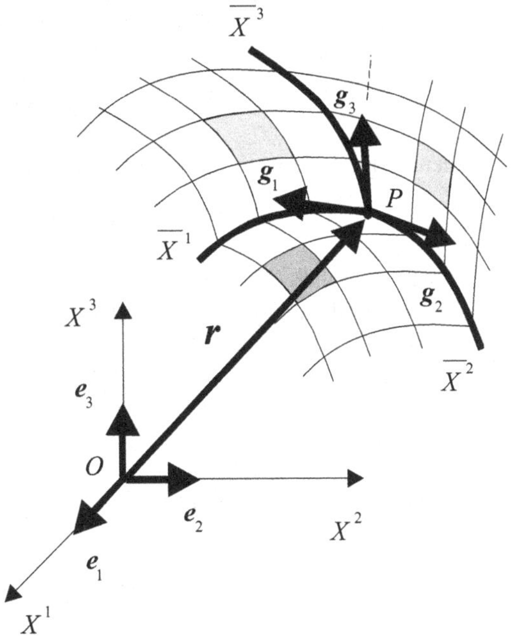

Fig. 2.2 Cartesian X^i and curvilinear $\overline{X}^i$ coordinates with basis vectors e_i and g_i, respectively

that only represents a contravariant vector if, and only if, $\overline{x}^i$ is a linear function of x^k. The first term on the right of this expression represents an ordinary differentiation of a vectorial function expressed in Cartesian coordinates, but the second term contains the derivatives curvilinear coordinates $\overline{x}^i$, relative to a coordinate system that varies as a function of the points of the space.

The study of this term is carried out considering the Cartesian coordinate system X^i and the curvilinear coordinate system $\overline{X}^i$, with unit vectors $\boldsymbol{e}_i$ and $\boldsymbol{g}_i$, respectively (Fig. 2.2). Whereby defining the position vector $\boldsymbol{r}$ of point P with respect to the coordinate system X^i by means of their contravariant components

$$\boldsymbol{r} = x^i \boldsymbol{e}_i$$

the differential total of this vector is given by

$$d\boldsymbol{r} = \frac{\partial \boldsymbol{r}}{\partial x^i} dx^i \tag{2.3.2}$$

As the basis vectors $\boldsymbol{e}_i$ do not depend on point P:

$$d\boldsymbol{r} = dx^i \boldsymbol{e}_i$$

so

$$\boldsymbol{e}_i = \frac{\partial \boldsymbol{r}}{\partial x^i} \tag{2.3.3}$$

With respect to the local coordinate system $\overline{X}^i$ the result is the differential total of the position vector $\boldsymbol{r}$:

$$d\boldsymbol{r} = \boldsymbol{r}_{,i}\, d\overline{x}^i \tag{2.3.4}$$

Whereby the base vectors of the curvilinear coordinate system results

$$\boldsymbol{g}_i = \boldsymbol{r}_{,i} \tag{2.3.5}$$

that shows that the unit vectors $\boldsymbol{g}_i$ are tangent to the curves that define the curvilinear coordinate system $\overline{X}^i$, that varies for each point of the vectorial space E_3, and as the unit vectors $\boldsymbol{e}_i$ do not vary

$$\frac{\partial \boldsymbol{r}}{\partial \overline{x}^k} = \frac{\partial x^i}{\partial \overline{x}^k} \boldsymbol{e}_i$$

Comparing with expression (2.3.5)

$$\boldsymbol{g}_k = \frac{\partial x^i}{\partial \overline{x}^k} \boldsymbol{e}_i \tag{2.3.6}$$

then

$$\boldsymbol{e}_i = \frac{\partial \overline{x}^j}{\partial x^i} \boldsymbol{g}_j \tag{2.3.7}$$

The covariant derivative of the base vector defined by the expression (2.3.6) is given by

$$\boldsymbol{g}_{k,\ell} = \frac{\partial^2 x^i}{\partial \overline{x}^k \partial \overline{x}^\ell} \boldsymbol{e}_i \tag{2.3.8}$$

and substituting expression (2.3.7) in this expression

$$\boldsymbol{g}_{k,\ell} = \frac{\partial \overline{x}^j}{\partial x^i} \frac{\partial^2 x^i}{\partial \overline{x}^k \partial \overline{x}^\ell} \boldsymbol{g}_j$$

Defining the variety

$$\Gamma^j_{k\ell} = \frac{\partial \overline{x}^j}{\partial x^i} \frac{\partial^2 x^i}{\partial \overline{x}^k \partial \overline{x}^\ell} \tag{2.3.9}$$

with which the covariant derivatives of the basis vectors of the curvilinear coordinate system can be written as linear combination of the base vector $\boldsymbol{g}_j$:

$$\boldsymbol{g}_{k,\ell} = \Gamma^j_{k\ell} \boldsymbol{g}_j \tag{2.3.10}$$

2.3.1 Christoffel Symbols

The coefficients determined by expression (2.3.9) can be expressed in terms of the derivatives of the metric tensor and its conjugated tensor. For the derivatives of the contravariant basis vectors considering another variety $\overline{\Gamma}^{i}_{jm}$:

$$g^{i}_{,j} = \overline{\Gamma}^{i}_{jm} g^{m} \tag{2.4.1}$$

Writing

$$\left(\boldsymbol{g}^{i}\cdot \boldsymbol{g}_{j}\right)_{,k} = \left(\delta^{i}_{j}\right)_{,k} = \boldsymbol{g}^{i}_{,k}\cdot \boldsymbol{g}_{j} + \boldsymbol{g}_{j,k}\cdot \boldsymbol{g}^{i} = 0 \tag{2.4.2}$$

and substituting the expressions (2.3.10) and (2.4.1) in expression (2.4.2)

$$\overline{\Gamma}^{i}_{jk}\boldsymbol{g}^{k}\cdot \boldsymbol{g}_{j} + \Gamma^{i}_{jk}\boldsymbol{g}_{j}\cdot \boldsymbol{g}^{i} = 0 \tag{2.4.3}$$

then

$$\overline{\Gamma}^{i}_{jk} = -\Gamma^{i}_{jk}$$

and with the expressions (2.3.9), (2.3.10), (2.4.1) and with the prior expression it follows that

$$\begin{aligned}
\Gamma^{m}_{ij} &= \frac{\partial \overline{x}^{m}}{\partial x^{k}}\frac{\partial^{2} x^{k}}{\partial^{2}\overline{x}^{i}\partial \overline{x}^{j}}\\
\boldsymbol{g}_{i,j} &= \Gamma^{m}_{ij}\boldsymbol{g}_{m}\\
\boldsymbol{g}^{i}_{,j} &= -\Gamma^{i}_{jm}\boldsymbol{g}^{m}
\end{aligned} \tag{2.4.4}$$

The relation between the covariant and contravariant unit vectors is defined by

$$\boldsymbol{g}_{i} = g_{ij}\boldsymbol{g}^{j}$$

and the derivative of this expression with respect to the coordinate x^{k} is given by

$$\boldsymbol{g}_{i,k} = g_{ij,k}\boldsymbol{g}^{j} + g_{ij}\boldsymbol{g}^{j}_{,k}$$

Replacing expressions (2.3.10) and (2.4.4) in this last expression

$$\Gamma^{m}_{ik}\boldsymbol{g}_{m} = g_{ij,k}\boldsymbol{g}^{j} - g_{ij}\Gamma^{j}_{km}\boldsymbol{g}^{m}$$

and with the multiplying by $\boldsymbol{g}^{n}$

$$\Gamma_{ik}^{m}\boldsymbol{g_m}\cdot\boldsymbol{g^n} = g_{ij,k}\boldsymbol{g^j}\cdot\boldsymbol{g^n} - g_{ij}\Gamma_{km}^{j}\boldsymbol{g^m}\cdot\boldsymbol{g^n}$$

$$\Gamma_{ik}^{m}\delta_m^n = g_{ij,k}g^{jn} - g_{ij}\Gamma_{km}^{j}g^{mn}$$

$$\Gamma_{ik}^{n} = g_{ij,k}g^{jn} - g_{ij}g^{mn}\Gamma_{km}^{j}$$

The multiplying of this last expression by g_{np} provides

$$g_{np}\Gamma_{ik}^{n} + g_{ij}g^{mn}g_{np}\Gamma_{km}^{j} = g_{ij,k}g^{jn}g_{np} \Rightarrow g_{np}\Gamma_{ik}^{n} + g_{ij}\delta_p^m\Gamma_{km}^{j} = g_{ij,k}\delta_p^j$$

whereby

$$g_{np}\Gamma_{ik}^{n} + g_{ij}\Gamma_{kp}^{j} = g_{ip,k} \tag{2.4.5}$$

and with the cyclic permutation of the free indexes i, p, k of expression (2.4.5)

$$g_{nk}\Gamma_{pi}^{n} + g_{pj}\Gamma_{ik}^{j} = g_{pk,i} \tag{2.4.6}$$

$$g_{ni}\Gamma_{kp}^{n} + g_{kj}\Gamma_{pi}^{j} = g_{ki,p} \tag{2.4.7}$$

Multiplying expression (2.4.5) by $-1/2$ and expressions (2.4.6) and (2.4.7) by 1/2 and adding

$$-\frac{1}{2}\left(g_{np}\Gamma_{ik}^{n} + g_{ij}\Gamma_{kp}^{j}\right) + \frac{1}{2}\left(g_{nk}\Gamma_{pi}^{n} + g_{pj}\Gamma_{ik}^{j}\right) + \frac{1}{2}\left(g_{ni}\Gamma_{kp}^{n} + g_{kj}\Gamma_{pi}^{j}\right)$$
$$= \frac{1}{2}\left(g_{pk,i} + g_{ki,p} - g_{ip,k}\right)$$

and with the change of the index n for the index j, and considering the symmetry of the metric tensor

$$g_{kj}\Gamma_{ip}^{j} = \frac{1}{2}\left(g_{pk,i} + g_{ki,p} - g_{ip,k}\right)$$

The term to the right of the expression shows the existence of coefficients that are functions only of the partial derivatives of the metric tensor that define the Christoffel symbol of first kind

$$[p,k] = \Gamma_{ip,k} = \frac{1}{2}\left(g_{pk,i} + g_{ki,p} - g_{ip,k}\right) = \frac{1}{2}\left(\frac{\partial g_{pk}}{\partial x^i} + \frac{\partial g_{ki}}{\partial x^p} - \frac{\partial g_{ip}}{\partial x^k}\right) \tag{2.4.8}$$

Multiplying expression (2.4.8) by g^{km}:

$$g^{km}g_{kj}\Gamma_{ip}^{j} = \frac{1}{2}g^{km}\left(g_{pk,i} + g_{ki,p} - g_{ip,k}\right) \Rightarrow \delta_j^m\Gamma_{ip}^{j} = \frac{1}{2}g^{km}\left(g_{pk,i} + g_{ki,p} - g_{ip,k}\right)$$

whereby

$$\left\{ \begin{matrix} m \\ ip \end{matrix} \right\} = \Gamma^m_{ip} = \frac{1}{2} g^{km} \left(g_{pk,i} + g_{ki,p} - g_{ip,k} \right)$$
$$= \frac{1}{2} g^{km} \left(\frac{\partial g_{pk}}{\partial x^i} + \frac{\partial g_{ki}}{\partial x^p} - \frac{\partial g_{ip}}{\partial x^k} \right) \tag{2.4.9}$$

The term to the right of this expression shows the existence of coefficients that depend on the partial derivatives of the metric tensor and the conjugate metric tensor. The coefficients represented by $\Gamma^i_{k\ell}$ given by expressions (2.3.9) and (2.4.9) define Christoffel symbol of second kind. Expression (2.4.9) is more convenient for calculating these coefficients than expression (2.3.9).

Multiplying the terms of expression (2.3.10) by $\mathbf{g}^i$:

$$\mathbf{g}^i \cdot \mathbf{g}_{k,\ell} = \Gamma^j_{k\ell} \mathbf{g}^i \cdot \mathbf{g}_j = \delta^i_j \Gamma^j_{k\ell}$$
$$\Gamma^i_{k\ell} = \mathbf{g}^i \cdot \mathbf{g}_{k,\ell} \tag{2.4.10}$$

2.3.2 Relation Between the Christoffel Symbols

Expressions (2.4.8) and (2.4.9) relate the two Christoffel symbols, i.e.:

$$\Gamma^m_{ij} = g^{km} \Gamma_{ij,k} \tag{2.4.11}$$

then the Christoffel symbol of second kind is the raising of the third index of the Christoffel symbol of first kind.

Expression (2.4.10) written in terms of the Christoffel symbol of first kind is given by

$$\Gamma^k_{ij} = g^{kp} \Gamma_{ij,p} = \mathbf{g}^k \cdot \mathbf{g}_{i,j}$$

and multiplying the members by g_{kp}

$$g_{kp} g^{kp} \Gamma_{ij,p} = g_{kp} \mathbf{g}^k \cdot \mathbf{g}_{i,j} \Rightarrow \Gamma_{ij,p} = g_{kp} \mathbf{g}^k \cdot \mathbf{g}_{i,j} \Rightarrow \Gamma_{ij,p} = \mathbf{g}_k \cdot \mathbf{g}_p \cdot \mathbf{g}^k \cdot \mathbf{g}_{i,j}$$

then

$$\Gamma_{ij,p} = \mathbf{g}_p \cdot \mathbf{g}_{i,j} \tag{2.4.12}$$

2.3.3 Symmetry

For the Christoffel symbol of first kind

$$\Gamma_{ij,k} = \frac{1}{2}\left(\frac{\partial g_{jk}}{\partial x^i} + \frac{\partial g_{ik}}{\partial x^j} - \frac{\partial g_{ij}}{\partial x^k}\right) \quad \Gamma_{ji,k} = \frac{1}{2}\left(\frac{\partial g_{ik}}{\partial x^j} + \frac{\partial g_{jk}}{\partial x^i} - \frac{\partial g_{ji}}{\partial x^k}\right)$$

and considering the symmetry of the metric tensor $g_{ik} = g_{ki}, g_{jk} = g_{kj}, g_{ij} = g_{ji}$ it results in

$$\Gamma_{ij,k} = \Gamma_{ji,k}$$

then the Christoffel symbol of first kind is symmetrical with respect to the first two indexes, and with considering this symmetry results to the Christoffel symbol of second kind

$$\Gamma_{ij}^m = g^{km}\Gamma_{ij,k} = g^{km}\Gamma_{ji,k} = \Gamma_{ji}^m$$

that is symmetrical in regard to a permutation of the lower indexes.

2.3.4 Cartesian Coordinate System

For the Cartesian coordinate systems the elements of the metric tensor are $g_{ij} = \delta_{ij}$, whereby for $p = 1, 2, \ldots, N$ it results in

$$\frac{\partial g_{ip}}{\partial x^p} = 0$$

By means of the definition of the Christoffel symbol of first kind

$$\Gamma_{ij,k} = \frac{1}{2}\left(\frac{\partial g_{jk}}{\partial x^i} + \frac{\partial g_{ik}}{\partial x^j} - \frac{\partial g_{ij}}{\partial x^k}\right) = 0$$

It is verified that the Christoffel symbol of second kind is cancelled, for

$$\Gamma_{ij}^p = g^{pk}\Gamma_{ij,k} = 0$$

then for the Cartesian, orthogonal, or oblique coordinate systems, all the terms of $\Gamma_{ij,k}$ and Γ_{ij}^p are null.

2.3.5 Notation

The oldest notations for the Christoffel symbols are $\begin{bmatrix} ij \\ k \end{bmatrix}$ for the symbol of first kind, and $\begin{Bmatrix} ij \\ k \end{Bmatrix}$ for the symbol of second kind. Improving this notations Levi-Civita adopted the spelling $[ij, k]$ and $\{ij, k\}$, which second symbol was later improved by various authors to $\begin{Bmatrix} k \\ ij \end{Bmatrix}$, where the indexes were placed in more adequate and logical positions. This notation is well adopted, using the representation $[ij, k]$ for the Christoffel symbol of first kind.

Hermann Weyl used the Greek letter Γ to denote these symbols, which positions of the indexes indicates the kind it represents: $\Gamma_{ij,k}$ and Γ_{ij}^{k}. This symbology is known as the notation of the Princeton School. A few authors invert the position of the indexes and write $\Gamma_{k,ij}$.

The argument for adopting the notations $[ij, k]$ and $\begin{Bmatrix} k \\ ij \end{Bmatrix}$ is that the Princeton School notation leads to confusing these coefficients with a tensor. However, this argument does not make its adoption valid, for the use of brackets or keys could also lead to confusion with a matrix or column matrix. That is not the case. The use of the notations $\Gamma_{ij,k}$ and Γ_{ij}^{k}, even not being universally accepted, has in its favor the economy of characters in a text with many expressions containing these symbols. Several authors do not use the comma for indicating the differentiation with respect to one of the indexes in the Christoffel symbol of first kind, and write Γ_{ijk}.

2.3.6 Number of Different Terms

For the tensorial space E_N where $i,j = 1, 2, \ldots, N$, it is verified that the metric tensor g_{ij} has N^2 terms

$$g_{ij} = \begin{bmatrix} g_{11} & g_{12} & \cdots & g_{1N} \\ g_{21} & g_{22} & \cdots & g_{2N} \\ \vdots & \vdots & \vdots & \vdots \\ g_{N1} & g_{N2} & \cdots & g_{NN} \end{bmatrix}_{N \times N}$$

This matrix has N diagonal terms g_{ii}, so $\left(N^2 - N\right)$ terms remain in their sides. As g_{ij} is symmetrical, the result is $\frac{1}{2}\left(N^2 - N\right)$ terms for $i \neq j$. The total of different terms in the metric tensor is $N^2 + \frac{1}{2}\left(N^2 - N\right) = \frac{N(N+1)}{2}$.

For each kind of Christoffel symbol N derivatives of g_{ij} are calculated, so the number of different terms for these coefficients is given by $\frac{N\left(N^2+1\right)}{2}$.

2.3.7 Transformation of the Christoffel Symbol of First Kind

Let $\Gamma_{pq,r}$ defined in the coordinate system X^i and $\overline{\Gamma}_{ij,k}$ expressed in the coordinate system $\overline{X}^i$, then using the expression that defines the Christoffel symbol of first kind, and adopting the notation $g_{ij,k} = \frac{\partial g_{ij}}{\partial x^k}$ it follows that

$$\begin{aligned} \overline{g}_{ij,k} &= \frac{\partial}{\partial \overline{x}^k}\left(g_{pq}\frac{\partial x^p}{\partial \overline{x}^i}\frac{\partial x^q}{\partial \overline{x}^j}\right) \\ &= \frac{\partial g_{pq}}{\partial \overline{x}^k}\frac{\partial x^p}{\partial \overline{x}^i}\frac{\partial x^q}{\partial \overline{x}^j} + g_{pq}\frac{\partial^2 x^p}{\partial \overline{x}^k \partial \overline{x}^i}\frac{\partial x^q}{\partial \overline{x}^j} + g_{pq}\frac{\partial x^p}{\partial \overline{x}^i}\frac{\partial^2 x^q}{\partial \overline{x}^k \partial \overline{x}^j} \end{aligned} \tag{2.4.13}$$

By the chain rule

$$\frac{\partial g_{pq}}{\partial \overline{x}^k} = \frac{\partial g_{pq}}{\partial \overline{x}^r}\frac{\partial x^r}{\partial \overline{x}^k} = g_{pq,r}\frac{\partial x^r}{\partial \overline{x}^k} \tag{2.4.14}$$

$$\overline{g}_{ij,k} = g_{pq,r}\frac{\partial x^p}{\partial \overline{x}^i}\frac{\partial x^q}{\partial \overline{x}^j}\frac{\partial x^r}{\partial \overline{x}^k} + g_{pq}\left(\frac{\partial^2 x^r}{\partial \overline{x}^k \partial \overline{x}^i}\frac{\partial x^s}{\partial \overline{x}^j}\right) + g_{pq}\left(\frac{\partial^2 x^s}{\partial \overline{x}^k \partial \overline{x}^j}\frac{\partial x^r}{\partial \overline{x}^i}\right) \tag{2.4.15}$$

and with cyclic permutation of the indexes in each term of the previous expression and with $g_{qp} = g_{pq}$

$$\overline{g}_{jk,i} = g_{qr,p}\frac{\partial x^q}{\partial \overline{x}^j}\frac{\partial x^r}{\partial \overline{x}^k}\frac{\partial x^p}{\partial \overline{x}^i} + g_{pq}\left(\frac{\partial^2 x^q}{\partial \overline{x}^i \partial \overline{x}^j}\frac{\partial x^p}{\partial \overline{x}^k}\right) + g_{pq}\left(\frac{\partial^2 x^p}{\partial \overline{x}^i \partial \overline{x}^k}\frac{\partial x^q}{\partial \overline{x}^j}\right) \tag{2.4.16}$$

$$\overline{g}_{ki,j} = g_{rp,q}\frac{\partial x^r}{\partial \overline{x}^k}\frac{\partial x^p}{\partial \overline{x}^i}\frac{\partial x^q}{\partial \overline{x}^j} + g_{pq}\left(\frac{\partial^2 x^p}{\partial \overline{x}^j \partial \overline{x}^k}\frac{\partial x^q}{\partial \overline{x}^i}\right) + g_{pq}\left(\frac{\partial^2 x^p}{\partial \overline{x}^j \partial \overline{x}^i}\frac{\partial x^q}{\partial \overline{x}^k}\right) \tag{2.4.17}$$

$$\overline{\Gamma}_{ij,k} = \frac{1}{2}\left(\frac{\partial \overline{g}_{jk}}{\partial \overline{x}^i} + \frac{\partial \overline{g}_{ki}}{\partial \overline{x}^j} - \frac{\partial \overline{g}_{ij}}{\partial \overline{x}^k}\right) \tag{2.4.18}$$

By substitution

$$\overline{\Gamma}_{ij,k} = \frac{\partial x^p}{\partial \overline{x}^i}\frac{\partial x^q}{\partial \overline{x}^j}\frac{\partial x^r}{\partial \overline{x}^k}\Gamma_{pq,r} + g_{pq}\frac{\partial^2 x^p}{\partial \overline{x}^i \partial \overline{x}^j}\frac{\partial x^q}{\partial \overline{x}^k} \tag{2.4.19}$$

that is the transformation law of the Christoffel symbol of first kind. The second term to the right of this expression shows that these coefficients are not the components of a tensor.

2.3.8 Transformation of the Christoffel Symbol of Second Kind

Writing the Christoffel symbol of second kind in terms of the components of a new coordinate system $\overline{X}^i$:

$$\overline{\Gamma}^i_{jk} = \overline{g}^{ip}\overline{\Gamma}_{jk,p} = \left(g^{qr}\frac{\partial \overline{x}^j}{\partial x^q}\frac{\partial \overline{x}^p}{\partial x^r}\right)\overline{\Gamma}_{jk,p} \tag{2.4.20}$$

As the Christoffel symbol of first kind transforms by mean of expression (2.4.19)

$$\overline{\Gamma}_{jk,p} = \Gamma_{\ell m,n}\frac{\partial x^\ell}{\partial \overline{x}^j}\frac{\partial x^m}{\partial \overline{x}^k}\frac{\partial x^n}{\partial \overline{x}^p} + g_{\ell m}\frac{\partial^2 x^\ell}{\partial \overline{x}^j \partial \overline{x}^k}\frac{\partial x^m}{\partial \overline{x}^p} \tag{2.4.21}$$

and substituting expression (2.4.21) in expression (2.4.20) it follows that

$$\overline{\Gamma}^i_{jk} = g^{qr}\Gamma_{rn}\left(\frac{\partial x^\ell}{\partial \overline{x}^j}\frac{\partial x^m}{\partial \overline{x}^k}\frac{\partial x^n}{\partial \overline{x}^p}\frac{\partial \overline{x}^i}{\partial x^q}\frac{\partial \overline{x}^p}{\partial x^r}\right) + g^{qr}g_{\ell m}\left(\frac{\partial^2 x^\ell}{\partial \overline{x}^j \partial \overline{x}^k}\frac{\partial x^m}{\partial \overline{x}^p}\frac{\partial \overline{x}^i}{\partial x^q}\frac{\partial \overline{x}^p}{\partial x^r}\right)$$

$$\overline{\Gamma}^i_{jk} = g^{qr}\Gamma_{\ell m,n}\delta^n_r\frac{\partial x^\ell}{\partial \overline{x}^j}\frac{\partial x^m}{\partial \overline{x}^k}\frac{\partial \overline{x}^i}{\partial x^q} + g^{qr}g_{\ell m}\delta^m_r\left(\frac{\partial^2 x^\ell}{\partial \overline{x}^j \partial \overline{x}^k}\frac{\partial \overline{x}^i}{\partial x^q}\right)$$

$$\overline{\Gamma}^i_{jk} = g^{qr}\Gamma_{\ell m,n}\delta^n_r\frac{\partial x^\ell}{\partial \overline{x}^j}\frac{\partial x^m}{\partial \overline{x}^k}\frac{\partial \overline{x}^i}{\partial x^q} + g^{qr}g_{\ell m}\delta^m_r\frac{\partial^2 x^\ell}{\partial \overline{x}^j \partial \overline{x}^k}\frac{\partial \overline{x}^i}{\partial x^q}$$

$$\overline{\Gamma}^i_{jk} = g^{qr}\Gamma_{\ell m,n}\delta^n_r\frac{\partial x^\ell}{\partial \overline{x}^j}\frac{\partial x^m}{\partial \overline{x}^k}\frac{\partial \overline{x}^i}{\partial x^q} + g^{qr}g_{\ell r}\frac{\partial^2 x^\ell}{\partial \overline{x}^j \partial \overline{x}^k}\frac{\partial \overline{x}^i}{\partial x^q}$$

$$g^{qr}\Gamma_{\ell m,r} = \Gamma^q_{\ell m}$$

$$g^{qr}g_{\ell r} = \delta^q_\ell$$

$$\overline{\Gamma}^i_{jk} = \Gamma^q_{\ell m}\frac{\partial x^\ell}{\partial \overline{x}^j}\frac{\partial x^m}{\partial \overline{x}^k}\frac{\partial \overline{x}^i}{\partial x^q} + \frac{\partial^2 x^\ell}{\partial \overline{x}^j \partial \overline{x}^k}\frac{\partial \overline{x}^i}{\partial x^q}$$

Replacing the dummy indexes $\ell \to q$, $q \to p$, $m \to r$ results in

$$\overline{\Gamma}^i_{jk} = \Gamma^p_{qr}\frac{\partial x^q}{\partial \overline{x}^j}\frac{\partial x^r}{\partial \overline{x}^k}\frac{\partial \overline{x}^i}{\partial x^p} + \frac{\partial^2 x^q}{\partial \overline{x}^j \partial \overline{x}^k}\frac{\partial \overline{x}^i}{\partial x^p} \tag{2.4.22}$$

Expression (2.4.22) is the transformation law of the Christoffel symbol of second kind. The second term to the right of this expression shows that these coefficients are not the components of tensor. The Christoffel symbols do not depend only on the coordinate system, but depend also on the rate with which this coordinate system varies in each point of the space. This variation rate is not present in the transformation law of tensors.

2.3.9 Linear Transformations

Consider the transformation of coordinates between two coordinate systems given by linear relation

$$\overline{x}^j = a_i^j x^i + b^j$$

where a_i^j and b^j are constants, and with the techniques of successive differentiation

$$\frac{\partial \overline{x}^j}{\partial x^i} = a_i^j \Rightarrow \frac{\partial^2 \overline{x}^j}{\partial x^i \partial x^k} = 0$$

then for this kind of transformation of coordinates the Christoffel symbols transform as tensors.

2.3.10 Orthogonal Coordinate Systems

In the orthogonal coordinate systems, the tensorial space E_N is defined by the metric tensor $g_{ij} \neq 0$ for $i = j$ and $g_{ij} = 0$ for $i \neq j$.

Putting $h_i^2 = g_{ii}$, where g_{ii} does not indicate the summation of the terms, with the Christoffel symbol of first kind

$$\Gamma_{ij,k} = \frac{1}{2}\left(\frac{\partial g_{jk}}{\partial x^i} + \frac{\partial g_{ik}}{\partial x^j} + \frac{\partial g_{ij}}{\partial x^k}\right)$$

and with the components of the tensor g^{ij} given by

$$g^{\ell k} = \begin{cases} 1 \rightarrow i = j \\ 0 \rightarrow i \neq j \end{cases}$$

$$g_{ij} = \frac{1}{g^{ij}}$$

it results for the relation between the Christoffel symbols

$$\Gamma_{ij}^{\ell} = g^{\ell k}\Gamma_{ij,k} = \begin{cases} g^{kk}\Gamma_{ij,k} \rightarrow \ell = k \\ 0 \rightarrow\rightarrow \ell \neq k \end{cases}$$

Varying the indexes:

- $i = j = k$

$$\Gamma_{ii,i} = \frac{1}{2}\left(\frac{\partial g_{ii}}{\partial x^j} + \frac{\partial g_{ii}}{\partial x^j} - \frac{\partial g_{ii}}{\partial x^j}\right) = \frac{1}{2}\frac{\partial g_{ii}}{\partial x^j} = h_i \frac{\partial h_i}{\partial x^j}$$

$$\Gamma_{ij}^k = \Gamma_{ii}^i = g^{ii}\Gamma_{ii,i} = \frac{1}{2g_{ii}}\frac{\partial g_{ii}}{\partial x^j} = \frac{1}{2}\frac{\partial(\ell n\, g_{ii})}{\partial x^j} = \frac{\partial\left(\ell n\sqrt{g_{ii}}\right)}{\partial x^j} = \frac{1}{h_i}\frac{\partial h_i}{\partial x^j}$$

– $i = j \neq k$

$$\Gamma_{ii,k} = \frac{1}{2}\left(\frac{\partial g_{ik}}{\partial x^i} + \frac{\partial g_{ik}}{\partial x^i} - \frac{\partial g_{ii}}{\partial x^k}\right)$$

as $i \neq k$ it implies by definition of the metric tensor that $g_{ik} = 0$, so

$$\Gamma_{ii,k} = -\frac{1}{2}\frac{\partial g_{ii}}{\partial x^k} = -h_i \frac{\partial h_i}{\partial x^k}$$

$$\Gamma_{ij}^k = \Gamma_{ii}^k = g^{kk}\Gamma_{ii,k} = -\frac{1}{2g_{kk}}\frac{\partial g_{ii}}{\partial x^k} = -\frac{h_i}{(h_k)^2}\frac{\partial h_i}{\partial x^k}$$

– $i = k \neq j$

$$\Gamma_{ij,i} = \frac{1}{2}\left(\frac{\partial g_{ji}}{\partial x^i} + \frac{\partial g_{ii}}{\partial x^j} + \frac{\partial g_{ij}}{\partial x^i}\right)$$

and in an analogous way to the previous case where $g_{ji} = g_{ij} = 0$, so

$$\Gamma_{ij,i} = \frac{1}{2}\frac{\partial g_{ii}}{\partial x^j} = h_i \frac{\partial h_i}{\partial x^j}$$

$$\Gamma_{ij}^k = \Gamma_{ij}^i = g^{ii}\Gamma_{ij,i} = \frac{1}{2g_{ii}}\frac{\partial g_{ii}}{\partial x^j} = \frac{\partial\left(\ell n\sqrt{g_{ii}}\right)}{\partial x^j} = \frac{1}{h_i}\frac{\partial h_i}{\partial x^j}$$

– for $i \neq j, j \neq k, i \neq k$ it results in $\Gamma_{ij,k} = 0$, for by the definition of the metric tensor it implies $g_{ij} = g^{ij} = 0$ if $i \neq j$, whereby $\Gamma_{ij}^k = 0$.

2.3.11 Contraction

The tensorial expressions at times contain Christoffel symbols. However, the calculation of their components can be avoided, for an expression can be obtained that relates the derivative of the natural logarithm of the metric tensor with these symbols.

Let the Christoffel symbol of second kind

$$\Gamma_{ik}^{j} = \frac{1}{2} g^{mj} \left(\frac{\partial g_{km}}{\partial x^i} + \frac{\partial g_{mi}}{\partial x^k} - \frac{\partial g_{ik}}{\partial x^m} \right)$$

and with contraction of the indexes j and k

$$\Gamma_{ij}^{j} = \frac{1}{2} g^{mj} \left(\frac{\partial g_{jm}}{\partial x^i} + \frac{\partial g_{mi}}{\partial x^j} - \frac{\partial g_{ij}}{\partial x^m} \right)$$

The symmetry of the metric tensor provides

$$g^{mj} \frac{\partial g_{mi}}{\partial x^j} = g^{jm} \frac{\partial g_{ji}}{\partial x^m} = g^{mj} \frac{\partial g_{ij}}{\partial x^m}$$

where the second equality was obtained by means of indexes interchanging the $m \leftrightarrow j$.

Substituting $g^{mj} \frac{\partial g_{mi}}{\partial x^j}$ in the expression of the contracted Christoffel symbol

$$\Gamma_{ij}^{j} = \frac{1}{2} g^{jm} \frac{\partial g_{jm}}{\partial x^i}$$

The conjugate metric tensor can be written as

$$g^{jm} = \frac{G^{jm}}{g} \Rightarrow g = G^{jm} g_{jm}$$

being G^{jm} the cofactor of this matrix and $g = \det g_{jm}$, it follows that

$$\Gamma_{ij}^{j} = \frac{1}{2g} G^{jm} \frac{\partial g_{jm}}{\partial x^i} = \frac{1}{2g} \frac{\partial g}{\partial x^i} = \frac{1}{2} \frac{\partial (\ell n g)}{\partial x^i} = \frac{\partial \left(\ell n \sqrt{g}\right)}{\partial x^i}$$

and with the contracted form of the Christoffel symbol of second kind

$$\Gamma_{ij}^{j} = \frac{1}{\sqrt{g}} \frac{\partial \left(\sqrt{g}\right)}{\partial x^i} \tag{2.4.23}$$

that is of great use in manipulations of tensorial expressions, for it reduces the algebrism in calculating the Christoffel symbol.

For $g = |g_{ij}| < 0$ the analysis is analogous, having only to change the sign of the determinant in the expression shown in the previous demonstration

$$\Gamma_{ip}^{i} = \frac{\partial \left(\ell n \sqrt{-g}\right)}{\partial x^p} = \frac{1}{\sqrt{-g}} \frac{\partial \left(\sqrt{-g}\right)}{\partial x^p} \tag{2.4.24}$$

2.3.12 Christoffel Relations

Consider the transformation of the Christoffel symbol of second kind from one coordinate system X^i to another coordinate system $\overline{X}^j$, whereby rewriting expression (2.4.22)

$$\overline{\Gamma}^r_{pq} = \Gamma^m_{ij} \frac{\partial \overline{x}^r}{\partial x^m} \frac{\partial x^i}{\partial \overline{x}^p} \frac{\partial x^j}{\partial \overline{x}^q} + \frac{\partial \overline{x}^r}{\partial x^j} \frac{\partial^2 x^j}{\partial \overline{x}^q \partial \overline{x}^p}$$

and multiplying by $\frac{\partial x^s}{\partial \overline{x}^r}$ it follows that

$$\frac{\partial x^s}{\partial \overline{x}^r} \overline{\Gamma}^r_{pq} = \Gamma^m_{ij} \frac{\partial x^s}{\partial \overline{x}^r} \frac{\partial \overline{x}^r}{\partial x^m} \frac{\partial x^i}{\partial \overline{x}^p} \frac{\partial x^j}{\partial \overline{x}^q} + \frac{\partial x^s}{\partial \overline{x}^r} \frac{\partial \overline{x}^r}{\partial x^j} \frac{\partial^2 x^j}{\partial \overline{x}^q \partial \overline{x}^p}$$

$$\frac{\partial x^s}{\partial \overline{x}^r} \frac{\partial \overline{x}^r}{\partial x^m} = \delta^s_m \qquad \frac{\partial x^s}{\partial \overline{x}^r} \frac{\partial \overline{x}^r}{\partial x^j} = \delta^s_j$$

$$\frac{\partial x^s}{\partial \overline{x}^r} \overline{\Gamma}^r_{pq} = \Gamma^m_{ij} \delta^s_m \frac{\partial x^i}{\partial \overline{x}^p} \frac{\partial x^j}{\partial \overline{x}^q} + \delta^s_j \frac{\partial^2 x^j}{\partial \overline{x}^q \partial \overline{x}^p} \Rightarrow \frac{\partial x^s}{\partial \overline{x}^r} \overline{\Gamma}^r_{pq} = \Gamma^s_{ij} \frac{\partial x^i}{\partial \overline{x}^p} \frac{\partial x^j}{\partial \overline{x}^q} + \frac{\partial^2 x^s}{\partial \overline{x}^q \partial \overline{x}^p}$$

$$\frac{\partial^2 x^s}{\partial \overline{x}^q \partial \overline{x}^p} = \frac{\partial x^s}{\partial \overline{x}^r} \overline{\Gamma}^r_{pq} - \Gamma^s_{ij} \frac{\partial x^i}{\partial \overline{x}^p} \frac{\partial x^j}{\partial \overline{x}^q} \tag{2.4.25}$$

Expression (2.4.25) shows that the second derivative of the coordinate x^s can be decomposed into terms with the first derivatives of this coordinate and the coordinates x^i, x^j, and with the Christoffel symbols of second kind. This important expression was deducted in 1869 by Elwin Bruno Christoffel.

Let an inverse transformation for the Christoffel symbol of second kind of the coordinate system $\overline{X}^j$ to another referential system X^i given by

$$\Gamma^j_{\ell k} = \overline{\Gamma}^r_{pq} \frac{\partial \overline{x}^p}{\partial x^\ell} \frac{\partial \overline{x}^q}{\partial x^k} \frac{\partial x^j}{\partial \overline{x}^r} + \frac{\partial x^j}{\partial \overline{x}^r} \frac{\partial^2 \overline{x}^r}{\partial x^\ell \partial x^k}$$

and multiplying both members by $\frac{\partial \overline{x}^m}{\partial x^j}$ and proceeding in a manner that is analogous to the previous one

$$\frac{\partial^2 \overline{x}^m}{\partial x^\ell \partial x^k} = \frac{\partial \overline{x}^m}{\partial x^j} \Gamma^j_{\ell k} - \overline{\Gamma}^m_{pq} \frac{\partial \overline{x}^p}{\partial x^\ell} \frac{\partial \overline{x}^q}{\partial x^k} \tag{2.4.26}$$

The transformation of the Christoffel symbols from one coordinate system X^i to another coordinate system $\overline{X}^j$, and from this one to a third coordinates system $\widetilde{X}^k$ is identical to the transformation from X^i directly to $\widetilde{X}^k$, so the transitive property is valid for the transformations of the Christoffel symbols. This shows that these symbols form a group.

The Christoffel relation given by expression (2.4.25) can be written as

$$\frac{\partial^2 x^k}{\partial \bar{x}^j \partial \bar{x}^m}\frac{\partial \bar{x}^s}{\partial x^k} = \bar{\Gamma}^s_{jm} - \frac{\partial \bar{x}^s}{\partial x^p}\frac{\partial x^k}{\partial \bar{x}^j}\frac{\partial x^r}{\partial \bar{x}^m}\Gamma^p_{kr}$$

and contracting the terms in the indexes s and m

$$\frac{\partial^2 x^k}{\partial \bar{x}^j \partial \bar{x}^m}\frac{\partial \bar{x}^m}{\partial x^k} = \bar{\Gamma}^m_{jm} - \frac{\partial \bar{x}^m}{\partial x^p}\frac{\partial x^k}{\partial \bar{x}^j}\frac{\partial x^r}{\partial \bar{x}^m}\Gamma^p_{kr}$$

$$\frac{\partial^2 x^k}{\partial \bar{x}^j \partial \bar{x}^m}\frac{\partial \bar{x}^m}{\partial x^k} = \bar{\Gamma}^m_{jm} - \delta^r_p\frac{\partial x^k}{\partial \bar{x}^j}\Gamma^p_{kr}$$

$$\bar{\Gamma}^m_{jm} = \frac{\partial x^k}{\partial \bar{x}^j}\Gamma^r_{kr} + \frac{\partial^2 x^k}{\partial \bar{x}^j \partial \bar{x}^m}\frac{\partial \bar{x}^m}{\partial x^k}$$

that is the transformation law of the contracted Christoffel symbol of second kind.

2.3.13 Ricci Identity

Another usual expression in Tensor Calculus is obtained by means of defining the Christoffel symbol of first kind

$$\Gamma_{ji,k} = \frac{1}{2}\left(\frac{\partial g_{ik}}{\partial x^j} + \frac{\partial g_{jk}}{\partial x^i} - \frac{\partial g_{ij}}{\partial x^k}\right) \quad \Gamma_{ki,j} = \frac{1}{2}\left(\frac{\partial g_{ij}}{\partial x^k} + \frac{\partial g_{kj}}{\partial x^i} - \frac{\partial g_{ik}}{\partial x^j}\right)$$

The sum of these two expressions provides the Ricci identity

$$\frac{\partial g_{jk}}{\partial x^i} = \Gamma_{ji,k} + \Gamma_{ki,j} \tag{2.4.27}$$

In an analogous way, subtracting the second expression from the first expression of the Christoffel symbol of first kind provides

$$\frac{\partial g_{ij}}{\partial x^k} - \frac{\partial g_{jk}}{\partial x^i} = \Gamma_{kj,i} - \Gamma_{ij,k} \tag{2.4.28}$$

Expressions (2.4.27) and (2.4.28) are very useful in manipulations of tensorial equations.

Exercise 2.3 If T^{ij} and g_{ik} are the components of a symmetric *tensor* and the metric tensor, respectively, show that $T^{jk}\Gamma_{ij,k} = \frac{1}{2}T^{jk}\frac{\partial g_{jk}}{\partial x^i}$.

With the Ricci identity

$$\frac{\partial g_{ik}}{\partial x^i} = \Gamma_{ji,k} + \Gamma_{ki,j}$$

$$\frac{1}{2}T^{jk}\frac{\partial g_{jk}}{\partial x^i} = \frac{1}{2}T^{jk}\left(\Gamma_{ji,k} + \Gamma_{ki,j}\right) = \frac{1}{2}\left(T^{jk}\Gamma_{ji,k} + T^{jk}\Gamma_{ki,j}\right)$$

Interchanging the indexes $j \leftrightarrow k$ in the last term to the right of the expression and considering the tensor's symmetry then

$$\frac{1}{2}T^{jk}\frac{\partial g_{jk}}{\partial x^i} = \frac{1}{2}T^{jk}\left(\Gamma_{ji,k} + \Gamma_{ki,j}\right) = \frac{1}{2}\left(T^{jk}\Gamma_{ji,k} + T^{kj}\Gamma_{ji,k}\right) = \frac{1}{2} \times 2T^{jk}\Gamma_{ji,k}$$

$$\frac{1}{2}T^{jk}\frac{\partial g_{jk}}{\partial x^i} = T^{jk}\Gamma_{ji,k} \qquad Q.E.D.$$

2.3.14 Fundamental Relations

The derivative of the metric tensor with respect to an arbitrary variable can be placed in terms of Christoffel symbols of second kind and the metric tensor, thus from the definition of this symbol

$$\Gamma^p_{ik} = g^{pj}\Gamma_{ik,j} \quad \Gamma^p_{jk} = g^{pi}\Gamma_{jk,i}$$

Multiplying these two expressions by g_{pj} and g_{pi}, respectively:

$$g_{pj}\Gamma^p_{ik} = g_{pj}g^{pj}\Gamma_{ik,j} \Rightarrow g_{pk}\Gamma^p_{ik} = \delta_{pj}\Gamma_{ik,j}$$

$$g_{pi}\Gamma^p_{jk} = g_{pi}g^{pi}\Gamma_{jk,i} \Rightarrow g_{pi}\Gamma^p_{jk} = \delta_{pi}\Gamma_{jk,i}$$

whereby

$$\Gamma_{ik,j} = g_{pj}\Gamma^p_{ik} \quad \Gamma_{jk,i} = g_{ip}\Gamma^p_{jk}$$

Adding these two expressions and considering the Ricci identity

$$\frac{\partial g_{ik}}{\partial x^i} = \Gamma_{ji,k} + \Gamma_{ki,j} \tag{2.4.29}$$

and as $g_{ip} = g_{pi}$ it results in

$$\frac{\partial g_{ij}}{\partial x^k} = g_{pj}\Gamma^p_{ik} + g_{ip}\Gamma^p_{jk} \tag{2.4.30}$$

With analogous analysis this derivative can be placed in terms of Christoffel symbols of the second kind and the conjugate metric tensor, and with

$$g^{ij}g_{kj} = \delta^i_k$$

the derivative is

$$\frac{\partial g^{ij}}{\partial x^p}g_{kj} + g^{ij}\frac{\partial g_{kj}}{\partial x^p} = 0 \Rightarrow \frac{\partial g^{ij}}{\partial x^p}g_{kj} = -g^{ij}\frac{\partial g_{kj}}{\partial x^p}$$

Multiplying both members of this last expression by g^{kq}:

$$g^{kq}g_{kj}\frac{\partial g^{ij}}{\partial x^p} = -g^{kq}g^{ij}\frac{\partial g_{kj}}{\partial x^p} \Rightarrow \delta^q_j\frac{\partial g^{ij}}{\partial x^p} = -g^{kq}g^{ij}\frac{\partial g_{kj}}{\partial x^p} \Rightarrow \frac{\partial g^{iq}}{\partial x^p} = -g^{ij}g^{kq}\frac{\partial g_{kj}}{\partial x^p}$$

and with the Ricci identity

$$\frac{\partial g_{kj}}{\partial x^p} = \Gamma_{kp,j} + \Gamma_{jp,k}$$

that substituted in the previous expression provides

$$\frac{\partial g^{iq}}{\partial x^p} = -g^{ij}g^{kq}\left(\Gamma_{kp,j} + \Gamma_{jp,k}\right) = -g^{ij}g^{kq}\Gamma_{kp,j} - g^{ij}g^{kq}\Gamma_{jp,k}$$

For the Christoffel symbol of second kind the result is the following relations

$$g^{kq}g^{ij}\Gamma_{kp,j} = g^{kq}\Gamma^i_{kp} \quad g^{ij}g^{kq}\Gamma_{jp,k} = g^{ij}\Gamma^q_{jp}$$

whereby

$$\frac{\partial g^{iq}}{\partial x^p} = -g^{kq}\Gamma^i_{kp} - g^{ij}\Gamma^q_{jp}$$

As $\Gamma^q_{jp} = \Gamma^q_{pj}$, the result is

$$\frac{\partial g^{iq}}{\partial x^p} = -g^{ij}\Gamma^q_{pj} - g^{kq}\Gamma^i_{kp} \tag{2.4.31}$$

Expressions (2.4.30) and (2.4.31) are well used in the development of tensorial expressions.

Exercise 2.4 Calculate the Christoffel symbols Γ_{ijk} and Γ^k_{ij} for the polar coordinates systems, which metric tensor is given by

$$g_{ij} = \begin{bmatrix} 1 & 0 \\ 0 & (x^1)^2 \end{bmatrix}$$

The Christoffel symbol of first kind is given by

$$\Gamma_{ij,k} = \frac{1}{2}\left(\frac{\partial g_{jk}}{\partial x^i} + \frac{\partial g_{ik}}{\partial x^j} - \frac{\partial g_{ij}}{\partial x^k}\right)$$

so

$$g_{11} = 1 \Rightarrow g_{11,1} = 0, g_{11,2} = 0$$

$$g_{22} = (x^1)^2 \Rightarrow g_{22,1} = 2x^1, g_{22,2} = 0$$

$$g_{12,1} = g_{12,2} = g_{21,1} = g_{21,2} = g_{22,2} = 0$$

It follows that

$$\Gamma_{11,1} = \Gamma_{11,2} = \Gamma_{12,1} = \Gamma_{21,1} = \Gamma_{22,2} = 0$$

$$\Gamma_{12,2} = \Gamma_{21,2} = x^1$$

$$\Gamma_{22,1} = -x^1$$

In matrix form the result is

$$\Gamma_{ij,1} = \begin{bmatrix} 0 & 0 \\ 0 & -x^1 \end{bmatrix} \quad \Gamma_{ij,2} = \begin{bmatrix} 0 & x^1 \\ x^1 & 0 \end{bmatrix}$$

For the Christoffel symbol of second kind it follows that

$$\left[g_{ij}\right]^{-1} = g^{ij} = \begin{bmatrix} 1 & 0 \\ 0 & \frac{1}{(x^1)^2} \end{bmatrix}$$

$$\Gamma^k_{12} = g^{k2}\Gamma_{12,2}$$

$$\Gamma^k_{22} = g^{k1}\Gamma_{22,1}$$

$$k = 1 \Rightarrow \begin{cases} \Gamma^1_{12} = g^{12}\Gamma_{12,2} = 0 \\ \Gamma^1_{22} = g^{11}(-x^1) = -x^1 \end{cases}$$

In matrix form the result is

$$\Gamma^1_{ij} = \begin{bmatrix} 0 & 0 \\ 0 & -x^1 \end{bmatrix} \quad \Gamma^2_{ij} = \begin{bmatrix} 0 & \frac{1}{x^1} \\ \frac{1}{x^1} & 0 \end{bmatrix}$$

Exercise 2.5 Calculate the Christoffel symbols for the *cylindrical* coordinates system, defined by $r \equiv x^1$, $\theta \equiv x^2$, $z \equiv x^3$, where $-\infty \leq r \leq \infty$, $0 \leq \theta \leq 2\pi$, $-\infty \leq z \leq \infty$, which metric tensor and its conjugated tensor are given, respectively, by

$$g_{ij} = \begin{bmatrix} 1 & 0 & 0 \\ 0 & r^2 & 0 \\ 0 & 0 & 1 \end{bmatrix} \quad g^{ij} = \begin{bmatrix} 1 & 0 & 0 \\ 0 & \frac{1}{r^2} & 0 \\ 0 & 0 & 1 \end{bmatrix}$$

Using the expressions deducted for the *orthogonal* coordinate systems:

- $i = j = k$

$$\Gamma_{ii,i} = \frac{1}{2}\frac{\partial g_{ii}}{\partial x^i}$$

$$\Gamma_{11,1} = \frac{1}{2}g_{11,1} = 0 \quad \Gamma_{22,2} = \frac{1}{2}g_{22,2} = 0 \quad \Gamma_{33,3} = \frac{1}{2}g_{33,3} = 0$$

- $i = j \neq k$

$$\Gamma_{ii,k} = -\frac{1}{2}\frac{\partial g_{ii}}{\partial x^k}$$

$$\Gamma_{11,2} = -\frac{1}{2}g_{11,2} = 0 \quad \Gamma_{11,3} = -\frac{1}{2}g_{11,3} = 0$$

$$\Gamma_{22,1} = -\frac{1}{2}g_{22,1} = -r \quad \Gamma_{22,3} = -\frac{1}{2}g_{22,3} = 0$$

$$\Gamma_{33,1} = -\frac{1}{2}g_{33,1} = 0 \quad \Gamma_{33,2} = -\frac{1}{2}g_{33,2} = 0$$

- $i = k \neq j$

$$\Gamma_{ij,i} = \frac{1}{2}\frac{\partial g_{ii}}{\partial x^j}$$

$$\Gamma_{12,1} = \frac{1}{2}g_{11,2} = 0 \quad \Gamma_{13,1} = \frac{1}{2}g_{11,3} = 0$$

$$\Gamma_{21,2} = \frac{1}{2}g_{22,1} = r \quad \Gamma_{23,2} = \frac{1}{2}g_{22,3} = 0$$

$$\Gamma_{31,3} = \frac{1}{2}g_{33,1} = 0 \quad \Gamma_{ij,i} = \frac{1}{2}g_{33,2} = 0$$

- $i \neq j, j \neq k, i \neq k$ all the Christoffel symbols are null.

Putting these symbols in matrix form, the result is

$$\Gamma_{ij,1} = \begin{bmatrix} 0 & 0 & 0 \\ 0 & -r & 0 \\ 0 & 0 & 0 \end{bmatrix} \quad \Gamma_{ij,2} = \begin{bmatrix} 0 & r & 0 \\ r & 0 & 0 \\ 0 & 0 & 0 \end{bmatrix} \quad \Gamma_{ij,3} = [0]$$

For the Christoffel symbols of second kind it follows that

$$\Gamma^p_{ij} = g^{pk}\Gamma_{ij,k}$$

$$\Gamma^p_{22} = g^{p1}\Gamma_{22,1} \quad \Gamma^p_{12} = g^{p2}\Gamma_{12,2}$$

- $p = 1$

$$\Gamma^1_{22} = g^{11}\Gamma_{22,1} = r \quad \Gamma^1_{12} = g^{12}\Gamma_{12,2} = 0$$

- $p = 2$

$$\Gamma^2_{22} = g^{21}\Gamma_{22,1} = 0 \quad \Gamma^2_{12} = g^{22}\Gamma_{12,2} = \frac{1}{r}$$

- $p = 3$

$$\Gamma^3_{22} = g^{31}\Gamma_{22,1} = 0 \quad \Gamma^3_{13} = g^{32}\Gamma_{12,2} = 0$$

Putting these symbols in matrix form, the result is

$$\Gamma^1_{ij} = \begin{bmatrix} 0 & 0 & 0 \\ 0 & -r & 0 \\ 0 & 0 & 0 \end{bmatrix} \quad \Gamma^2_{ij} = \begin{bmatrix} 0 & \frac{1}{r} & 0 \\ \frac{1}{r} & 0 & 0 \\ 0 & 0 & 0 \end{bmatrix} \quad \Gamma^3_{ij} = [0]$$

Exercise 2.6 Calculate the Christoffel symbols for the spherical coordinates system $r \equiv x^1, \varphi \equiv x^2, \theta \equiv x^3$, where $-\infty \le r \le \infty, 0 \le \varphi \le \pi, 0 \le \theta \le 2\pi$, which metric tensor and its conjugated tensor are given, respectively, by

$$g_{ij} = \begin{bmatrix} 1 & 0 & 0 \\ 0 & r^2 & 0 \\ 0 & 0 & r^2 \sin^2\varphi \end{bmatrix} \quad g^{ij} = \begin{bmatrix} 1 & 0 & 0 \\ 0 & \frac{1}{r^2} & 0 \\ 0 & 0 & \frac{1}{r^2 \sin^2\varphi} \end{bmatrix}$$

Using the expressions deduced for the orthogonal coordinate systems the result is:
- $i = j = k$

$$\Gamma_{ii,i} = \frac{1}{2}\frac{\partial g_{ii}}{\partial x^i}$$

$$\Gamma_{11,1} = \frac{1}{2}g_{11,1} = 0 \quad \Gamma_{22,2} = \frac{1}{2}g_{22,2} = 0 \quad \Gamma_{33,3} = \frac{1}{2}g_{33,3} = 0$$

- $i = j \neq k$

$$\Gamma_{ii,k} = -\frac{1}{2}\frac{\partial g_{ii}}{\partial x^k}$$

$$\Gamma_{11,2} = -\frac{1}{2}g_{11,2} = 0 \quad \Gamma_{11,3} = -\frac{1}{2}g_{11,3} = 0$$

$$\Gamma_{22,1} = -\frac{1}{2}g_{22,1} = -r \quad \Gamma_{22,3} = -\frac{1}{2}g_{22,3} = 0$$

$$\Gamma_{33,1} = -\frac{1}{2}g_{33,1} = -r\sin^2\varphi \quad \Gamma_{33,2} = -\frac{1}{2}g_{33,2} = -r^2\sin\varphi\cos\varphi$$

- $i = k \neq j$

$$\Gamma_{ij,i} = \frac{1}{2}\frac{\partial g_{ii}}{\partial x^j}$$

$$\Gamma_{12,1} = \frac{1}{2}g_{11,2} = 0 \quad \Gamma_{13,1} = \frac{1}{2}g_{11,3} = 0$$

$$\Gamma_{21,2} = \frac{1}{2}g_{22,1} = r \quad \Gamma_{23,2} = \frac{1}{2}g_{22,3} = 0$$

$$\Gamma_{31,3} = \frac{1}{2}g_{33,1} = r\sin^2\varphi \quad \Gamma_{ij,i} = \frac{1}{2}g_{33,2} = r^2\sin\varphi\cos\varphi$$

- $i \neq j, j \neq k, i \neq k$ all the Christoffel symbols are null.
  Putting these symbols in matrix form, the result is

$$\Gamma_{ij,1} = \begin{bmatrix} 0 & 0 & 0 \\ 0 & -r & 0 \\ 0 & 0 & r\sin^2\varphi \end{bmatrix} \quad \Gamma_{ij,2} = \begin{bmatrix} 0 & r & 0 \\ r & 0 & 0 \\ 0 & 0 & -r^2\sin\varphi\cos\varphi \end{bmatrix}$$

$$\Gamma_{ij,3} = \begin{bmatrix} 0 & 0 & r\sin^2\varphi \\ 0 & 0 & r^2\sin\varphi\cos\varphi \\ r\sin^2\varphi & r^2\sin\varphi\cos\varphi & 0 \end{bmatrix}$$

For the Christoffel symbols of second kind it follows that

$$\Gamma_{ij}^p = g^{pk}\Gamma_{j,k}$$

- $p = k = 1$

$$\Gamma^1_{ij} = g^{11}\Gamma_{ij,1}$$

$$\Gamma^1_{22} = g^{11}\Gamma_{22,1} = -r \quad \Gamma^1_{33} = g^{11}\Gamma_{33,1} = -r\sin^2\varphi$$

- $p = k = 2$

$$\Gamma^2_{ij} = g^{22}\Gamma_{ij,2}$$

$$\Gamma^2_{12} = g^{22}\Gamma_{12,2} = \frac{1}{r} \quad \Gamma^2_{33} = g^{22}\Gamma_{33,2} = -\sin\varphi\cos\varphi$$

- $p = k = 3$

$$\Gamma^3_{ij} = g^{33}\Gamma_{ij,3}$$

$$\Gamma^3_{13} = g^{33}\Gamma_{13,3} = \frac{1}{r} \quad \Gamma^3_{23} = g^{33}\Gamma_{23,3} = \cot\varphi$$

Putting these symbols in matrix form, the result is

$$\Gamma^1_{ij} = \begin{bmatrix} 0 & 0 & 0 \\ 0 & -r & 0 \\ 0 & 0 & -r\sin^2\varphi \end{bmatrix} \quad \Gamma^2_{ij} = \begin{bmatrix} 0 & \frac{1}{r} & 0 \\ \frac{1}{r} & 0 & 0 \\ 0 & 0 & \sin\varphi\cos\varphi \end{bmatrix}$$

$$\Gamma^3_{ij} = \begin{bmatrix} 0 & 0 & \frac{1}{r} \\ 0 & 0 & \cot\varphi \\ \frac{1}{r} & \cot\varphi & 0 \end{bmatrix}$$

Exercise 2.7 For the antisymmetric tensor A^{ijk}, show that $A^{ijk}\Gamma^p_{ij} = A^{ijk}\Gamma^p_{jk} = A^{ijk}\Gamma^p_{ik} = 0$.

The symmetry of the Christoffel symbol of second kind allows writing

$$A^{ijk}\Gamma^p_{ij} = A^{ijk}\Gamma^p_{ji} = 0$$

and replacing the indexes $i \rightarrow j$ it follows that

$$A^{ijk}\Gamma^p_{ij} = -A^{ijk}\Gamma^p_{ij}$$

the result is

$$A^{ijk}\Gamma^p_{ij} = 0$$

Proceeding in an analogous way for $A^{ijk}\Gamma^p_{jk}$ and $A^{ijk}\Gamma^p_{ik}$ the equalities of what was enunciated are verified.

Exercise 2.8 Given the expression $\overline{\Gamma}^i_{jk} = \Gamma^i_{jk} + \delta^i_j u_k + \delta^i_k u_j$, where u_i is a covariant vector and A^{ij} is an antisymmetric tensor, show that $A^{jk}\overline{\Gamma}^i_{jk} = 0$.

The symmetry of the Christoffel symbol of second kind allows writing

$$\overline{\Gamma}^i_{jk} = \overline{\Gamma}^i_{kj}$$

$$\overline{\Gamma}^i_{jk} = \Gamma^i_{kj} + \delta^i_k u_j + \delta^i_j u_k$$

$$A^{jk}\overline{\Gamma}^i_{jk} = A^{kj}\overline{\Gamma}^i_{kj}$$

and with the consideration of the *anti-symmetry* $A^{jk} = -A_{kj}$ *verifies that*

$$A^{jk}\overline{\Gamma}^i_{jk} = 0 \qquad Q.E.D.$$

2.4 Covariant Derivative

The basic problem treated by the Tensorial Analysis is to research if the derivatives of tensors generate new tensors, which, in general, does not occur. For the case of Cartesian coordinate systems the variation rates of the tensors are expressed by partial derivatives. For instance, the variation rates of a vector's components indicate the variation of this vector. However, for the curvilinear coordinate systems, the expressions for these variation rates are not expressed only by partial derivatives. Figure 2.3a shows this coordinate systems for the case in which the vector $\boldsymbol{u}$ has constant modulus and directions (Fig. 2.3a), but their components u^1 vary. Figure 2.3b shows the behavior of the vectors $\boldsymbol{u}$ with constant modulus and different directions, whereby the three vectors are different, but their radial u^1 and tangential $u^2 = 0$ components remain constant. This example indicates the need for

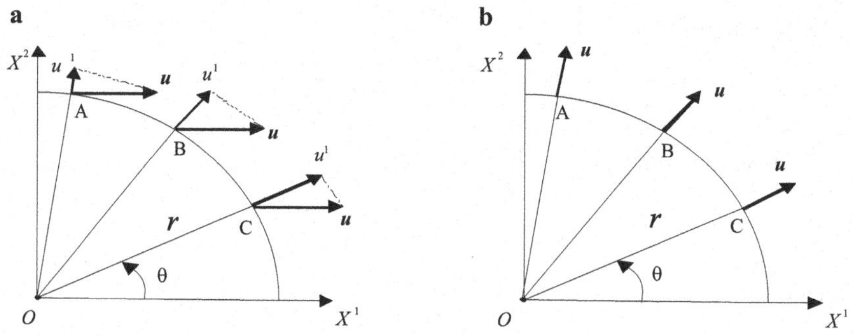

Fig. 2.3 Polar coordinates: (**a**) vector $\boldsymbol{u}$ with constant modulus and direction and (**b**) vector $\boldsymbol{u}$ with constant modulus and variable direction

researching the variation rates of vectors for the curvilinear coordinate systems, because the variation rates of their components do not represent the variation of these vectors.

To exemplify this fact let the scalar function $\phi = -mx$, where m is a scalar, which generates the potential $\boldsymbol{u} = -\text{grad}\phi$, with Cartesian components $u_1 = m$, $u_2 = 0$. This scalar function in polar coordinates is defined by $\phi = -mr\cos\theta$, which covariant components of its gradient are given by

$$\frac{\partial\phi}{\partial r} = -m\cos\theta \quad \frac{\partial\phi}{\partial\theta} = mr\sin\theta$$

and its physical components are $u_1^* = \frac{\partial\phi}{\partial r} = -m\cos\theta$ and $u_2^* = \frac{1}{r}\frac{\partial\phi}{\partial\theta} = m\sin\theta$. These components are not constant. The interpretation of this variation is carried out admitting a polar coordinates point $P(r;\theta)$ being displaced to another point nearby $P'(r + dr;\, \theta + d\theta)$, so the covariant components of the vector $\boldsymbol{u}$ initially given by

$$u_1 = \frac{\partial\phi}{\partial r} = -m\cos\theta \quad u_2 = \frac{\partial\phi}{\partial\theta} = mr\sin\theta$$

stay for this new point

$$\delta u_1 = -m\sin\theta d\theta \quad \delta u_2 = m\sin\theta dr + mr\cos\theta d\theta$$

The elemental variations of these new components are due to the change of coordinates, and not to the change of vector. This particular indicates the need of defining a kind of derivative that translates the vector's variation in an invariant manner, and leads to the definition of the covariant derivative.

The covariant derivative defines variation rate of parameters that are not dependent on the coordinate systems, and because of that it is of extreme importance in the expression of physical models, for it generates a new tensor. The denomination covariant derivative was adopted by Bruno Ricci-Curbastro when conceiving the Tensor Calculus. The term covariant denotes a kind of partial differentiation of tensors that generates new tensors with variance one order above the original tensors. The adjective covariant is used to indicate the tensorial characteristics of the differentiation of tensors, in which the set of Christoffel symbols Γ_{ij}^k are the coefficients of connections of the tensorial space E_N.

2.4.1 Contravariant Tensor

2.4.1.1 Contravariant Vector

Let the vector $\boldsymbol{u}$ defined by its contravariant components u^j:

$$\boldsymbol{u} = u^j \boldsymbol{g}_j \tag{2.5.1}$$

where the unit vectors $\boldsymbol{g}_j = \boldsymbol{g}_j(x^j)$ of the curvilinear coordinate system are functions of the coordinates that define this referential system.

Differentiating the expression (2.5.1) with respect to an arbitrary coordinate x^k results in

$$\frac{\partial \boldsymbol{u}}{\partial x^k} = \frac{\partial u^j}{\partial x^k}\boldsymbol{g}_j + u^j \frac{\partial \boldsymbol{g}_j}{\partial x^k}$$

and using expression (2.4.1)

$$\frac{\partial \boldsymbol{g}_j}{\partial x^k} = \Gamma^m_{jk}\boldsymbol{g}_m$$

then

$$\frac{\partial \boldsymbol{u}}{\partial x^k} = \frac{\partial u^j}{\partial x^k}\boldsymbol{g}_j + u^j\Gamma^m_{jk}\boldsymbol{g}_m \tag{2.5.2}$$

As j is a dummy index in the first term to the right, it can be changed for the index m:

$$\frac{\partial \boldsymbol{u}}{\partial x^k} = \frac{\partial u^m}{\partial x^k}\boldsymbol{g}_m + u^j\Gamma^m_{jk}\boldsymbol{g}_m = \left(\frac{\partial u^m}{\partial x^k} + u^j\Gamma^m_{jk}\right)\boldsymbol{g}_m$$

This expression shows that the covariant derivative of a contravariant vector is given by the N^2 functions

$$\partial_k u^m = \frac{\partial u^m}{\partial x^k} + u^j\Gamma^m_{jk} \tag{2.5.3}$$

whereby

$$\frac{\partial \boldsymbol{u}}{\partial x^k} = (\partial_k u^m)\boldsymbol{g}_m \tag{2.5.4}$$

For the Cartesian systems the Christoffel symbols are null, so the covariant derivative coincides with the partial derivative $\frac{\partial u^m}{\partial x^k}$.

In expression (2.5.3) the result is the variation rate of the vector $\boldsymbol{u}$ along the axes of the curvilinear coordinate system is given by $\frac{\partial u^j}{\partial x^k}$, and the variation of the unit vectors $\boldsymbol{g}_j$ along the axes of this coordinate system is expressed by $\frac{\partial \boldsymbol{g}_j}{\partial x^k}$. This physical interpretation of the covariant derivative is associated to the Christoffel symbols Γ^m_{jk} that are the connection coefficients of the tensorial space.

Various notations are found in the literature for the term to the left of expression (2.5.3), the most usual being: $\partial_k u^m = D_k u^m = \nabla_k u^m = u^m|_k = u^m|_k = u^m{}_{;k}$.

Expressions (2.5.1) and (2.5.4) are analogous, for $\partial_k u^m$ has the aspect of a vector. The transformation law of contravariant vectors is admitted to demonstrate that expression (2.5.4) is a tensor, thus

$$\overline{u}^i = \frac{\partial \overline{x}^i}{\partial x^p} u^p$$

which differentiated with respect to the coordinate $\overline{x}^j$ provides

$$\frac{\partial \overline{u}^i}{\partial \overline{x}^j} = \frac{\partial \overline{x}^i}{\partial x^p} \left(\frac{\partial u^p}{\partial x^q} \frac{\partial x^q}{\partial \overline{x}^j} \right) + u^p \left(\frac{\partial^2 \overline{x}^i}{\partial x^q \partial x^p} \frac{\partial x^q}{\partial \overline{x}^j} \right) \tag{2.5.5}$$

and with expression (2.4.26)

$$\frac{\partial^2 \overline{x}^i}{\partial x^q \partial x^p} = \Gamma^m_{pq} \frac{\partial \overline{x}^i}{\partial x^n} - \overline{\Gamma}^i_{\ell m} \frac{\partial \overline{x}^\ell}{\partial x^p} \frac{\partial \overline{x}^m}{\partial x^q} \tag{2.5.6}$$

Substituting expression (2.5.6) in expression (2.5.5)

$$\frac{\partial \overline{u}^i}{\partial \overline{x}^j} = \frac{\partial \overline{x}^i}{\partial x^p} \left(\frac{\partial u^p}{\partial x^q} \frac{\partial x^q}{\partial \overline{x}^j} \right) + u^p \Gamma^n_{pq} \frac{\partial \overline{x}^i}{\partial x^n} \frac{\partial x^q}{\partial \overline{x}^j} - u^p \overline{\Gamma}^i_{\ell m} \frac{\partial \overline{x}^\ell}{\partial x^p} \frac{\partial \overline{x}^m}{\partial x^q} \frac{\partial x^q}{\partial \overline{x}^j}$$

$$\frac{\partial \overline{u}^i}{\partial \overline{x}^j} + u^p \overline{\Gamma}^i_{\ell m} \frac{\partial \overline{x}^\ell}{\partial x^p} \frac{\partial \overline{x}^m}{\partial x^q} \frac{\partial x^q}{\partial \overline{x}^j} = \frac{\partial u^p}{\partial x^q} \frac{\partial \overline{x}^i}{\partial x^p} \frac{\partial x^q}{\partial \overline{x}^j} + u^p \Gamma^n_{pq} \frac{\partial \overline{x}^i}{\partial x^n} \frac{\partial x^q}{\partial \overline{x}^j}$$

The dummy index p in the first term to the right can be changed by the index n:

$$\frac{\partial \overline{u}^i}{\partial \overline{x}^j} + u^p \overline{\Gamma}^i_{\ell m} \frac{\partial \overline{x}^\ell}{\partial x^p} \frac{\partial \overline{x}^m}{\partial x^q} \frac{\partial x^q}{\partial \overline{x}^j} = \frac{\partial u^n}{\partial x^q} \frac{\partial \overline{x}^i}{\partial x^n} \frac{\partial x^q}{\partial \overline{x}^j} + u^p \Gamma^n_{pq} \frac{\partial \overline{x}^i}{\partial x^n} \frac{\partial x^q}{\partial \overline{x}^j}$$

and with expression

$$\frac{\partial \overline{x}^m}{\partial \overline{x}^j} = \delta^m_j$$

results in

$$\frac{\partial \overline{u}^i}{\partial \overline{x}^j} + u^p \overline{\Gamma}^i_{\ell m} \frac{\partial \overline{x}^\ell}{\partial x^p} \delta^m_j = \left(\frac{\partial u^n}{\partial x^q} + u^p \Gamma^n_{pq} \right) \frac{\partial \overline{x}^i}{\partial x^n} \frac{\partial x^q}{\partial \overline{x}^j}$$

With the transformation law of contravariant vectors

$$\overline{u}^\ell = u^p \frac{\partial \overline{x}^\ell}{\partial x^p}$$

the above expression becomes

$$\frac{\partial \overline{u}^i}{\partial \overline{x}^j} + \overline{u}^\ell \overline{\Gamma}^i_{\ell j} = \left(\frac{\partial u^n}{\partial x^q} + u^p \Gamma^n_{pq} \right) \frac{\partial \overline{x}^i}{\partial x^n} \frac{\partial x^q}{\partial \overline{x}^j} \tag{2.5.7}$$

It is verified that in expression (2.5.7) the variety in parenthesis transforms as a mixed second-order tensor, then the covariant derivative of a contravariant vector is a mixed second-order tensor, i.e., of variance $(1, 1)$.

2.4.2 Contravariant Tensor of the Second-Order

The transformation law of contravariant tensors of the second-order is given by

$$\overline{T}^{pq} = T^{ij} \frac{\partial \overline{x}^p}{\partial x^i} \frac{\partial \overline{x}^q}{\partial x^j}$$

The derivative of this expression with respect to coordinate x^ℓ is given by

$$\frac{\partial \overline{T}^{pq}}{\partial x^\ell} = \frac{\partial T^{ij}}{\partial x^k} \frac{\partial x^k}{\partial \overline{x}^\ell} \frac{\partial \overline{x}^p}{\partial x^i} \frac{\partial \overline{x}^q}{\partial x^j} + T^{ij} \frac{\partial^2 \overline{x}^p}{\partial x^k \partial x^i} \frac{\partial x^k}{\partial \overline{x}^\ell} \frac{\partial \overline{x}^q}{\partial x^j} + T^{ij} \frac{\partial \overline{x}^p}{\partial x^i} \frac{\partial^2 \overline{x}^q}{\partial x^k \partial x^j} \frac{\partial x^k}{\partial \overline{x}^\ell}$$

and with expression (2.4.26)

$$\frac{\partial^2 \overline{x}^p}{\partial x^k \partial x^i} = \Gamma^r_{ki} \frac{\partial \overline{x}^p}{\partial x^r} - \overline{\Gamma}^p_{\ell m} \frac{\partial \overline{x}^\ell}{\partial x^i} \frac{\partial \overline{x}^m}{\partial x^k} \qquad \frac{\partial^2 \overline{x}^q}{\partial x^k \partial x^j} = \Gamma^r_{kj} \frac{\partial \overline{x}^q}{\partial x^r} - \overline{\Gamma}^q_{\ell m} \frac{\partial \overline{x}^\ell}{\partial x^j} \frac{\partial \overline{x}^m}{\partial x^k}$$

then

$$\begin{aligned}\frac{\partial \overline{T}^{pq}}{\partial x^\ell} = & \frac{\partial T^{ij}}{\partial x^k} \frac{\partial x^k}{\partial \overline{x}^\ell} \frac{\partial \overline{x}^p}{\partial x^i} \frac{\partial \overline{x}^q}{\partial x^j} + T^{ij} \left(\Gamma^r_{ki} \frac{\partial \overline{x}^p}{\partial x^r} - \overline{\Gamma}^p_{\ell m} \frac{\partial \overline{x}^\ell}{\partial x^i} \frac{\partial \overline{x}^m}{\partial x^k} \right) \frac{\partial x^k}{\partial \overline{x}^\ell} \frac{\partial \overline{x}^q}{\partial x^j} \\ & + T^{ij} \left(\Gamma^r_{kj} \frac{\partial \overline{x}^q}{\partial x^r} - \overline{\Gamma}^q_{\ell m} \frac{\partial \overline{x}^\ell}{\partial x^j} \frac{\partial \overline{x}^m}{\partial x^k} \right) \frac{\partial x^k}{\partial \overline{x}^\ell} \frac{\partial \overline{x}^p}{\partial x^i}\end{aligned}$$

$$\begin{aligned}\frac{\partial \overline{T}^{pq}}{\partial x^\ell} = & \frac{\partial T^{ij}}{\partial x^k} \frac{\partial x^k}{\partial \overline{x}^\ell} \frac{\partial \overline{x}^p}{\partial x^i} \frac{\partial \overline{x}^q}{\partial x^j} + T^{ij} \Gamma^r_{ki} \frac{\partial \overline{x}^p}{\partial x^r} \frac{\partial x^k}{\partial \overline{x}^\ell} \frac{\partial \overline{x}^q}{\partial x^j} - T^{ij} \overline{\Gamma}^p_{\ell m} \frac{\partial \overline{x}^\ell}{\partial x^i} \frac{\partial \overline{x}^m}{\partial x^k} \frac{\partial x^k}{\partial \overline{x}^\ell} \frac{\partial \overline{x}^q}{\partial x^j} \\ & + T^{ij} \Gamma^r_{kj} \frac{\partial \overline{x}^q}{\partial x^r} \frac{\partial x^k}{\partial \overline{x}^\ell} \frac{\partial \overline{x}^p}{\partial x^i} - T^{ij} \overline{\Gamma}^q_{\ell m} \frac{\partial \overline{x}^\ell}{\partial x^j} \frac{\partial \overline{x}^m}{\partial x^k} \frac{\partial x^k}{\partial \overline{x}^\ell} \frac{\partial \overline{x}^p}{\partial x^i}\end{aligned}$$

With

$$\delta^m_\ell = \frac{\partial \overline{x}^m}{\partial x^k} \frac{\partial x^k}{\partial \overline{x}^\ell}$$

it follows that

$$\frac{\partial \overline{T}^{pq}}{\partial \overline{x}^{\ell}} = \frac{\partial T^{ij}}{\partial x^k}\frac{\partial x^k}{\partial \overline{x}^{\ell}}\frac{\partial \overline{x}^p}{\partial x^i}\frac{\partial \overline{x}^q}{\partial x^j} + T^{ij}\Gamma^r_{ki}\frac{\partial \overline{x}^p}{\partial x^r}\frac{\partial x^k}{\partial \overline{x}^{\ell}}\frac{\partial \overline{x}^q}{\partial x^j} - T^{ij}\overline{\Gamma}^p_{\ell m}\frac{\partial \overline{x}^{\ell}}{\partial x^i}\frac{\partial \overline{x}^q}{\partial x^j}$$
$$+ T^{ij}\Gamma^r_{kj}\frac{\partial \overline{x}^q}{\partial x^r}\frac{\partial x^k}{\partial \overline{x}^{\ell}}\frac{\partial \overline{x}^p}{\partial x^i} - T^{ij}\overline{\Gamma}^q_{\ell m}\frac{\partial \overline{x}^{\ell}}{\partial x^j}\frac{\partial \overline{x}^p}{\partial x^i}$$

In the second term to the right interchanging the indexes $i \leftrightarrow r$ and, likewise, in the same fourth term with the permutation of the indexes $j \leftrightarrow r$, it results in

$$\frac{\partial \overline{T}^{pq}}{\partial \overline{x}^{\ell}} = \frac{\partial T^{ij}}{\partial x^k}\frac{\partial x^k}{\partial \overline{x}^{\ell}}\frac{\partial \overline{x}^p}{\partial x^i}\frac{\partial \overline{x}^q}{\partial x^j} + T^{rj}\frac{\partial \overline{x}^p}{\partial x^i}\frac{\partial x^k}{\partial \overline{x}^{\ell}}\frac{\partial \overline{x}^q}{\partial x^j}\Gamma^i_{kr} - T^{ij}\frac{\partial \overline{x}^{\ell}}{\partial x^i}\frac{\partial \overline{x}^q}{\partial x^j}\overline{\Gamma}^p_{\ell m}$$
$$+ T^{ir}\frac{\partial \overline{x}^q}{\partial x^j}\frac{\partial x^k}{\partial \overline{x}^{\ell}}\frac{\partial \overline{x}^p}{\partial x^i}\Gamma^j_{kr} - T^{ij}\frac{\partial \overline{x}^{\ell}}{\partial x^j}\frac{\partial \overline{x}^p}{\partial x^i}\overline{\Gamma}^q_{\ell m}$$

and with the expressions

$$\overline{T}^{\ell q} = T^{ij}\frac{\partial \overline{x}^{\ell}}{\partial x^i}\frac{\partial \overline{x}^q}{\partial x^j} \qquad \overline{T}^{\ell p} = T^{ij}\frac{\partial \overline{x}^{\ell}}{\partial x^j}\frac{\partial \overline{x}^p}{\partial x^i}$$

it follows that

$$\frac{\partial \overline{T}^{pq}}{\partial \overline{x}^{\ell}} = \left(\frac{\partial T^{ij}}{\partial x^k} + T^{rj}\Gamma^i_{kr} + T^{ir}\Gamma^j_{kr}\right)\frac{\partial x^k}{\partial \overline{x}^{\ell}}\frac{\partial \overline{x}^p}{\partial x^i}\frac{\partial \overline{x}^q}{\partial x^j} - \overline{T}^{\ell q}\overline{\Gamma}^p_{\ell m} - \overline{T}^{\ell p}\overline{\Gamma}^q_{\ell m}$$

$$\frac{\partial \overline{T}^{pq}}{\partial \overline{x}^{\ell}} + \overline{T}^{\ell q}\overline{\Gamma}^p_{\ell m} + \overline{T}^{\ell p}\overline{\Gamma}^q_{\ell m} = \left(\frac{\partial T^{ij}}{\partial x^k} + T^{rj}\Gamma^i_{kr} + T^{ir}\Gamma^j_{kr}\right)\frac{\partial x^k}{\partial \overline{x}^{\ell}}\frac{\partial \overline{x}^p}{\partial x^i}\frac{\partial \overline{x}^q}{\partial x^j}$$

$$\partial_{\ell}\overline{T}^{pq} = \partial_k T^{ij}\frac{\partial x^k}{\partial \overline{x}^{\ell}}\frac{\partial \overline{x}^p}{\partial x^i}\frac{\partial \overline{x}^q}{\partial x^j} \tag{2.5.8}$$

where

$$\partial_k T^{ij} = \frac{\partial T^{ij}}{\partial x^k} + T^{rj}\Gamma^i_{kr} + T^{ir}\Gamma^j_{kr} \tag{2.5.9}$$

is the covariant derivative of the contravariant tensor of the second order.

Expression (2.5.8) indicates that the covariant derivative of a contravariant tensor of the second order is a mixed tensor of the third order, twice contravariant and once covariant, i.e., variance $(2,1)$. For the Cartesian coordinates the Christoffel symbols are null, so the covariant derivative coincides with the partial derivative $\frac{\partial T^{ij}}{\partial x^k}$.

2.4.2.1 Contravariant Tensor of Order Above Two

To generalize expression (2.5.9) for tensors of order above two, i.e., for instance, the covariant derivative of the contravariant tensor of the third order, which expression may be developed by means of the following steps:

(a) The basic structure of its expression is written considering the expression obtained for the covariant derivative of a contravariant tensor of the second order

$$\partial_p T^{ijk} = \frac{\partial T^{ijk}}{\partial x^p} + T^{\bullet\bullet\bullet}\Gamma^{\bullet}_{\bullet\bullet} + T^{\bullet\bullet\bullet}\Gamma^{\bullet}_{\bullet\bullet} + T^{\bullet\bullet\bullet}\Gamma^{\bullet}_{\bullet\bullet}$$

(b) The indexes of the Christoffel symbols corresponding to the coordinate with respect to which the differentiation is being carried out are placed

$$\partial_p T^{ijk} = \frac{\partial T^{ijk}}{\partial x^p} + T^{\bullet\bullet\bullet}\Gamma^{\bullet}_{\bullet p} + T^{\bullet\bullet\bullet}\Gamma^{\bullet}_{\bullet p} + T^{\bullet\bullet\bullet}\Gamma^{\bullet}_{\bullet p}$$

(c) The tensor indexes sequence must be obeyed on placing the contravariant indexes of the Christoffel symbol

$$\partial_p T^{ijk} = \frac{\partial T^{ijk}}{\partial x^p} + T^{\bullet\bullet\bullet}\Gamma^{i}_{\bullet p} + T^{\bullet\bullet\bullet}\Gamma^{j}_{\bullet p} + T^{\bullet\bullet\bullet}\Gamma^{k}_{\bullet p}$$

(d) The dummy index q is placed on the Christoffel symbol and in sequential form in the tensors

$$\partial_p T^{ijk} = \frac{\partial T^{ijk}}{\partial x^p} + T^{q\bullet\bullet}\Gamma^{i}_{qp} + T^{\bullet q\bullet}\Gamma^{j}_{qp} + T^{\bullet\bullet q}\Gamma^{k}_{qp}$$

(e) The remaining indexes are placed in the same sequence in which they appear on the tensor that is being differentiated

$$\partial_p T^{ijk} = \frac{\partial T^{ijk}}{\partial x^p} + T^{qjk}\Gamma^{i}_{qp} + T^{iqk}\Gamma^{j}_{qp} + T^{ijq}\Gamma^{k}_{qp}$$

This tensor generated by the differentiation of a variance tensor (3, 0) has a variance $(3, 1)$. Expression (2.5.9) can be generalized by adopting this indexes placement systematic for a contravariant tensor of order $p > 3$, then the variance of this new tensor will always be $(p, 1)$.

Exercise 2.9 Calculate the covariant derivative of the contravariant components of vector $\boldsymbol{u}$ expressed in polar coordinates.

In Exercise 2.4 the Christoffel symbols were calculated for the polar coordinates, given by

$$\Gamma_{ij,1} = \begin{bmatrix} 0 & 0 \\ 0 & -x^1 \end{bmatrix} \quad \Gamma_{ij,2} = \begin{bmatrix} 0 & x^1 \\ x^1 & 0 \end{bmatrix} \quad \Gamma^1_{ij} = \begin{bmatrix} 0 & 0 \\ 0 & -x^1 \end{bmatrix} \quad \Gamma^2_{ij} = \begin{bmatrix} 0 & \dfrac{1}{x^1} \\ \dfrac{1}{x^1} & 0 \end{bmatrix}$$

The expression for the derivative of the contravariant components of vector $\boldsymbol{u}$ is:

$$\partial_k u^m = \frac{\partial u^m}{\partial x^k} + u^j \Gamma^m_{jk}$$

- $m = 1$

$$\partial_k u^1 = \frac{\partial u^1}{\partial x^k} + u^j \Gamma^1_{jk}$$

$$k = 1 \Rightarrow \partial_1 u^1 = \frac{\partial u^1}{\partial x^1} + u^j \Gamma^1_{j1} \Rightarrow \partial_1 u^1 = \frac{\partial u^1}{\partial x^1} + u^1 \Gamma^1_{11} + u^2 \Gamma^1_{21}$$

$$\partial_1 u^1 = \frac{\partial u^1}{\partial x^1} + 0 + 0 = \frac{\partial u^1}{\partial x^1}$$

$$k = 2 \Rightarrow \partial_2 u^1 = \frac{\partial u^1}{\partial x^2} + u^j \Gamma^1_{j1} \Rightarrow \partial_2 u^1 = \frac{\partial u^1}{\partial x^2} + u^1 \Gamma^1_{12} + u^2 \Gamma^1_{22}$$

$$\partial_2 u^1 = \frac{\partial u^1}{\partial x^2} + \frac{u^1}{x^1} + 0 = \frac{\partial u^1}{\partial x^2} + \frac{u^1}{x^1}$$

- $m = 2$

$$\partial_k u^2 = \frac{\partial u^2}{\partial x^k} + u^j \Gamma^2_{jk}$$

$$k = 1 \Rightarrow \partial_1 u^2 = \frac{\partial u^2}{\partial x^1} + u^j \Gamma^2_{j1} \Rightarrow \partial_1 u^2 = \frac{\partial u^2}{\partial x^1} + u^1 \Gamma^2_{11} + u^2 \Gamma^2_{21}$$

$$\partial_1 u^2 = \frac{\partial u^2}{\partial x^1} + 0 + \frac{u^2}{x^1} = \frac{\partial u^2}{\partial x^1} + \frac{u^2}{x^1}$$

$$k = 2 \Rightarrow \partial_2 u^2 = \frac{\partial u^2}{\partial x^2} + u^j \Gamma^2_{j2} \Rightarrow \partial_2 u^2 = \frac{\partial u^2}{\partial x^2} + u^1 \Gamma^2_{12} + u^2 \Gamma^2_{22}$$

$$\partial_2 u^2 = \frac{\partial u^2}{\partial x^2} + \frac{u^1}{x^1} + 0 = \frac{\partial u^2}{\partial x^2} + \frac{u^1}{x^1}$$

Exercise 2.10 Show that $\partial_j T^{ij} = \frac{1}{\sqrt{g}} \frac{\partial\left(T^{ij}\sqrt{g}\right)}{\partial x^j} + T^{jp}\Gamma^i_{jp}$.

The expression of the covariant derivative of *a* contravariant tensor of the second order is given by

$$\partial_j T^{ij} = \frac{\partial T^{ij}}{\partial x^k} + T^{mj}\Gamma^i_{km} + T^{im}\Gamma^j_{mk}$$

and assuming $k = j$

$$\partial_j T^{ij} = \frac{\partial T^{ij}}{\partial x^k} + T^{mj}\Gamma^i_{jm} + T^{im}\Gamma^j_{mj}$$

In the study of the contraction of the Christoffel symbol, it was verified that

$$\Gamma^i_{mj} = \frac{\partial\left(\ell n\sqrt{g}\right)}{\partial x^r}$$

Substituting this expression in the previous expression

$$\partial_j T^{ij} = \frac{\partial T^{ij}}{\partial x^k} + T^{mj}\Gamma^i_{jm} + T^{im}\frac{\partial\left(\ell n\sqrt{g}\right)}{\partial x^m}$$

As m is a dummy index, it can be changed by the index j in the third term to the right

$$\partial_j T^{ij} = \frac{\partial T^{ij}}{\partial x^k} + T^{mj}\Gamma^i_{jm} + T^{ij}\frac{\partial\left(\ell n\sqrt{g}\right)}{\partial x^j} \Rightarrow \partial_j T^{ij} = \left[\frac{\partial T^{ij}}{\partial x^k} + T^{ij}\frac{\partial\left(\frac{1}{2}\ell n\, g\right)}{\partial x^j}\right] + T^{mj}\Gamma^i_{jm}$$

and multiplying and dividing the two terms between brackets by $\sqrt{g}$

$$\partial_j T^{ij} = \frac{1}{\sqrt{g}}\left(\sqrt{g}\frac{\partial T^{ij}}{\partial x^k} + T^{ij}\frac{1}{2\sqrt{g}}\frac{\partial g}{\partial x^j}\right) + T^{mj}\Gamma^i_{jm}$$

Changing the indexes $j \to p$ and $m \to j$ in the last term

$$\partial_j T^{ij} = \frac{1}{\sqrt{g}}\left(\sqrt{g}\frac{\partial T^{ij}}{\partial x^k} + T^{ij}\frac{1}{2\sqrt{g}}\frac{\partial g}{\partial x^j}\right) + T^{jp}\Gamma^i_{pj} \Rightarrow \partial_j T^{ij} = \frac{1}{\sqrt{g}}\frac{\partial\left(T^{ij}\sqrt{g}\right)}{\partial x^j} + T^{jp}\Gamma^i_{pj}$$

By means of the symmetry of the Christoffel symbol it results

$$\partial_j T^{ij} = \frac{1}{\sqrt{g}}\frac{\partial\left(T^{ij}\sqrt{g}\right)}{\partial x^j} + T^{jp}\Gamma^i_{pj} \qquad Q.E.D.$$

2.4.3 Covariant Tensor

2.4.3.1 Covariant Vector

Let the vector $\boldsymbol{u}$ defined by their covariant components u_j:

$$\boldsymbol{u} = u_i \boldsymbol{g}^j \tag{2.5.10}$$

where $\boldsymbol{g}^j = \boldsymbol{g}^j(x_j)$ are the basis vectors of the curvilinear coordinate system, which are functions of the coordinates that define this referential system.

Differentiating the expression (2.5.10) with respect to an arbitrary coordinate x^k:

$$\frac{\partial \boldsymbol{u}}{\partial x^k} = \frac{\partial u_i}{\partial x^k}\boldsymbol{g}^i + u_i \frac{\partial \boldsymbol{g}^i}{\partial x^k} \tag{2.5.11}$$

and substituting expression (2.4.4)

$$\frac{\partial \boldsymbol{g}^i}{\partial x^k} = -\Gamma^i_{kj}\boldsymbol{g}^j$$

in expression (2.5.11) the result is

$$\frac{\partial \boldsymbol{u}}{\partial x^k} = \frac{\partial u_i}{\partial x^k}\boldsymbol{g}^i - u^i \Gamma^i_{kj}\boldsymbol{g}^j \tag{2.5.12}$$

As i is a dummy index in the first term to the right of expression (2.5.12), it can be changed by j:

$$\frac{\partial \boldsymbol{u}}{\partial x^k} = \left(\frac{\partial u_j}{\partial x^k} - u^i \Gamma^i_{kj}\right)\boldsymbol{g}^j \tag{2.5.13}$$

thus the covariant derivative of a covariant vector is given by the N^2 functions

$$\partial_k u_j = \frac{\partial u_j}{\partial x^k} - u^i \Gamma^i_{kj} \tag{2.5.14}$$

whereby

$$\frac{\partial \boldsymbol{u}}{\partial x^k} = (\partial_k u_j)\boldsymbol{g}^j \tag{2.5.15}$$

For the Cartesian coordinate systems the Christoffel symbols are null, so in these referential systems the covariant derivative of a covariant vector coincides with the partial derivative $\frac{\partial u_j}{\partial x^k}$.

Expression (2.5.15) has the aspect of a vector, and to demonstrate that this expression is a tensor let the transformation law of covariant vectors

$$\overline{u}_p = \frac{\partial x^i}{\partial \overline{x}^p} u_i$$

that differentiated with respect to the coordinate $\overline{x}^q$ provides

$$\frac{\partial \overline{u}_p}{\partial \overline{x}^p} = \frac{\partial u_i}{\partial x^k} \frac{\partial x^k}{\partial \overline{x}^q} \frac{\partial x^i}{\partial \overline{x}^q} + u_i \frac{\partial^2 x^i}{\partial \overline{x}^q \partial \overline{x}^p} \tag{2.5.16}$$

Expression (2.4.25) can be written as

$$\frac{\partial^2 x^i}{\partial \overline{x}^q \partial \overline{x}^p} = \frac{\partial x^i}{\partial \overline{x}^s} \overline{\Gamma}^s_{pq} - \Gamma^i_{jk} \frac{\partial x^j}{\partial \overline{x}^p} \frac{\partial x^k}{\partial \overline{x}^q}$$

and substituting this expression in expression (2.5.15)

$$\frac{\partial \overline{u}_p}{\partial \overline{x}^q} = \frac{\partial u_i}{\partial x^k} \frac{\partial x^k}{\partial \overline{x}^q} \frac{\partial x^i}{\partial \overline{x}^p} + u_i \left(\frac{\partial x^i}{\partial \overline{x}^s} \overline{\Gamma}^s_{pq} - \Gamma^i_{jk} \frac{\partial x^j}{\partial \overline{x}^p} \frac{\partial x^k}{\partial \overline{x}^q} \right)$$

$$\frac{\partial \overline{u}_p}{\partial \overline{x}^q} - u_i \frac{\partial x^i}{\partial \overline{x}^s} \overline{\Gamma}^s_{pq} = \frac{\partial u_i}{\partial x^k} \frac{\partial x^k}{\partial \overline{x}^q} \frac{\partial x^i}{\partial \overline{x}^p} - u_i \frac{\partial x^j}{\partial \overline{x}^p} \frac{\partial x^k}{\partial \overline{x}^q} \Gamma^i_{jk}$$

Replacing the indexes $i \to \ell, j \to i$ in the second term to the right of the expression, and with

$$\overline{u}_s = u_i \frac{\partial x^i}{\partial \overline{x}^s}$$

this expression becomes

$$\frac{\partial \overline{u}_p}{\partial \overline{x}^q} - \overline{u}_s \overline{\Gamma}^s_{pq} = \left(\frac{\partial u_i}{\partial x^k} - u_\ell \Gamma^\ell_{ik} \right) \frac{\partial x^i}{\partial \overline{x}^p} \frac{\partial x^k}{\partial \overline{x}^q}$$

Putting

$$\partial_q \overline{u}_p = \frac{\partial \overline{u}_p}{\partial \overline{x}^q} - \overline{u}_s \overline{\Gamma}^s_{pq}$$

the result is

$$\partial_q \overline{u}_p = (\partial_k u_i) \frac{\partial x^i}{\partial \overline{x}^p} \frac{\partial x^k}{\partial \overline{x}^q} \tag{2.5.17}$$

Then the covariant derivative of a covariant vector is a covariant tensor of the second order, i.e., of variance $(0, 2)$.

Various notations are found in the literature for the covariant derivative. For the covariant vector, the most usual ones are: $\partial_k u_m = D_k u_m = \nabla_k u_m = u_{m|k} = u_{m;k}$.

2.4.3.2 Covariant Tensor of the Second Order

The transformation law of covariant tensors of the second order is given by

$$\overline{T}_{pq} = T_{ij} \frac{\partial x^i}{\partial \overline{x}^p} \frac{\partial x^j}{\partial \overline{x}^q}$$

and differentiating with respect to the coordinate $\overline{x}^r$

$$\frac{\partial \overline{T}_{pq}}{\partial \overline{x}^r} = \frac{\partial^2 x^i}{\partial \overline{x}^r \partial \overline{x}^p} \frac{\partial x^j}{\partial \overline{x}^q} T_{ij} + \frac{\partial x^i}{\partial \overline{x}^p} \frac{\partial^2 x^j}{\partial \overline{x}^r \partial \overline{x}^q} T_{ij} + \frac{\partial x^i}{\partial \overline{x}^p} \frac{\partial x^j}{\partial \overline{x}^q} \frac{\partial T_{ij}}{\partial x^k} \frac{\partial x^k}{\partial \overline{x}^r}$$

$$\frac{\partial^2 x^i}{\partial \overline{x}^r \partial \overline{x}^p} = \frac{\partial x^i}{\partial \overline{x}^s} \overline{\Gamma}^s_{rp} - \Gamma^i_{\ell m} \frac{\partial x^\ell}{\partial \overline{x}^p} \frac{\partial x^m}{\partial \overline{x}^r}$$

$$\frac{\partial^2 x^j}{\partial \overline{x}^r \partial \overline{x}^p} = \frac{\partial x^j}{\partial \overline{x}^s} \overline{\Gamma}^s_{rq} - \Gamma^j_{\ell m} \frac{\partial x^\ell}{\partial \overline{x}^r} \frac{\partial x^m}{\partial \overline{x}^q}$$

Substituting these two expressions in the expression of the covariant derivative

$$\frac{\partial \overline{T}_{pq}}{\partial \overline{x}^r} = \left(\frac{\partial x^i}{\partial \overline{x}^s} \overline{\Gamma}^s_{rp} - \Gamma^i_{\ell m} \frac{\partial x^\ell}{\partial \overline{x}^p} \frac{\partial x^m}{\partial \overline{x}^r} \right) \frac{\partial x^j}{\partial \overline{x}^q} T_{ij} + \frac{\partial x^i}{\partial \overline{x}^p} \left(\frac{\partial x^j}{\partial \overline{x}^s} \overline{\Gamma}^s_{rq} - \Gamma^j_{\ell m} \frac{\partial x^\ell}{\partial \overline{x}^r} \frac{\partial x^m}{\partial \overline{x}^q} \right) T_{ij}$$
$$+ \frac{\partial x^i}{\partial \overline{x}^p} \frac{\partial x^j}{\partial \overline{x}^q} \frac{\partial T_{ij}}{\partial x^k} \frac{\partial x^k}{\partial \overline{x}^r}$$

$$\frac{\partial \overline{T}_{pq}}{\partial \overline{x}^r} = T_{ij} \frac{\partial x^i}{\partial \overline{x}^s} \frac{\partial x^j}{\partial \overline{x}^q} \overline{\Gamma}^s_{rp} - T_{ij} \frac{\partial x^j}{\partial \overline{x}^q} \frac{\partial x^\ell}{\partial \overline{x}^p} \frac{\partial x^m}{\partial \overline{x}^r} \Gamma^i_{\ell m} + T_{ij} \frac{\partial x^j}{\partial \overline{x}^s} \frac{\partial x^i}{\partial \overline{x}^p} \overline{\Gamma}^s_{rq}$$
$$- T_{ij} \frac{\partial x^i}{\partial \overline{x}^p} \frac{\partial x^\ell}{\partial \overline{x}^r} \frac{\partial x^m}{\partial \overline{x}^q} \Gamma^j_{\ell m} + \frac{\partial T_{ij}}{\partial x^k} \frac{\partial x^i}{\partial \overline{x}^p} \frac{\partial x^j}{\partial \overline{x}^q} \frac{\partial x^k}{\partial \overline{x}^r}$$

In the second term to the right replacing the dummy index $m \to k$, and interchanging the indexes $i \leftrightarrow \ell$, and in the fourth term replacing the indexes $\ell \to k$ and interchanging the indexes $j \leftrightarrow m$ results in

$$\frac{\partial \overline{T}_{pq}}{\partial \overline{x}^r} = T_{ij} \frac{\partial x^i}{\partial \overline{x}^s} \frac{\partial x^j}{\partial \overline{x}^q} \overline{\Gamma}^s_{rp} - T_{\ell j} \frac{\partial x^i}{\partial \overline{x}^p} \frac{\partial x^j}{\partial \overline{x}^q} \frac{\partial x^k}{\partial \overline{x}^r} \Gamma^\ell_{ik} + T_{ij} \frac{\partial x^i}{\partial \overline{x}^p} \frac{\partial x^j}{\partial \overline{x}^s} \overline{\Gamma}^s_{rq}$$
$$- T_{im} \frac{\partial x^i}{\partial \overline{x}^p} \frac{\partial x^j}{\partial \overline{x}^q} \frac{\partial x^k}{\partial \overline{x}^r} \Gamma^m_{kj} + \frac{\partial T_{ij}}{\partial x^k} \frac{\partial x^i}{\partial \overline{x}^p} \frac{\partial x^j}{\partial \overline{x}^q} \frac{\partial x^k}{\partial \overline{x}^r}$$

and with the transformation law of covariant tensors of the second order

$$\overline{T}_{sq} = T_{ij}\frac{\partial x^i}{\partial \overline{x}^s}\frac{\partial x^j}{\partial \overline{x}^q} \quad \overline{T}_{ps} = T_{ij}\frac{\partial x^i}{\partial \overline{x}^p}\frac{\partial x^j}{\partial \overline{x}^s}$$

$$\frac{\partial \overline{T}_{pq}}{\partial \overline{x}^r} - \overline{T}_{sq}\overline{\Gamma}^s_{rp} - \overline{T}_{ps}\overline{\Gamma}^s_{rq} = \left(\frac{\partial T_{ij}}{\partial x^k} - T_{\ell j}\Gamma^\ell_{ik} - T_{im}\Gamma^m_{kj}\right)\frac{\partial x^i}{\partial \overline{x}^p}\frac{\partial x^j}{\partial \overline{x}^q}\frac{\partial x^k}{\partial \overline{x}^r}$$

Replacing the dummy indexes $m \to \ell$:

$$\frac{\partial \overline{T}_{pq}}{\partial \overline{x}^r} - \overline{T}_{sq}\overline{\Gamma}^s_{rp} - \overline{T}_{ps}\overline{\Gamma}^s_{rq} = \left(\frac{\partial T_{ij}}{\partial x^k} - T_{\ell j}\Gamma^\ell_{ik} - T_{i\ell}\Gamma^\ell_{kj}\right)\frac{\partial x^i}{\partial \overline{x}^p}\frac{\partial x^j}{\partial \overline{x}^q}\frac{\partial x^k}{\partial \overline{x}^r}$$

whereby

$$\partial_k\overline{T}_{pq} = (\partial_k T_{ij})\frac{\partial x^i}{\partial \overline{x}^p}\frac{\partial x^j}{\partial \overline{x}^q}\frac{\partial x^k}{\partial \overline{x}^r}$$

therefore the covariant derivative of a covariant tensor of the second order is a covariant tensor of the third order, i.e., of variance $(0,3)$. Whereby the covariant derivative of a covariant tensor of the second order is given by

$$\partial_k T_{ij} = \frac{\partial T_{ij}}{\partial x^k} - T_{\ell j}\Gamma^\ell_{ik} - T_{i\ell}\Gamma^\ell_{kj} \tag{2.5.18}$$

For the Cartesian coordinates the Christoffel symbols are null, so in these referential systems the covariant derivative of the tensor T_{ij} coincides with the partial derivative $\frac{\partial T_{ij}}{\partial x^k}$.

2.4.3.3 Covariant Tensor of Order Above Two

To generalize expression (2.5.18) for tensors of order above two, i.e., for instance, the covariant derivative of the covariant tensor of the third order, which expression may be developed by means of the following steps:

(a) The basic structure of its expression is written considering the expression obtained for the covariant derivative of a covariant tensor of the second order

$$\partial_p T_{ijk} = \frac{\partial T_{ijk}}{\partial x^p} + T_{\bullet\bullet\bullet}\Gamma^\bullet_{\bullet\bullet} + T_{\bullet\bullet\bullet}\Gamma^\bullet_{\bullet\bullet} + T_{\bullet\bullet\bullet}\Gamma^\bullet_{\bullet\bullet}$$

(b) The indexes of the Christoffel symbols corresponding to the coordinate with respect to which the differentiation is being carried out are placed

$$\partial_p T_{ijk} = \frac{\partial T_{ijk}}{\partial x^p} + T_{\bullet\bullet\bullet}\Gamma^\bullet_{\bullet p} + T_{\bullet\bullet\bullet}\Gamma^\bullet_{\bullet p} + T_{\bullet\bullet\bullet}\Gamma^\bullet_{\bullet p}$$

(c) The covariant indexes of the Christoffel symbols must be completed obeying the sequence of the indexes of the tensor that is being differentiated

$$\partial_p T_{ijk} = \frac{\partial T_{ijk}}{\partial x^p} + T_{\bullet\bullet\bullet}\Gamma^{\bullet}_{ip} + T_{\bullet\bullet\bullet}\Gamma^{\bullet}_{jp} + T_{\bullet\bullet\bullet}\Gamma^{\bullet}_{kp}$$

(d) The dummy index q is placed on the Christoffel symbols and in sequential form in the tensors

$$\partial_p T_{ijk} = \frac{\partial T_{ijk}}{\partial x^p} + T_{q\bullet\bullet}\Gamma^{q}_{ip} + T_{\bullet q\bullet}\Gamma^{q}_{jp} + T_{\bullet\bullet q}\Gamma^{q}_{kp}$$

(e) The remaining indexes are placed in the same sequence in which they appear on the tensor that is being differentiated

$$\partial_p T_{ijk} = \frac{\partial T_{ijk}}{\partial x^p} + T_{qjk}\Gamma^{q}_{ip} + T_{iqk}\Gamma^{q}_{jp} + T_{ijq}\Gamma^{q}_{kp}$$

This tensor generated by the differentiation of a variance tensor $(0,4)$. Expression (2.5.18) can be generalized by adopting this indexes placement systematic for a covariant tensor of order $q > 3$, and the variance of this new tensor will always be $(0, q+1)$.

2.4.4 Mixed Tensor

Consider the transformation law of the mixed tensors of the second

$$\overline{T}^m_n = T^i_j \frac{\partial \overline{x}^m}{\partial x^i} \frac{\partial x^j}{\partial \overline{x}^n}$$

that can be written as

$$\overline{T}^m_n \frac{\partial x^i}{\partial \overline{x}^m} = T^i_j \frac{\partial x^j}{\partial \overline{x}^n}$$

which derivative with respect to coordinate $\overline{x}^r$ is given by

$$\frac{\partial \overline{T}^m_n}{\partial \overline{x}^r} \frac{\partial x^i}{\partial \overline{x}^m} + \overline{T}^m_n \frac{\partial^2 x^i}{\partial \overline{x}^r \partial \overline{x}^m} = \frac{\partial T^i_j}{\partial x^k} \frac{\partial x^k}{\partial \overline{x}^r} \frac{\partial x^j}{\partial \overline{x}^n} + T^i_j \frac{\partial^2 x^j}{\partial \overline{x}^r \partial \overline{x}^n}$$

and with the following expressions

$$\frac{\partial^2 x^i}{\partial \overline{x}^r \partial \overline{x}^m} = \frac{\partial x^i}{\partial \overline{x}^s} \overline{\Gamma}^s_{rm} - \Gamma^i_{\ell j} \frac{\partial x^\ell}{\partial \overline{x}^m} \frac{\partial x^j}{\partial \overline{x}^r} \qquad \frac{\partial^2 x^j}{\partial \overline{x}^r \partial \overline{x}^m} = \frac{\partial x^j}{\partial \overline{x}^s} \overline{\Gamma}^s_{mr} - \Gamma^j_{\ell p} \frac{\partial x^\ell}{\partial \overline{x}^n} \frac{\partial x^p}{\partial \overline{x}^r}$$

this expression becomes

$$\frac{\partial \overline{T}_{n}^{m}}{\partial \overline{x}^{r}}\frac{\partial x^{i}}{\partial \overline{x}^{m}}+\overline{T}_{n}^{m}\left(\frac{\partial x^{i}}{\partial \overline{x}^{s}}\overline{\Gamma}_{rm}^{s}-\Gamma_{\ell j}^{i}\frac{\partial x^{\ell}}{\partial \overline{x}^{m}}\frac{\partial x^{j}}{\partial \overline{x}^{r}}\right)$$
$$=\frac{\partial T_{j}^{i}}{\partial x^{k}}\frac{\partial x^{k}}{\partial \overline{x}^{r}}\frac{\partial x^{j}}{\partial \overline{x}^{n}}+T_{j}^{i}\left(\frac{\partial x^{j}}{\partial \overline{x}^{s}}\overline{\Gamma}_{mr}^{s}-\Gamma_{\ell p}^{j}\frac{\partial x^{\ell}}{\partial \overline{x}^{n}}\frac{\partial x^{p}}{\partial \overline{x}^{r}}\right)$$

As

$$\overline{T}_{n}^{m}=T_{q}^{p}\frac{\partial \overline{x}^{m}}{\partial x^{p}}\frac{\partial x^{q}}{\partial \overline{x}^{n}}\qquad T_{j}^{i}=\overline{T}_{e}^{m}\frac{\partial x^{i}}{\partial \overline{x}^{m}}\frac{\partial \overline{x}^{e}}{\partial x^{j}}$$

it follows that

$$\frac{\partial \overline{T}_{n}^{m}}{\partial \overline{x}^{r}}\frac{\partial x^{i}}{\partial \overline{x}^{m}}+\overline{T}_{n}^{m}\frac{\partial x^{i}}{\partial \overline{x}^{s}}\overline{\Gamma}_{rm}^{s}-T_{q}^{p}\frac{\partial \overline{x}^{m}}{\partial x^{p}}\frac{\partial x^{q}}{\partial \overline{x}^{n}}\frac{\partial x^{\ell}}{\partial \overline{x}^{m}}\frac{\partial x^{j}}{\partial \overline{x}^{r}}\Gamma_{\ell j}^{i}$$
$$=\frac{\partial T_{j}^{i}}{\partial x^{k}}\frac{\partial x^{k}}{\partial \overline{x}^{r}}\frac{\partial x^{j}}{\partial \overline{x}^{n}}+\overline{T}_{e}^{m}\frac{\partial x^{i}}{\partial \overline{x}^{m}}\frac{\partial \overline{x}^{e}}{\partial x^{j}}\frac{\partial x^{j}}{\partial \overline{x}^{s}}\overline{\Gamma}_{mr}^{s}-T_{j}^{i}\frac{\partial x^{\ell}}{\partial \overline{x}^{n}}\frac{\partial x^{p}}{\partial \overline{x}^{r}}\Gamma_{\ell p}^{j}$$

$$\frac{\partial \overline{T}_{n}^{m}}{\partial \overline{x}^{r}}\frac{\partial x^{i}}{\partial \overline{x}^{m}}+\overline{T}_{n}^{m}\frac{\partial x^{i}}{\partial \overline{x}^{s}}\overline{\Gamma}_{rm}^{s}-T_{q}^{p}\delta_{p}^{\ell}\frac{\partial x^{q}}{\partial \overline{x}^{n}}\frac{\partial x^{j}}{\partial \overline{x}^{r}}\Gamma_{\ell j}^{i}$$
$$=\frac{\partial T_{j}^{i}}{\partial x^{k}}\frac{\partial x^{k}}{\partial \overline{x}^{r}}\frac{\partial x^{j}}{\partial \overline{x}^{n}}+\overline{T}_{s}^{m}\frac{\partial x^{i}}{\partial \overline{x}^{m}}\delta_{s}^{e}\overline{\Gamma}_{mr}^{s}-T_{j}^{i}\frac{\partial x^{\ell}}{\partial \overline{x}^{n}}\frac{\partial x^{p}}{\partial \overline{x}^{r}}\Gamma_{\ell p}^{j}$$

$$\frac{\partial \overline{T}_{n}^{m}}{\partial \overline{x}^{r}}\frac{\partial x^{i}}{\partial \overline{x}^{m}}+\overline{T}_{n}^{m}\frac{\partial x^{i}}{\partial \overline{x}^{s}}\overline{\Gamma}_{rm}^{s}-T_{q}^{\ell}\frac{\partial x^{q}}{\partial \overline{x}^{n}}\frac{\partial x^{j}}{\partial \overline{x}^{r}}\Gamma_{pj}^{i}$$
$$=\frac{\partial T_{j}^{i}}{\partial x^{k}}\frac{\partial x^{k}}{\partial \overline{x}^{r}}\frac{\partial x^{j}}{\partial \overline{x}^{n}}+\overline{T}_{s}^{m}\frac{\partial x^{i}}{\partial \overline{x}^{m}}\overline{\Gamma}_{mr}^{s}-T_{j}^{i}\frac{\partial x^{\ell}}{\partial \overline{x}^{n}}\frac{\partial x^{p}}{\partial \overline{x}^{r}}\Gamma_{\ell p}^{j}$$

Interchanging the indexes in the second term on the left $m \leftrightarrow s$, in the last term on the right, interchanging the indexes $j \leftrightarrow \ell$ and replacing the indexes $p \rightarrow k$ results in

$$\frac{\partial \overline{T}_{n}^{m}}{\partial \overline{x}^{r}}\frac{\partial x^{i}}{\partial \overline{x}^{m}}+\overline{T}_{n}^{s}\frac{\partial x^{i}}{\partial \overline{x}^{m}}\overline{\Gamma}_{rs}^{m}-T_{q}^{\ell}\frac{\partial x^{q}}{\partial \overline{x}^{n}}\frac{\partial x^{j}}{\partial \overline{x}^{r}}\Gamma_{\ell j}^{i}=\frac{\partial T_{j}^{i}}{\partial x^{k}}\frac{\partial x^{k}}{\partial \overline{x}^{r}}\frac{\partial x^{j}}{\partial \overline{x}^{n}}+\overline{T}_{s}^{m}\frac{\partial x^{j}}{\partial \overline{x}^{m}}\overline{\Gamma}_{mr}^{s}$$
$$-T_{\ell}^{i}\frac{\partial x^{j}}{\partial \overline{x}^{n}}\frac{\partial x^{k}}{\partial \overline{x}^{r}}\Gamma_{jk}^{\ell}$$

and replacing the indexes $j \rightarrow k$ and $q \rightarrow j$ in the last term on the left

$$\frac{\partial \overline{T}_{n}^{m}}{\partial \overline{x}^{r}}\frac{\partial x^{i}}{\partial \overline{x}^{m}}+\overline{T}_{n}^{s}\frac{\partial x^{i}}{\partial \overline{x}^{m}}\overline{\Gamma}_{rs}^{m}-T_{j}^{\ell}\frac{\partial x^{j}}{\partial \overline{x}^{n}}\frac{\partial x^{k}}{\partial \overline{x}^{r}}\Gamma_{\ell k}^{i}=\frac{\partial T_{j}^{i}}{\partial x^{k}}\frac{\partial x^{k}}{\partial \overline{x}^{r}}\frac{\partial x^{j}}{\partial \overline{x}^{n}}$$

$$+\overline{T}_{s}^{m}\frac{\partial x^{j}}{\partial \overline{x}^{m}}\overline{\Gamma}_{mr}^{s}-T_{\ell}^{i}\frac{\partial x^{j}}{\partial \overline{x}^{n}}\frac{\partial x^{k}}{\partial \overline{x}^{r}}\Gamma_{jk}^{\ell}$$

that can be written as

$$\left(\frac{\partial \overline{T}_{n}^{m}}{\partial \overline{x}^{r}}+\overline{T}_{n}^{s}\overline{\Gamma}_{rs}^{m}-\overline{T}_{s}^{m}\overline{\Gamma}_{mr}^{s}\right)\frac{\partial x^{i}}{\partial \overline{x}^{m}}=\left(\frac{\partial T_{j}^{i}}{\partial x^{k}}+T_{j}^{\ell}\Gamma_{\ell k}^{i}-T_{\ell}^{i}\Gamma_{jk}^{\ell}\right)\frac{\partial x^{j}}{\partial \overline{x}^{n}}\frac{\partial x^{k}}{\partial \overline{x}^{r}}$$

then

$$\frac{\partial \overline{T}_{n}^{m}}{\partial \overline{x}^{r}}+\overline{T}_{n}^{s}\overline{\Gamma}_{rs}^{m}-\overline{T}_{s}^{m}\overline{\Gamma}_{mr}^{s}=\left(\frac{\partial T_{j}^{i}}{\partial x^{k}}+T_{j}^{\ell}\Gamma_{\ell k}^{i}-T_{\ell}^{i}\Gamma_{jk}^{\ell}\right)\frac{\partial \overline{x}^{m}}{\partial x^{i}}\frac{\partial x^{j}}{\partial \overline{x}^{n}}\frac{\partial x^{k}}{\partial \overline{x}^{r}} \tag{2.5.19}$$

Putting

$$\partial_{r}\overline{T}_{n}^{m}=\frac{\partial \overline{T}_{n}^{m}}{\partial \overline{x}^{r}}+\overline{T}_{n}^{s}\overline{\Gamma}_{rs}^{m}-\overline{T}_{s}^{m}\overline{\Gamma}_{mr}^{s} \tag{2.5.20}$$

$$\partial_{r}T_{j}^{i}=\frac{\partial T_{j}^{i}}{\partial x^{r}}+T_{j}^{\ell}\Gamma_{\ell k}^{i}-T_{\ell}^{i}\Gamma_{jk}^{\ell} \tag{2.5.21}$$

the result is the expressions that represent the covariant derivative of the mixed tensors of the second-order $\overline{T}_{n}^{m}$ and T_{j}^{i}, whereby

$$\partial_{r}\overline{T}_{n}^{m}=\partial_{r}T_{j}^{i}\frac{\partial \overline{x}^{m}}{\partial x^{i}}\frac{\partial x^{j}}{\partial \overline{x}^{n}}\frac{\partial x^{k}}{\partial \overline{x}^{r}} \tag{2.5.22}$$

Expression (2.5.22) shows that the derivative of a mixed tensor of the second order is a mixed tensor of the third order, once contravariant and twice covariant, i.e., of variance $(1, 2)$.

The covariant derivative of a mixed tensor of variance (p, q) generates a variance tensor $(p, q+1)$. To generalize expression (2.5.22) for mixed tensors of order above two, assume as an example the covariant derivatives of a mixed tensor of the third order of variance $(1, 2)$ and of a mixed tensor of fifth order of variance $(3, 2)$, which are given, respectively, by the expressions

$$\partial_{k}T_{p\ell}^{j}=\frac{\partial T_{p\ell}^{j}}{\partial x^{k}}-T_{q\ell}^{j}\Gamma_{pk}^{q}-T_{pq}^{j}\Gamma_{\ell k}^{q}+T_{p\ell}^{q}\Gamma_{kq}^{j}$$

$$\partial_{k}T_{rs}^{j\ell m}=\frac{\partial T_{rs}^{j\ell m}}{\partial x^{k}}-T_{qs}^{j\ell m}\Gamma_{rk}^{q}-T_{rq}^{j\ell m}\Gamma_{sk}^{q}+T_{rs}^{q\ell m}\Gamma_{kq}^{j}+T_{rs}^{jqm}\Gamma_{kq}^{\ell}+T_{rs}^{j\ell q}\Gamma_{kq}^{m}$$

2.4.5 Covariant Derivative of the Addition, Subtraction, and Product of Tensors

Expression (2.5.21) shows that the covariant derivative of a mixed tensor comprises a partial derivative of this tensor and the terms containing Christoffel symbols, which are always linear in the components of the original tensor. This characteristic indicates that the covariant differentiation follows the same rules of the ordinary differentiation of Differential Calculus.

To stress the properties of the covariant derivative let the scalar $\phi(x^i)$ which ordinary derivative is equal to its covariant derivative, that can be written as the dot product of the vectors u^i and v_i expressed in Cartesian coordinates

$$\phi(x^i) = u^i v_i$$

and differentiating

$$\partial_k \phi(x^i) = \partial_k(u^i v_i) = \frac{d(u^i v_i)}{dx^k} = \frac{du^i}{dx^k} v_i + u^i \frac{dv_i}{dx^k}$$

As the covariant and ordinary derivatives are equal, it results in

$$\partial_k \phi(x^i) = \partial_k(u^i)\, v_i + u^i \partial_k(v_i)$$

Substituting the expressions of the covariant derivatives of contravariant and covariant vectors

$$\partial_k \phi(x^i) = \left(\frac{\partial u^i}{\partial x^k} + u^i \Gamma^i_{kj}\right) v_i + u^i \left(\frac{\partial v_i}{\partial x^k} - v_i \Gamma^i_{kj}\right) = \frac{\partial u^i}{\partial x^k} v_i + u^i \frac{\partial v_i}{\partial x^k}$$

This expression suggests that the covariant derivative of an inner product of tensors behaves in a manner that is similar to the ordinary derivative. To prove this assumption, let, for instance, the tensors A_{ij} and B_{ij} for which the following properties of the covariant derivative are admitted a priori as valid:

(a) $\partial_k(A_{ij} + B_{ij}) = \partial_k A_{ij} + \partial_k B_{ij}$;
(b) $\partial_k(A_{ij} - B_{ij}) = \partial_k A_{ij} - \partial_k B_{ij}$;
(c) $\partial_k(A_{ij} B_{ij}) = (\partial_k A_{ij}) B_{ij} + A_{ij}(\partial_k B_{ij})$.

To demonstrate property (a) let the tensor $C_{ij} = A_{ij} + B_{ij}$, so

$$\partial_k(A_{ij}+B_{ij}) = \partial_k C_{ij} = \frac{\partial C_{ij}}{\partial x^k} - C_{\ell j}\Gamma^{\ell}_{ik} - C_{i\ell}\Gamma^{\ell}_{kj}$$
$$= \frac{\partial(A_{ij}+B_{ij})}{\partial x^k} - (A_{\ell j}+B_{\ell j})\Gamma^{\ell}_{ik} - (A_{i\ell}+B_{i\ell})\Gamma^{\ell}_{kj}$$
$$= \left(\frac{\partial A_{ij}}{\partial x^k} - A_{\ell j}\Gamma^{\ell}_{ik} - A_{i\ell}\Gamma^{\ell}_{kj}\right) + \frac{\partial B_{ij}}{\partial x^k} - B_{\ell j}\Gamma^{\ell}_{ik} - B_{i\ell}\Gamma^{\ell}_{kj}$$
$$= \partial_k A_{ij} + \partial_k B_{ij}$$

In an analogous way, it is possible to prove property (b), replacing only the addition sign for the subtraction sign in the previous demonstration.

To demonstrate property (c) let the inner product $A_{ij}B_{\ell m} = C_{ij\ell m}$ that generates a covariant tensor of the fourth order

$$\partial_k(A_{ij}B_{\ell m}) = \partial_k C_{ij\ell m} = \frac{\partial C_{ij\ell m}}{\partial x^k} - C_{pj\ell m}\Gamma^{p}_{ki} - C_{ip\ell m}\Gamma^{p}_{kj} - C_{ijpm}\Gamma^{p}_{k\ell} - C_{ij\ell p}\Gamma^{p}_{km}$$

Substituting the expressions of the tensor of the fourth order in terms of the inner product

$$\partial_k(A_{ij}B_{\ell m}) = \frac{\partial(A_{ij}B_{\ell m})}{\partial x^k} - A_{pj}B_{\ell m}\Gamma^{p}_{ki} - A_{ip}B_{\ell m}\Gamma^{p}_{k\ell} - A_{ij}B_{pm}\Gamma^{p}_{k\ell} - A_{ij}B_{\ell p}\Gamma^{p}_{km}$$
$$= \left(\frac{\partial A_{ij}}{\partial x^k} - A_{pj}\Gamma^{p}_{ki} - A_{ip}\Gamma^{p}_{kj}\right)B_{\ell m} + A_{ij}\left(\frac{\partial B_{\ell m}}{\partial x^k} - B_{pm}\Gamma^{p}_{k\ell} - B_{\ell p}\Gamma^{p}_{km}\right)$$

The terms in parenthesis are the covariant derivatives of the covariant tensors of the second-order, whereby

$$\partial_k(A_{ij}B_{\ell m}) = (\partial_k A_{ij})B_{\ell m} + A_{ij}(\partial_k B_{\ell m})$$

thus the covariant derivative of an inner product of tensors follows the same rule as the derivative of the product of functions in Differential Calculus.

2.4.6 Covariant Derivative of Tensors g_{ij}, g^{ij}, δ^{i}_{j}

Ricci's Lemma

The metric tensor behaves as a constant when calculating the covariant derivative.

The covariant derivative of the metric tensor g_{ij} is calculated to demonstrate this lemma, thus

$$\partial_k g_{ij} = \frac{\partial g_{ij}}{\partial x^k} - g_{pj}\Gamma^p_{ik} - g_{ip}\Gamma^p_{kj}$$

$$\partial_k g_{ij} = \frac{\partial g_{ij}}{\partial x^k} - \left(\Gamma_{ik,j} + \Gamma_{kj,i}\right)$$

and with the Ricci identity

$$\frac{\partial g_{ij}}{\partial x^k} = \Gamma_{ik,j} + \Gamma_{jk,i}$$

and by the symmetry $\Gamma_{jk,i} = \Gamma_{kj,i}$

$$\partial_k g_{ij} = \frac{\partial g_{ij}}{\partial x^k} - \frac{\partial g_{ij}}{\partial x^k} = 0$$

In an analogous way the conjugate metric tensor g^{ij} is given by

$$\partial_k g^{ij} = \frac{\partial g^{ij}}{\partial x^k} + g^{pj}\Gamma^i_{kp} + g^{ip}\Gamma^j_{kp} \tag{2.5.23}$$

Since

$$g_{ij}g^{jp} = \delta^p_i \Rightarrow \frac{\partial\left(g_{ij}g^{jp}\right)}{\partial x^k} = 0 \Rightarrow \frac{\partial g_{ij}}{\partial x^k} g^{jp} + g_{ij}\frac{\partial g^{jp}}{\partial x^k} = 0$$

and multiplying by g^{iq}

$$g^{iq}g^{jp}\frac{\partial g_{ij}}{\partial x^k} + g^{iq}g_{ij}\frac{\partial g^{jp}}{\partial x^k} = 0 \Rightarrow g^{iq}g^{jp}\frac{\partial g_{ij}}{\partial x^k} + \delta^q_j\frac{\partial g^{jp}}{\partial x^k} = 0 \Rightarrow \frac{\partial g^{qp}}{\partial x^k} = -g^{iq}g^{jp}\frac{\partial g_{ij}}{\partial x^k}$$

it follows that

$$\begin{aligned}\frac{\partial g^{qp}}{\partial x^k} &= -g^{iq}g^{jp}\left(\Gamma_{ik,j} + \Gamma_{jk,i}\right) = -g^{iq}g^{jp}\Gamma_{ik,j} + -g^{iq}g^{jp}\Gamma_{jk,i} \\ &= -g^{iq}\Gamma^p_{ik} - g^{iq}\Gamma^q_{jk}\end{aligned}$$

Replacing the indexes $i \to p$, $q \to i$, and $p \to j$:

$$\frac{\partial g^{qp}}{\partial x^k} = -g^{pi}\Gamma^j_{pk} - g^{pj}\Gamma^i_{pk}$$

and substituting this expression in expression (2.5.23)

$$\partial_k g^{ij} = \left(-g^{pi}\Gamma^j_{pk} - g^{pj}\Gamma^i_{pk}\right) + g^{pj}\Gamma^i_{kp} + g^{ip}\Gamma^j_{kp}$$

and with the symmetry of g^{ij} and the Christoffel symbol of second kind

$$\partial_k g^{ij} = \left(-g^{ip}\Gamma^j_{kp} - g^{pj}\Gamma^i_{kp}\right) + g^{pj}\Gamma^i_{kp} + g^{ip}\Gamma^j_{kp} = 0$$

Following the same systematic it implies for the covariant derivative of the Kronecker delta

$$\partial_k \delta^i_j = \frac{\partial \delta^i_j}{\partial x^k} + \delta^p_j \Gamma^i_{pk} - \delta^i_p \Gamma^p_{jk} = 0 + \Gamma^i_{jk} - \Gamma^i_{jk} = 0$$

These deductions show that the conjugate metric tensor g^{ij} and the Kronecker delta δ^i_j also behave as constants in calculating the covariant derivative.

Exercise 2.11 Show that $\partial_k T^i_j = g^{im}\left(\partial_k T^i_j\right)$.

Expressing the mixed tensor by

$$T^i_j = g^{im} T_{mj}$$

the result for its covariant derivative is

$$\partial_k T^i_j = \left(\partial_k g^{im}\right) T^i_j + g^{im}\left(\partial_k T^i_j\right)$$

As $\partial_k g^{im} = 0$, it results in

$$\partial_k T^i_j = g^{im}\left(\partial_k T^i_j\right) \qquad Q.E.D.$$

Exercise 2.12 Show that $\left(\frac{\partial u_i}{\partial x^j} - \frac{\partial u_j}{\partial x^i}\right)$ is a covariant tensor of the second order, being u_i a covariant vector.

The covariant derivative of a covariant vector is given by

$$\partial_j u_i = \frac{\partial u_i}{\partial x^j} - u_p \Gamma^p_{ij} \Rightarrow \frac{\partial u_i}{\partial x^j} = \partial_j u_i + u_p \Gamma^p_{ij}$$

and replacing the indexes $i \to j$ results in

$$\frac{\partial u_j}{\partial x^i} = \partial_i u_j + u_p \Gamma^p_{ji}$$

Carrying out the subtraction presented in the enunciation

$$\left(\frac{\partial u_i}{\partial x^j}-\frac{\partial u_j}{\partial x^i}\right)=\left(\partial_j u_i+u_p\Gamma^p_{ji}\right)-\left(\partial_j u_i+u_p\Gamma^p_{ij}\right)$$

and with the symmetry $\Gamma^p_{ij}=\Gamma^p_{ji}$

$$\left(\frac{\partial u_i}{\partial x^j}-\frac{\partial u_j}{\partial x^i}\right)=\partial_j u_i-\partial_i u_j$$

As the covariant derivative of a covariant vector is a tensor of the second order, then this expression represents a tensor of variance $(0,2)$.

Exercise 2.13 Show that $\Gamma^p_{ij}=\frac{1}{2}T^{pq}\left(\frac{\partial T_{ik}}{\partial x^j}+\frac{\partial T_{jk}}{\partial x^i}-\frac{\partial T_{ij}}{\partial x^k}\right)$, being T_{ij} a symmetric tensor and $\det T_{ij}\neq 0$, and with covariant derivative $\partial_k T_{ij}=0$.

The tensor T^{pk} can be written under the form

$$T^{pk}=g^{ip}g^{jk}T_{ij}$$

For the tensor T_{ij} the covariant derivative is given by

$$\partial_k T_{ij}=\frac{\partial T_{ij}}{\partial x^k}-T_{pj}\Gamma^p_{ik}-T_{ip}\Gamma^p_{jk}=0\Rightarrow\frac{\partial T_{ij}}{\partial x^k}=T_{pj}\Gamma^p_{ik}+T_{ip}\Gamma^p_{jk}$$

Interchanging the indexes i, j, k cyclically

$$\frac{\partial T_{jk}}{\partial x^i}=T_{pk}\Gamma^p_{ji}+T_{jp}\Gamma^p_{ki}\qquad\frac{\partial T_{ki}}{\partial x^j}=T_{pi}\Gamma^p_{kj}+T_{kp}\Gamma^p_{ij}$$

and adding these two expressions and subtracting the one that comes before them, and considering the tensor's symmetry

$$\begin{aligned}\frac{\partial T_{jk}}{\partial x^i}+\frac{\partial T_{ki}}{\partial x^j}-\frac{\partial T_{ij}}{\partial x^k}&=\left(T_{pk}\Gamma^p_{ji}+T_{jp}\Gamma^p_{ki}\right)+\left(T_{pi}\Gamma^p_{kj}+T_{kp}\Gamma^p_{ij}\right)-\left(T_{pj}\Gamma^p_{ik}+T_{ip}\Gamma^p_{jk}\right)\\&=2T_{kp}\Gamma^p_{ij}\end{aligned}$$

The dummy index p can be changed by the index q, *so*

$$\frac{\partial T_{jk}}{\partial x^i}+\frac{\partial T_{ki}}{\partial x^j}-\frac{\partial T_{ij}}{\partial x^k}=2T_{kq}\Gamma^q_{ij}\Rightarrow\frac{1}{2}\left(\frac{\partial T_{jk}}{\partial x^i}+\frac{\partial T_{ki}}{\partial x^j}-\frac{\partial T_{ij}}{\partial x^k}\right)=T_{kq}\Gamma^q_{ij}$$

and multiplying by T^{pk}

$$\frac{1}{2}T^{pq}\left(\frac{\partial T_{jk}}{\partial x^i}+\frac{\partial T_{ki}}{\partial x^j}-\frac{\partial T_{ij}}{\partial x^k}\right)=T^{pk}T_{kq}\Gamma^q_{ij}$$

and with the contraction

$$T^{pk}T_{kq} = \delta_q^p$$

it follows that

$$\frac{1}{2}T^{pq}\left(\frac{\partial T_{jk}}{\partial x^i} + \frac{\partial T_{ki}}{\partial x^j} - \frac{\partial T_{ij}}{\partial x^k}\right) = \delta_q^p \Gamma_{ij}^q = \Gamma_{ij}^p \qquad \text{Q.E.D.}$$

2.4.7 Particularities of the Covariant Derivative

To exemplify a particularity of the covariant derivative let the vector $\boldsymbol{u}$ defined by its covariant components $u_j = g_{ij}u^i$, then

$$\partial_k u_j = \partial_k\left(g_{ij}u^i\right) = \left(\partial_k g_{ij}\right) u^i + g_{ij}\partial_k u^i$$

and with Ricci's lemma

$$\partial_k\left(g_{ij}u^i\right) = g_{ij}\partial_k u^i$$

The covariant derivative of the contravariant vector is given by

$$\partial_k u^i = \frac{\partial u^i}{\partial x^k} + u^\ell \Gamma_{\ell k}^i$$

so by substitution

$$\partial_k\left(g_{ij}u^i\right) = g_{ij}\left(\frac{\partial u^i}{\partial x^k} + u^\ell \Gamma_{\ell k}^i\right)$$

The contravariant components of the vector can be expressed in terms of their covariant components

$$\begin{aligned}\partial_k\left(g_{ij}u^i\right) &= g_{ij}\frac{\partial\left(g^{i\ell}u_\ell\right)}{\partial x^k} + g_{ij}u^\ell\Gamma_{\ell k}^i = g_{ij}\frac{\partial\left(g^{i\ell}u_\ell\right)}{\partial x^k} + u^\ell\Gamma_{\ell k,j} \\ &= g_{ij}\frac{\partial g^{i\ell}}{\partial x^k}u_\ell + g_{ij}g^{i\ell}\frac{\partial u_\ell}{\partial x^k} + u^\ell\Gamma_{\ell k,j}\end{aligned}$$

Rewriting expression (2.4.31)

$$\frac{\partial g^{i\ell}}{\partial x^k} = -g^{\ell m}\Gamma_{mk}^i - g^{im}\Gamma_{mk}^\ell$$

which substituted in the previous expression provides

$$\begin{aligned}\partial_k\left(g_{ij}u^i\right) &= g_{ij}\left(-g^{\ell m}\Gamma^i_{mk} - g^{im}\Gamma^\ell_{mk}\right)u_\ell + g_{ij}g^{i\ell}\frac{\partial u_\ell}{\partial x^k} + u^\ell\Gamma_{\ell k,j}\\ &= g_{ij}\left(-g^{\ell m}u_\ell\Gamma^i_{mk} - g^{im}u_\ell\Gamma^\ell_{mk}\right) + \delta^\ell_j\frac{\partial u_\ell}{\partial x^k} + u^\ell\Gamma_{\ell k,j}\\ &= -g_{ij}u^m\Gamma^i_{mk} - \delta^m_j u_\ell\Gamma^\ell_{mk} + \frac{\partial u_j}{\partial x^k} + u^\ell\Gamma_{\ell k,j}\\ &= u^m\Gamma_{mk,j} - u_\ell\Gamma^\ell_{mk} + \frac{\partial u_j}{\partial x^k} + u^\ell\Gamma_{\ell k,j}\end{aligned}$$

Replacing the dummy indexes $\ell \to m$:

$$\begin{aligned}\partial_k\left(g_{ij}u^i\right) &= -u^m\Gamma_{mk,j} - u_\ell\Gamma^\ell_{mk} + \frac{\partial u_j}{\partial x^k} + u^m\Gamma_{mk,j} \Rightarrow \partial_k u_j = \partial_k\left(g_{ij}u^i\right)\\ &= \frac{\partial u_j}{\partial x^k} - u_\ell\Gamma^\ell_{mk}\end{aligned}$$

then the covariant derivative of a covariant vector is equal to the covariant derivative of the product of the metric tensor by the contravariant components of this vector. This characteristic of the covariant derivative can be generalized for tensors of order above one, for instance, for a contravariant tensor of the second order the result is

$$\partial_k\left(g_{ip}g_{jq}T^{pq}\right) = \partial_k T_{ij}$$

Another particularity of the covariant derivative is its successive differentiation of a scalar function. Let a scalar function ϕ that represents an invariant, so its derivative with respect to its coordinate x^i is a covariant vector given by

$$\phi_{,\,i} = \frac{\partial\phi}{\partial x^i} = \partial_i\phi$$

Taking the derivative of this function again, now with respect to the coordinate x^j:

$$\phi_{,\,ij} = \frac{\partial^2\phi}{\partial x^j\partial x^i} = \partial_j(\partial_i\phi) = \frac{\partial^2\phi}{\partial x^j\partial x^i} - \frac{\partial\phi}{\partial x^m}\Gamma^m_{ij}$$

The dummy index m can be changed, and as the Christoffel symbol is symmetric, it results in

$$\partial_j(\partial_i\phi) = \partial_i\left(\partial_j\phi\right)$$

Then the covariant derivative of an invariant is commutative.

2.5 Covariant Derivative of Relative Tensors

The covariant derivative of relative tensors has characteristics that differ from the covariant derivative of absolute tensors. For studying the derivatives of these varieties in a progressive manner, a scalar density of weight W with respect to the coordinate system $\overline{X}^i$ is admitted, given by $J^W\overline{\phi}(\overline{x}^i)$. Taking the derivative of this function

$$\frac{\partial\left(J^W\overline{\phi}\right)}{\partial\overline{x}^j}=J^W\frac{\partial\phi}{\partial x^k}\frac{\partial x^k}{\partial\overline{x}^j}+WJ^{W-1}\frac{\partial J}{\partial\overline{x}^j}\phi \tag{2.6.1}$$

The second parcel on the right shows that the gradient of a scalar density is not a vector. It is verified that for $W=0$ the result is a scalar function and

$$\frac{\partial\overline{\phi}}{\partial\overline{x}^j}=\frac{\partial\phi}{\partial x^k}\frac{\partial x^k}{\partial\overline{x}^j}$$

is the transformation law of the vectors.

Let the Jacobian cofactor

$$C_k^m=\frac{\partial x^k}{\partial\overline{x}^m}$$

or

$$\frac{\partial x^r}{\partial\overline{x}^j}C_r^m=J\delta_r^r\Rightarrow C_r^m=J\frac{\partial\overline{x}^m}{\partial x^k}$$

it follows that

$$\frac{\partial J}{\partial\overline{x}^j}=\left[\frac{\partial}{\partial\overline{x}^j}\left(\frac{\partial x^k}{\partial\overline{x}^m}\right)\right]C_k^m\Rightarrow\frac{\partial J}{\partial\overline{x}^j}=J\frac{\partial^2x^k}{\partial\overline{x}^j\partial\overline{x}^m}\frac{\partial\overline{x}^m}{\partial x^k}$$

The substitution of this expression in expression (2.6.1) provides

$$\frac{\partial\left(J^W\overline{\phi}\right)}{\partial\overline{x}^j}=J^W\left(\frac{\partial\phi}{\partial x^k}\frac{\partial x^k}{\partial\overline{x}^j}+W\frac{\partial^2x^k}{\partial\overline{x}^j\partial\overline{x}^m}\frac{\partial\overline{x}^m}{\partial x^k}\phi\right) \tag{2.6.2}$$

that is the transformation law of the pseudoscalar $J^W\phi(x^i)$. Using expression (2.4.25) the second term in parenthesis can be written as

$$\frac{\partial^2x^k}{\partial\overline{x}^j\partial\overline{x}^m}\frac{\partial\overline{x}^m}{\partial x^k}=\overline{\Gamma}_{j\ell}^m-\frac{\partial\overline{x}^m}{\partial x^p}\frac{\partial x^k}{\partial\overline{x}^j}\frac{\partial x^q}{\partial\overline{x}^\ell}\Gamma_{kq}^p$$

The contraction in the indexes m and ℓ provides

$$\frac{\partial^2 x^k}{\partial \overline{x}^j \partial \overline{x}^m}\frac{\partial \overline{x}^m}{\partial x^k} = \overline{\Gamma}^m_{j\ell} - \frac{\partial \overline{x}^\ell}{\partial x^p}\frac{\partial x^k}{\partial \overline{x}^j}\frac{\partial x^q}{\partial \overline{x}^\ell}\Gamma^p_{kq}$$

and with

$$\delta^q_p = \frac{\partial \overline{x}^\ell}{\partial x^p}\frac{\partial x^q}{\partial \overline{x}^\ell}$$

the result is

$$\frac{\partial^2 x^k}{\partial \overline{x}^j \partial \overline{x}^m}\frac{\partial \overline{x}^m}{\partial x^k} = \overline{\Gamma}^m_{j\ell} - \frac{\partial x^k}{\partial \overline{x}^j}\Gamma^q_{kq}$$

The substitution of this expression in expression (2.6.2) provides

$$\frac{\partial (J^W \overline{\phi})}{\partial \overline{x}^j} = J^W \frac{\partial \phi}{\partial x^k}\frac{\partial x^k}{\partial \overline{x}^j} + WJ^W \overline{\Gamma}^m_{j\ell}\phi - WJ^W \frac{\partial x^k}{\partial \overline{x}^j}\Gamma^q_{kq}\phi$$

Let a scalar density which transformation law is given by

$$\overline{\phi} = J^W \phi$$

it results in

$$\frac{\partial (J^W \overline{\phi})}{\partial \overline{x}^j} - W\overline{\Gamma}^m_{j\ell}\overline{\phi} = J^W \frac{\partial x^k}{\partial \overline{x}^j}\left(\frac{\partial \phi}{\partial x^k} - W\Gamma^q_{kq}\phi\right)$$

The term in parenthesis to the right represents a covariant pseudovector of weight W. This expression shows that the covariant derivative of a scalar density presents an additional term in its expression, in which the factor multiplies the contracted Christoffel symbol. For $W = 0$ this expression is reduced to the gradient expression of the scalar function $\phi(x^i)$

$$\frac{\partial \phi}{\partial \overline{x}^j} = \frac{\partial \phi}{\partial x^k}\frac{\partial x^k}{\partial \overline{x}^j}$$

For a contravariant pseudovector of weight W it follows by means of this expression that is analogous to the one shown for a scalar density, the next expression

$$\partial_k u^j = \frac{\partial u^j}{\partial x^k} + u^q \Gamma^j_{kq} - W u^j \Gamma^q_{kq} \tag{2.6.3}$$

and the contraction of the indexes j and k provides

$$\partial_j u^j = \frac{\partial u^j}{\partial x^j} + u^q \Gamma^j_{jq} - W u^j \Gamma^q_{jq}$$

The dummy index j in the third term to the right can be changed by the index q:

$$\partial_j u^j = \frac{\partial u^j}{\partial x^j} + (1 - W)\, u^q \Gamma^q_{jq}$$

If the pseudovector has weight $W = 1$ this expression is simplified for $\partial_j u^j = \frac{\partial u^j}{\partial x^j}$, which represents the divergence of vector u^j.

The generalization of expression (2.6.3) for a relative tensor of weight W and variance $(1,1)$ is given by

$$\partial_r T^i_j = \frac{\partial T^i_j}{\partial x^r} + T^\ell_j \Gamma^j_{\ell k} - T^i_\ell \Gamma^\ell_{jk} - W T^i_j \Gamma^q_{rq} \tag{2.6.4}$$

For a relative tensor $T^{i\cdots}_{j\cdots}$ of weight W and variance (p, q) it results in

$$\partial_r T^{i\cdots}_{j\cdots} = \frac{\partial T^{i\cdots}_{j\cdots}}{\partial x^r} + T^\ell_j \Gamma^j_{\ell k} + \underbrace{\cdots\cdots\cdots}_{\substack{\text{terms relative to} \\ \text{the contravariance}}} - T^i_\ell \Gamma^\ell_{jk} - \underbrace{\cdots\cdots\cdots}_{\substack{\text{terms relative to} \\ \text{the covariance}}} - W T^i_j \Gamma^q_{rq} \tag{2.6.5}$$

By means of the considerations presented in the first paragraph of item (2.5.4), and adding that the parcel $W T^i_j \Gamma^q_{rq}$ in expression (2.6.5) is linear in terms of the original tensor, it implies that the rules of ordinary differentiation of Differential Calculus are applicable to the covariant differentiation of relative tensors.

2.5.1 Covariant Derivative of the Ricci Pseudotensor

The covariant derivative of the Ricci pseudotensor in its contravariant form is given by

$$\partial_i e^{jk\ell} = \partial_i \left(\frac{e^{jk\ell}}{\sqrt{g}} \right) = \frac{\partial \left(\frac{e^{ijk}}{\sqrt{g}} \right)}{\partial x^i} + \Gamma^j_{ip} \frac{e^{pk\ell}}{\sqrt{g}} + \Gamma^k_{ip} \frac{e^{jp\ell}}{\sqrt{g}} + \Gamma^\ell_{ip} \frac{e^{jkp}}{\sqrt{g}}$$

$$\frac{\partial \left(\frac{e^{jk\ell}}{\sqrt{g}} \right)}{\partial x^i} = \frac{1}{\sqrt{g}} \frac{\partial e^{jk\ell}}{\partial x^i} + e^{jk\ell} \frac{\partial \left(\frac{1}{\sqrt{g}} \right)}{\partial x^i} = e^{jk\ell} \frac{\partial \left(\frac{1}{\sqrt{g}} \right)}{\partial x^i} = -\frac{e^{jk\ell}}{2 g^{\frac{2}{3}}} \frac{\partial g}{\partial x^i}$$

The contraction of the Christoffel symbol provides

$$\frac{\partial g}{\partial x^i} = 2g\Gamma^p_{pi}$$

whereby

$$\frac{\partial\left(\frac{e^{jk\ell}}{\sqrt{g}}\right)}{\partial x^i} = -\frac{e^{jk\ell}}{\sqrt{g}}\Gamma^p_{pi}$$

Substituting this expression in the expression of covariant derivative

$$\partial_i \varepsilon^{jk\ell} = \partial_i\left(\frac{e^{jk\ell}}{\sqrt{g}}\right) = -\frac{e^{jk\ell}}{\sqrt{g}}\Gamma^p_{pi} + \Gamma^j_{ip}\frac{e^{pk\ell}}{\sqrt{g}} + \Gamma^k_{ip}\frac{e^{jp\ell}}{\sqrt{g}} + \Gamma^\ell_{ip}\frac{e^{jkp}}{\sqrt{g}}$$

The conditions for which this pseudotensor is non-null are that the first three indexes be different, i.e., $j \neq k \neq \ell$, then for $j = 1$, $k = 2$, $\ell = 3$:

$$\partial_i \varepsilon^{jk\ell} = \partial_i\left(\frac{e^{jk\ell}}{\sqrt{g}}\right) = -\frac{e^{123}}{\sqrt{g}}\Gamma^p_{pi} + \Gamma^j_{ip}\frac{e^{p23}}{\sqrt{g}} + \Gamma^k_{ip}\frac{e^{1p3}}{\sqrt{g}} + \Gamma^\ell_{ip}\frac{e^{12p}}{\sqrt{g}}$$

With $p = 1, 2, 3$:

$$\partial_i \varepsilon^{jk\ell} = \partial_i\left(\frac{e^{jk\ell}}{\sqrt{g}}\right) = -\frac{e^{123}}{\sqrt{g}}\left(\Gamma^1_{1i} + \Gamma^2_{2i} + \Gamma^3_{3i}\right) + \Gamma^j_{i1}\frac{e^{123}}{\sqrt{g}} + \Gamma^k_{i2}\frac{e^{123}}{\sqrt{g}} + \Gamma^\ell_{i3}\frac{e^{123}}{\sqrt{g}}$$

and with the symmetry of the Christoffel symbol it results in

$$\partial_i \varepsilon^{jk\ell} = \partial_i\left(\frac{e^{jk\ell}}{\sqrt{g}}\right) = 0$$

With an analogous expression for the covariant form of the Ricci pseudotensor $\varepsilon_{ijk} = \sqrt{g}e_{ijk}$ it results for its covariant derivative

$$\partial_i \varepsilon_{ijk} = \partial_i\left(\sqrt{g}e_{ijk}\right) = \frac{\partial\left(\sqrt{g}e_{ijk}\right)}{\partial x^i} - \sqrt{g}e_{pjk}\Gamma^p_{i\ell} - \sqrt{g}e_{ipk}\Gamma^p_{j\ell} - \sqrt{g}e_{ijp}\Gamma^p_{k\ell}$$

The partial derivative referent to the first term to the right is given by

$$\frac{\partial\left(\sqrt{g}e_{ijk}\right)}{\partial x^\ell} = \frac{\partial\left(\sqrt{g}\right)}{\partial x^\ell}e_{ijk} + \sqrt{g}\frac{\partial\left(e_{ijk}\right)}{\partial x^\ell}$$

but

$$\frac{\partial\left(e_{ijk}\right)}{\partial x^\ell} = 0$$

it results in

$$\frac{\partial\left(\sqrt{g}e_{ijk}\right)}{\partial x^{\ell}} = \frac{\partial\left(\sqrt{g}\right)}{\partial x^{\ell}} e_{ijk}$$

Expression (2.4.23) can be written as

$$\frac{\partial\left(\sqrt{g}\right)}{\partial x^{\ell}} = \sqrt{g}\Gamma^{p}_{p\ell} \Rightarrow \frac{\partial\left(\sqrt{g}e_{ijk}\right)}{\partial x^{\ell}} = \sqrt{g}e_{ijk}\Gamma^{p}_{p\ell}$$

Substituting this expression in the expression of the covariant derivative

$$\partial_i \varepsilon_{ijk} = \partial_i\left(\sqrt{g}e_{ijk}\right) = \sqrt{g}e_{ijk}\Gamma^{p}_{p\ell} - \sqrt{g}e_{pjk}\Gamma^{p}_{i\ell} - \sqrt{g}e_{ipk}\Gamma^{p}_{j\ell} - \sqrt{g}e_{ijp}\Gamma^{p}_{k\ell}$$

The conditions for which this pseudotensor is non-null are that the first three indexes be different, i.e., $j \neq k \neq \ell$, then for $j = 1$, $k = 2$, $\ell = 3$:

$$\partial_i \varepsilon_{ijk} = \partial_i\left(\sqrt{g}e_{ijk}\right) = \sqrt{g}e_{123}\Gamma^{p}_{p\ell} - \sqrt{g}e_{p23}\Gamma^{p}_{1\ell} - \sqrt{g}e_{1p3}\Gamma^{p}_{2\ell} - \sqrt{g}e_{12p}\Gamma^{p}_{3\ell}$$

With $p = 1, 2, 3$:

$$\begin{aligned}\partial_i \varepsilon_{ijk} &= \partial_i\left(\sqrt{g}e_{ijk}\right)\\ &= \sqrt{g}e_{123}\left(\Gamma^{1}_{1\ell} + \Gamma^{2}_{2\ell} + \Gamma^{3}_{3\ell}\right) - \sqrt{g}e_{123}\Gamma^{1}_{1\ell} - \sqrt{g}e_{123}\Gamma^{2}_{2\ell} - \sqrt{g}e_{123}\Gamma^{3}_{3\ell}\end{aligned}$$

whereby

$$\partial_i \varepsilon_{ijk} = \partial_i\left(\sqrt{g}e_{ijk}\right) = 0$$

These derivatives show that

$$\partial_i\left(\delta^{ijk}_{pqr}\right) = \partial_i\left(\varepsilon^{ijk}\varepsilon_{pqr}\right) = \partial_i\left(\varepsilon^{ijk}\right)\varepsilon_{pqr} + \varepsilon^{ijk}\partial_i\left(\varepsilon_{pqr}\right) = 0$$

The covariant derivatives of the Ricci pseudotensors ε^{ijk}, ε_{pqr} and the generalized Kronecker delta δ^{ijk}_{pqr} being null, it implies that these varieties behave as constants in the calculation of the covariant derivative.

As an example of an application of this characteristic, let the tensorial expression $\varepsilon^{ijk}\partial_j u_k$, which covariant derivative is given by

$$\partial_i\left(\varepsilon^{ijk}\partial_j u_k\right) = \varepsilon^{ijk}\partial_i\left(\partial_j u_k\right) + \partial_j u_k \partial_i\left(\partial_j \varepsilon^{ijk}\right)$$

but with

$$\partial_i \varepsilon^{ijk} = 0$$

this expression becomes

$$\partial_i\left(\varepsilon^{ijk}\partial_j u_k\right) = \varepsilon^{ijk}\partial_i\left(\partial_j u_k\right)$$

2.6 Intrinsic or Absolute Derivative

The absolute derivative of a variety is calculated when the coordinates x^i vary as a function of time, i.e., $x^i = x^i(t)$.

A covariant derivative of an invariant $\phi(x^k)$ is given by

$$\partial_k\phi\left(x^k\right) = \frac{\partial\phi\left(x^k\right)}{\partial x^k}$$

which is equal to its partial derivative.

For the absolute derivative

$$\frac{\delta\phi\left(x^k\right)}{\delta t} = \frac{\partial\phi\left(x^k\right)}{\partial t} + \partial_k\phi\left(x^k\right)\frac{dx^k}{dt} = \frac{\partial\phi\left(x^k\right)}{\partial t} + \frac{\partial\phi\left(x^k\right)}{\partial x^k}\frac{dx^k}{dt} = \frac{d\phi\left(x^k\right)}{dt}$$

then this derivative is equal to its total derivative.

For the vector $\boldsymbol{u}(x^i)$ where x^i varies as a function of time, which is expressed by means of its contravariant coordinates, or

$$\boldsymbol{u} = u^k\left[x^i(t), t\right]\boldsymbol{g_k}\left[x^i(t)\right]$$

The derivative with respect to time is given by

$$\frac{d\boldsymbol{u}}{dt} = \frac{d\left(u^k\boldsymbol{g_k}\right)}{dt} = \frac{du^k}{dt}\boldsymbol{g_k} + u^k\frac{\partial\boldsymbol{g_k}}{\partial t}\frac{dx^i}{dt} \tag{2.7.1}$$

and with

$$\frac{du^k}{dt} = \frac{\partial u^k}{\partial t} + \frac{\partial u^k}{\partial x^i}\frac{dx^i}{dt} \tag{2.7.2}$$

The following expression (item 2.3)

$$\frac{\partial\boldsymbol{g_k}}{\partial x^i} = \boldsymbol{g_m}\Gamma^m_{ki}$$

substituted in expression (2.7.1) provides

$$\frac{d\boldsymbol{u}}{dt} = \frac{du^k}{dt}\boldsymbol{g_k} + u^k\Gamma^m_{ki}\frac{dx^i}{dt}\boldsymbol{g_m}$$

Replacing the indexes $k \rightarrow m$ in the first term to the right

$$\frac{d\boldsymbol{u}}{dt} = \left(\frac{du^m}{dt} + u^k \Gamma^m_{ki} \frac{dx^i}{dt}\right) \boldsymbol{g_m}$$

thus the absolute derivative of a vector generates a vector.

The covariant derivative of the contravariant vector is written as

$$\frac{\delta u^m}{\delta t} = \frac{du^m}{dt} + u^k \Gamma^m_{ki} \frac{dx^i}{dt} \tag{2.7.3}$$

and substituting expression (2.7.2) in expression (2.7.3)

$$\frac{\delta u^m}{\delta t} = \frac{\partial u^k}{\partial t} + \frac{\partial u^k}{\partial x^i} \frac{dx^i}{dt} + u^k \Gamma^m_{ki} \frac{dx^i}{dt} \Rightarrow \frac{\delta u^m}{\delta t} = \frac{\partial u^k}{\partial t} + \left(\frac{\partial u^k}{\partial x^i} + u^k \Gamma^m_{ki}\right) \frac{dx^i}{dt}$$

The covariant derivative of the contravariant vector is given by

$$\partial_i u^k = \frac{\partial u^k}{\partial x^i} + u^k \Gamma^m_{ki}$$

or in vectorial form

$$\frac{\partial \boldsymbol{u}}{\partial x^i} = \left(\partial_i u^k\right) \boldsymbol{g_k}$$

whereby for the absolute derivative of vector $\boldsymbol{u}$ it results that

$$\frac{\delta u^m}{\delta t} = \frac{\partial u^k}{\partial t} + \partial_i u^k \frac{dx^i}{dt}$$

or in vectorial form

$$\frac{d\boldsymbol{u}}{dt} = \left(\frac{\delta u^m}{\delta t}\right) \boldsymbol{g_m}$$

The vector $\boldsymbol{u}$ in terms of their covariant components is given by

$$\boldsymbol{u} = u_k \left[x^i(t), t\right] \boldsymbol{g^k} \left[x^i(t)\right]$$

and with an analogous analysis to the one shown for the contravariant vectors, and with

$$\boldsymbol{g^k_{,i}} = -\Gamma^k_{im} \boldsymbol{g^m}$$

it results for the absolute derivative of vector $\boldsymbol{u}$

$$\frac{\delta u_k}{\delta t} = \frac{\partial u_m}{\partial t} + \partial_i u_k \frac{dx^i}{dt}$$

where $\partial_i u_k$ is the covariant derivative of the covariant vector.

These expressions can be generalized for the tensors

$$\frac{\delta T^{ij}}{\delta t} = \frac{\partial T^{ij}}{\partial t} + \partial_k T^{ij} \frac{dx^k}{dt} \tag{2.7.4}$$

$$\frac{\delta T_{ij}}{\delta t} = \frac{\partial T_{ij}}{\partial t} + \partial_k T_{ij} \frac{dx^k}{dt} \tag{2.7.5}$$

$$\frac{\delta T^{ij}_{\ \ m}}{\delta t} = \frac{\partial T^{ij}_{\ \ m}}{\partial t} + \partial_k T^{ij}_{\ \ m} \frac{dx^k}{dt} \tag{2.7.6}$$

The differentiation rules of Differential Calculus are applicable to absolute differentiation, which can be proven, for instance, for two tensors A_{ij} and B_{ij}, which algebraic addition generates the tensors $C_{ij} = A_{ij} \pm B_{ij}$, and which product results in $A_{ij}B_{ij}$. Calculating the absolute derivative of this sum

$$\frac{\delta C_{ij}}{\delta t} = \partial_k C_{ij} \frac{dx^k}{dt} = \partial_k \left(A_{ij} + B_{ij}\right) \frac{dx^k}{dt} = \partial_k A_{ij} \frac{dx^k}{dt} + \partial_k B_{ij} \frac{dx^k}{dt} = \frac{\delta A_{ij}}{\delta t} + \frac{\delta B_{ij}}{\delta t}$$

Calculating the absolute derivative of the product of the tensors

$$\begin{aligned} \frac{\delta\left(A_{ij}B_{ij}\right)}{\delta t} &= \partial_k \left(A_{ij}B_{ij}\right) \frac{dx^k}{dt} = \left(\partial_k A_{ij}\right) B_{ij} \frac{dx^k}{dt} + A_{ij}\left(\partial_k B_{ij}\right) \frac{dx^k}{dt} \\ &= \partial_k A_{ij} \frac{dx^k}{dt} B_{ij} + A_{ij} \partial_k B_{ij} \frac{dx^k}{dt} = \frac{\delta A_{ij}}{\delta t} B_{ij} + A_{ij} \frac{\delta B_{ij}}{\delta t} \end{aligned}$$

The absolute derivative of vector $\boldsymbol{u}$ calculated along the curve $x^i = x^i(t)$ can be defined by means of the inner product of its covariant derivative by the tangent vector to this curve $\frac{dx^i}{dt}$. For a tensor of order above the unit, and with an analogous way, the absolute derivative is the inner product of this tensor by the vector tangent to a curve, then

$$\frac{\delta T^{ij}_{\ \ pqr}}{\delta t} = \partial_k T^{ij}_{\ \ pqr} \frac{dx^k}{dt}$$

This definition in conjunction with the considerations made in the first paragraph of item 2.5.4 indicates that the absolute derivative follows the rules of Differential Calculus, such as shown for the addition and product of two tensors.

The derivative of the metric tensor g_{ij} is given by

$$\frac{\delta g_{ij}}{\delta t} = \frac{\partial g_{ij}}{\partial t} + \partial_k g_{ij} \frac{dx^k}{dt}$$

Ricci's lemma shows that $\partial_k g_{ij} = 0$, then

$$\frac{\delta g_{ij}}{\delta t} = \frac{\partial g_{ij}}{\partial t}$$

As the metric tensor is independent of time it implies that $\frac{\partial g_{ij}}{\partial t} = 0$, whereby it results that $\frac{\delta g_{ij}}{\delta t} = 0$, i.e., its absolute derivative is null.

For the tensors g^{ij} and δ^i_j, which have the same characteristics of the metric tensor, developing an analysis analogous to the one shown for this tensor it results in

$$\frac{\delta g^{ij}}{\delta t} = \frac{\partial g^{ij}}{\partial t} + \partial_k g^{ij} \frac{dx^k}{dt} = \frac{\partial g^{ij}}{\partial t} = 0$$

$$\frac{\delta \delta^i_j}{\delta t} = \frac{\partial \delta^i_j}{\partial t} + \partial_k \delta^i_j \frac{dx^k}{dt} = \frac{\partial \delta^i_j}{\partial t} = 0$$

2.6.1 Uniqueness of the Absolute Derivative

The covariant derivative of a Cartesian tensor coincides with its partial derivative, then the absolute derivative of this variety, calculated along a curve $x^i = x^i(t)$, can be defined by means of the scalar product of this derivative by the vector tangent to this curve $\frac{dx^i}{dt}$. For instance, for a Cartesian tensor of variance $(2, 3)$ it results in

$$\frac{\delta T^{ij}_{pqr}}{\delta t} = \frac{\partial T^{ij}_{pqr}}{\partial t} \frac{dx^k}{dt}$$

As the partial derivative of a Cartesian tensor is unique, and the scalar product that defines the absolute derivative generates an invariant, it is possible to conclude that this derivative is also unique. This analysis can be generalized for arbitrary tensors.

Exercise 2.14 Calculate the absolute derivative of: (a) $g_{ij}u^i v^j$; *(b)* $g_{ij}u^i u^j$; (c) vector u_i knowing that $\frac{\delta u^i}{\delta t} = 0$.
(a) The expression $g_{ij}u^i v^j$ represents a scalar, and taking the derivative

$$\frac{\delta\left(g_{ij}u^iv^j\right)}{\delta t} = \frac{d\left(g_{ij}u^iv^j\right)}{dt}$$

$$\frac{\delta\left(g_{ij}u^iv^j\right)}{\delta t} = \frac{\delta\left(g_{ij}\right)}{\delta t}u^iv^j + g_{ij}\frac{\delta(u^i)}{\delta t}v^j + g_{ij}u^i\frac{\delta(v^j)}{\delta t} = g_{ij}\frac{\delta(u^i)}{\delta t}v^j + g_{ij}u^i\frac{\delta(v^j)}{\delta t}$$

(b) The change of vector v^j by vector u^j in the expression calculated in the previous item provides

$$\frac{\delta\left(g_{ij}u^iu^j\right)}{\delta t} = g_{ij}\frac{\delta(u^i)}{\delta t}u^j + g_{ij}u^i\frac{\delta(u^j)}{\delta t}$$

Interchanging the indexes $i \leftrightarrow j$ in the first term to the right, and with the symmetry of the metric tensor results in

$$\frac{\delta\left(g_{ij}u^iu^j\right)}{\delta t} = g_{ji}\frac{\delta(u^j)}{\delta t}u^i + g_{ij}u^i\frac{\delta(u^j)}{\delta t} = 2g_{ij}u^i\frac{\delta(u^j)}{\delta t}$$

As $g_{ij}u^iu^j = \|\boldsymbol{u}\|^2$, it implies that $\frac{\delta\left(g_{ij}u^iu^j\right)}{\delta t} = 0$, which indicates that $\frac{\delta u^j}{\delta t} = 0$.

(c) The covariant components of the vector are given by

$$u_j = g_{ij}u^i$$

whereby differentiating

$$\frac{\delta u_j}{\delta t} = \frac{\delta\left(g_{ij}u^i\right)}{\delta t} = \frac{\delta g_{ij}}{\delta t}u^i + g_{ij}\frac{\delta u^i}{\delta t} = g_{ij}\frac{\delta u^i}{\delta t} = 0$$

Exercise 2.15 Show that $\frac{\delta}{\delta t}\left(\frac{dx^i}{dt}\right) = \frac{d^2x^i}{dt^2} + \Gamma^i_{jk}\frac{dx^j}{dt}\frac{dx^k}{dt}$.

Putting

$$u^i = \frac{dx^i}{dt}$$

results for the absolute derivative of this vector

$$\frac{\delta u^i}{\delta t} = \partial_k u_i\frac{dx^k}{dt} = \left(\frac{\partial u^i}{\partial x^k} + u^j\Gamma^i_{jk}\right)\frac{dx^k}{dt} = \frac{\partial u^i}{\partial x^k}\frac{dx^k}{dt} + u^j\Gamma^i_{jk}\frac{dx^k}{dt}$$

and with

$$u^j = \frac{dx^j}{dt}$$

it implies

$$\frac{\delta u^i}{\delta t} = \frac{\partial u^i}{\partial x^k}\frac{dx^k}{dt} + \Gamma^i_{jk}\frac{dx^j}{dt}\frac{dx^k}{dt}$$

It follows that

$$\frac{\delta}{\delta t}\left(\frac{dx^i}{dt}\right) = \frac{d}{dt}\left(\frac{dx^i}{dt}\right) + \Gamma^i_{jk}\frac{dx^j}{dt}\frac{dx^k}{dt}$$

$$\frac{\delta}{\delta t}\left(\frac{dx^i}{dt}\right) = \frac{d^2x^i}{dt^2} + \Gamma^i_{jk}\frac{dx^j}{dt}\frac{dx^k}{dt} \qquad \text{Q.E.D.}$$

2.7 Contravariant Derivative

The contravariant derivative is defined considering the tensorial nature of the covariant derivative, for the raising of the index of tensor $\partial_k \ldots$ the result is

$$\partial^\ell \ldots = g^{k\ell}\partial_k \ldots \tag{2.8.1}$$

It is promptly verified with Ricci's lemma that $\partial^k g_{ij} = 0$, as well as $\partial^k g^{ij} = 0$ and $\partial^k \delta^i_j = 0$. These relations show that the tensors $g_{ij}, g^{ij}, \delta^i_j$ behave as constants in the calculation of the contravariant derivative.

For the variance tensors (p, q) the result by means of the expression (2.8.1) is

$$\partial^k T^{\cdots}_{\cdots} = g^{kj}\partial_j T^{\cdots}_{\cdots} = \partial_j\left(g^{kj}T^{\cdots}_{\cdots}\right) \tag{2.8.2}$$

Then the contravariant derivative is equivalent to the raising of the indexes of tensor $\partial_k \ldots$, or the covariant derivative of tensor $g^{kj}T^{\cdots}_{\cdots}$.

For instance, for the covariant vector u_k:

$$\partial^k u_k = g^{kj}\partial_j u_k = \partial_j\left(g^{kj}u_k\right) = \partial_j u^j$$

Problems

2.1 Calculate the Christoffel symbols for the coordinates X^i which metric tensor is given by

$$g_{ij} = \begin{bmatrix} 1 & 0 \\ 0 & \dfrac{1}{(x^2)^2} \end{bmatrix}$$

Answer:

$$\Gamma_{ij,1} = 0 \quad \text{for} \quad i,j = 1,2 \quad \Gamma_{ij,2} = \begin{bmatrix} 0 & 0 \\ 0 & -\dfrac{1}{(x^2)^2} \end{bmatrix}$$

$$\Gamma^1_{ij} = 0 \quad \text{for} \quad i,j = 1,2 \quad \Gamma^2_{ij} = \begin{bmatrix} 0 & 0 \\ 0 & -1 \end{bmatrix}$$

2.2 Calculate the Christoffel symbols for the coordinate system X^i which metric tensor and its conjugated metric tensor are given by

$$g_{ij} = \begin{bmatrix} 1 & 0 & 0 \\ 0 & (x^1)^2 & 0 \\ 0 & 0 & (x^1 \sin x^2)^2 \end{bmatrix} \quad g^{ij} = \begin{bmatrix} 1 & 0 & 0 \\ 0 & \dfrac{1}{(x^1)^2} & 0 \\ 0 & 0 & \dfrac{1}{(x^1 \sin x^2)^2} \end{bmatrix}$$

Answer:

$$\Gamma_{ij,1} = \begin{bmatrix} 0 & 0 & 0 \\ 0 & -x^1 & 0 \\ 0 & 0 & -x^1(\sin x^2)^2 \end{bmatrix} \quad \Gamma_{ij,2} = \begin{bmatrix} 0 & x^1 & 0 \\ x^1 & 0 & 0 \\ 0 & 0 & -(x^1)^2 \sin x^2 \cos x^2 \end{bmatrix}$$

$$\Gamma_{ij,3} = \begin{bmatrix} 0 & 0 & x^1(\sin x^2)^2 \\ 0 & 0 & (x^1)^2 \sin x^2 \cos x^2 \\ x^1(\sin x^2)^2 & (x^1)^2 \sin x^2 \cos x^2 & 0 \end{bmatrix}$$

$$\Gamma^1_{ij} = \begin{bmatrix} 0 & 0 & 0 \\ 0 & -x^1 & 0 \\ 0 & 0 & -x^1(\sin x^2)^2 \end{bmatrix}$$

$$\Gamma^2_{ij} = \begin{bmatrix} 0 & \dfrac{1}{x^1} & 0 \\ \dfrac{1}{x^1} & 0 & 0 \\ 0 & 0 & -(x^1)^2 \sin x^2 \cos x^2 \end{bmatrix} \quad \Gamma^3_{ij} = \begin{bmatrix} 0 & 0 & \dfrac{1}{x^1} \\ 0 & 0 & \cot x^2 \\ \dfrac{1}{x^1} & \cot x^2 & 0 \end{bmatrix}$$

2.3 Calculate the Christoffel symbols of the second kind, where $F(x^1;x^2)$ is a function of the coordinates, for the referential system which metric tensor is

$$g_{ij} = \begin{bmatrix} 1 & 0 \\ 0 & F(x^1;x^2) \end{bmatrix}$$

Answer:

$$\Gamma_{ij}^1 = \begin{bmatrix} 1 & 0 \\ 0 & -\frac{1}{2}\frac{\partial F}{\partial x^1} \end{bmatrix} \quad \Gamma_{ij}^2 = \begin{bmatrix} 1 & \frac{1}{2F}\frac{\partial F}{\partial x^1} \\ \frac{1}{2F}\frac{\partial F}{\partial x^1} & \frac{1}{2F}\frac{\partial F}{\partial x^2} \end{bmatrix}$$

2.4 Calculate the Christoffel symbols for the space defined by the metric tensor

$$g_{ij} = \begin{bmatrix} -1 & 0 & 0 & 0 \\ 0 & -1 & 0 & 0 \\ 0 & 0 & -1 & 0 \\ 0 & 0 & 0 & e^{-x^4} \end{bmatrix}$$

Answer: $\Gamma_{44,4} = -\frac{1}{2}e^{-x^4}$; $\Gamma_{44}^4 = -\frac{1}{2}$.

2.5 Calculate the covariant derivative of the inner product of the tensors A_k^j and $B_n^{\ell m}$ with respect to coordinate x^p.

Answer:

$$\left(\partial_p A_k^j\right) B_n^{\ell m} + A_k^j\left(\partial_p B_n^{\ell m}\right)$$

2.6 Show that $\frac{\partial\left(\sqrt{g}g^{ij}\right)}{\partial x^i} + \Gamma_{pq}^j\sqrt{g}g^{pq} = 0$.

Chapter 3
Integral Theorems

3.1 Basic Concepts

The integral theorems and the concepts presented in this chapter are treated in Differential and Integral Calculus of multiple variables.

The approach of this subject is carried in a concise and direct manner, and seeks solely to provide theoretical subsides so that the gradient, divergence, and curl differential operators can be physically interpreted.

3.1.1 Smooth Surface

The surface S, open or closed, with upward normal $\boldsymbol{n}$ unique in each point, which direction is a continuous function of its points, is classified as a smooth surface. For instance, the surface of a sphere is closed smooth, and the surface of a cube is closed smooth by parts, for it can be decomposed into six smooth surfaces.

3.1.2 Simply Connected Domain

For every closed curve C defined in the domain D, the region formed by C and its interior is fully contained in D. This curve defines a region $R \subset D$, and D is called simply connected domain (Fig. 3.1a).

The interior of a circle and the interior of a sphere are simply connected regions. Two concentric spheres define a simply connected region.

© Springer International Publishing Switzerland 2016

E. de Souza Sánchez Filho, *Tensor Calculus for Engineers and Physicists*,
DOI 10.1007/978-3-319-31520-1_3

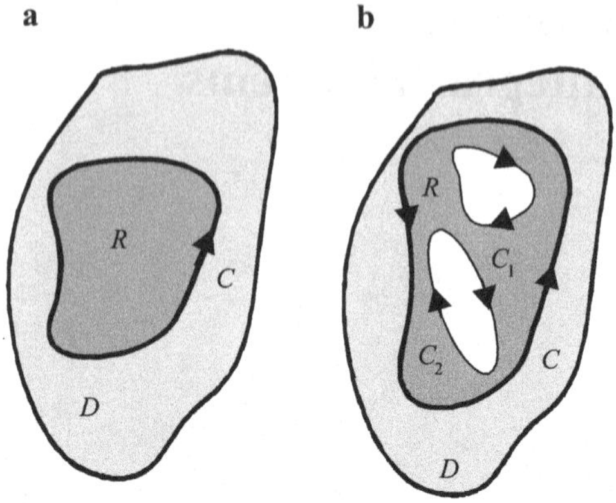

Fig. 3.1 Domain: (**a**) simply connected and (**b**) multiply connected

3.1.3 Multiply Connected Domain

Multiply Connected Domain is the domain D that contains a region R with N "holes" (Fig. 3.1b). A circle excluded its center defines a simply connected domain, and the "hole" is reduced to a point, but the region between two coaxial cylinders is multiply connected.

3.1.4 Oriented Curve

The closed smooth curve C that limits a region R is counterclockwise oriented if this region stays to its left, i.e., this curve is positively oriented.

3.1.5 Surface Integral

Consider S a smooth surface by parts with upward unit normal vector $\boldsymbol{n}$, and $\phi(x^i)$ a function that represents a smooth curve C over this surface (Fig. 3.2).

Dividing this finite area surface, defined by the function $\phi(x^i)$ in N elementary areas dS_i, $i = 1, 2, \ldots, N$, where the elementary area contains the point $P(x^i)$, and carrying out the sum $\sum_{i=1}^{N} \phi\left(x^i\right) dS_i$ and for $N \to \infty$, thus $dS_i \to 0$, implies the limit $\iint_S \phi\left(x^i\right) dS$ that represents the integral of surface S.

This limit exists and is independent of the number of divisions made. For a vectorial function, it results in a similar way $\iint_S \boldsymbol{u} dS$.

Fig. 3.2 Smooth surface

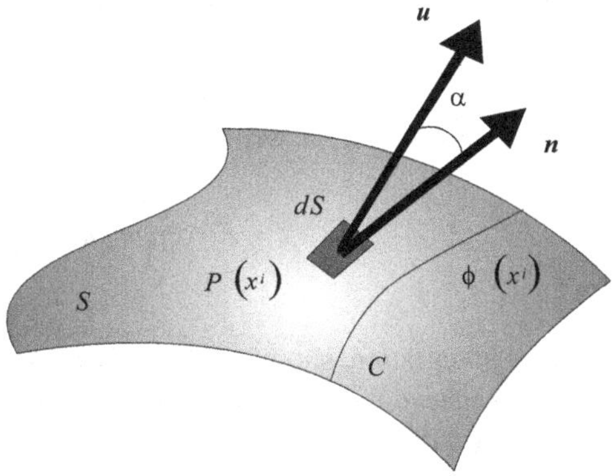

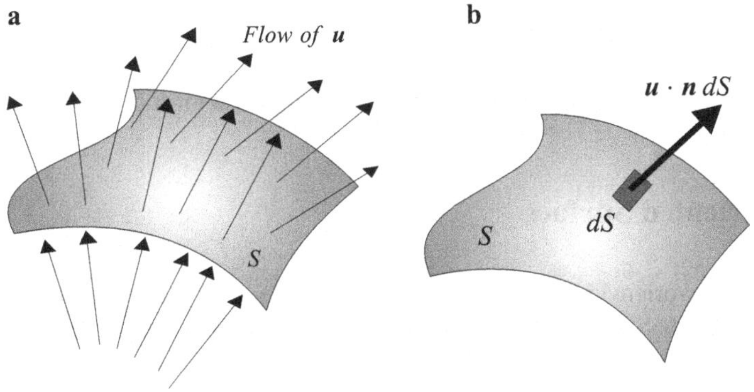

Fig. 3.3 Flow: (**a**) through the surface S and (**b**) component of the vectorial function $\boldsymbol{u}$ in the direction normal to the surface S

3.1.6 Flow

Let the vectorial function $\boldsymbol{u}$ dependent on point $P(x^i)$ located on the surface S. The component of $\boldsymbol{u}$ in the direction of the unit normal vector to the surface in this point is given by the scalar product $\boldsymbol{u} \cdot \boldsymbol{n}$. With this dot product for all the points located in the surface elements dS, and carrying out the sum $\sum_{i=1}^{N} \boldsymbol{u} \cdot \boldsymbol{n} dS$, and for $N \to \infty$, and $dS \to 0$ implies the integral

$$F = \iint_S \boldsymbol{u} \cdot \boldsymbol{n} dS \tag{3.1.1}$$

that defines the flow of the vectorial function $\boldsymbol{u}$ on the surface S (Fig. 3.3).

The surface area element dS is associated to the area vector $d\boldsymbol{S}$, with modulus dS and same direction of $\boldsymbol{n}$, then

$$d\boldsymbol{S} = \boldsymbol{n}dS \tag{3.1.2}$$

Expression (3.1.1) is written as

$$F = \iint_S \boldsymbol{u} \cdot \boldsymbol{n}dS = \iint_S \boldsymbol{u} \cdot d\boldsymbol{S} \tag{3.1.3}$$

and the integration shown in this expression is independent of the coordinate system, because the dot product $\boldsymbol{u} \cdot \boldsymbol{n}$ is invariant. In terms of the components of $\boldsymbol{u}$, it follows that

$$F = \iint_S u_i n_i dS \tag{3.1.4}$$

where n_i are the direction cosines of the unit normal vector $\boldsymbol{n}$.

3.2 Oriented Surface

Let S a surface oriented by means of its upward unit normal vector $\boldsymbol{n}$, then its outline C is oriented positively if S stays to its left, thus this curve is anticlockwise oriented.

Figure 3.4 shows a smooth surface S with upward unit normal vector $\boldsymbol{n}$, defined in a Cartesian coordinate system. This surface is expressed by the function $z = \phi(x; y)$, which orthogonal projection in plane OX^3 determines the region $R = S_{12}$. The unit

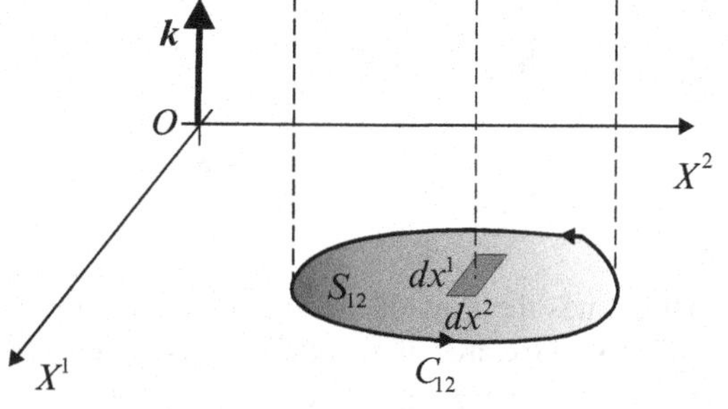

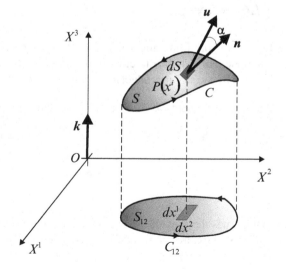

Fig. 3.4 Smooth surface S with upward unit normal vector $\boldsymbol{n}$ which outline is a curve closed smooth C

normal vector $\boldsymbol{n}$ forms an angle α with the axis OX^3, being $\cos\alpha$ its direction cosine. The orthogonal projection of the area element dS is given by

$$dS = \frac{dx^1 dx^2}{\cos\,\alpha}$$

The dot product of the unit vectors $\boldsymbol{n}$ and $\boldsymbol{k}$ is given by

$$\boldsymbol{n}\cdot\boldsymbol{k} = \|\boldsymbol{n}\|\,\|\boldsymbol{k}\|\cos\alpha$$

so

$$\|\boldsymbol{n}\cdot\boldsymbol{k}\| = \cos\alpha$$

and therefore

$$dS = \frac{dx^1 dx^2}{\|\boldsymbol{n}\cdot\boldsymbol{k}\|}$$

Substituting this expression in expression (3.1.3) results in

$$F = \iint\limits_{S} \boldsymbol{u}\cdot\boldsymbol{n}dS = \iint\limits_{S} \boldsymbol{u}\cdot\boldsymbol{n}\frac{dx^1 dx^2}{\|\boldsymbol{n}\cdot\boldsymbol{k}\|} \tag{3.1.5}$$

then the surface integral can be calculated as a double integral defined in the region R. The algebraic value of the flow depends on the field's orientation. If $\alpha < \frac{\pi}{2}$ then $F > 0$, i.e., the flow "is outward," and if $\alpha > \frac{\pi}{2}$ then $F < 0$, i.e., "the flow is inward."

3.2.1 Volume Integral

Consider the closed smooth surface S that contains a volume V, and $\phi(x^i)$ a function of position defined on this volume. Dividing V into elementary volumes dV_i, then for the point $P(x^i)$ situated over S implies $\phi[P(x^i)] = \phi(x^i)$. Carrying out the sum of elementary volumes $\sum_{i=1}^{N}\phi(x^i)\,dV_i$ and for $N\to\infty$, thus $dV_i\to 0$, results the limit $\iiint\limits_{V}\phi(x^i)\,dV$ that represents the volume integral. This limit exists and is independent on the number of divisions. If the function is vectorial, it results in a similar way $\iiint\limits_{V}\boldsymbol{u}\,dV$.

3.3 Green's Theorem

Consider R a region in the plane OX^1X^2 involved by the closed smooth curve C with R to its left. Let the real continuous functions $F_1(x^1;x^2)$ and $F_2(x^1;x^2)$, with continuous partial derivatives in $R \cup C$. Then

$$\iint_R \left(\frac{\partial F_2}{\partial x^1} - \frac{\partial F_1}{\partial x^2} \right) dx^1 dx^2 = \oint_C \left(F_1 dx^1 + F_2 dx^2 \right) \qquad (3.2.1)$$

This theorem is due to George Green (1793–1841) and deals with a generalization of the fundamental theorem of Integral Calculus for two dimensions.

Figure 3.5 shows the region R involved by the closed smooth curve C, in which there are lines parallel to the coordinate axes that are tangent to this curve. It is assumed as a premise that C is intersect by straight lines parallel to the coordinate axes in a maximum of two points.

The region R is defined by

$$\begin{cases} a \le x^1 \le b, & f(x^1) \le x^2 \le g(x^1) \\ c \le x^2 \le d, & p(x^2) \le x^1 \le q(x^2) \end{cases}$$

Let $C = AEB \cup BFA$, with AEB given by $x^2 = f(x^1)$, and BFA by $x^2 = g(x^1)$. In an analogous way results $C = FAE \cup EBF$, with FAE given by $x^1 = p(x^2)$, and EBF by $x^1 = q(x^2)$. With

$$\iint_R \frac{\partial F_1}{\partial x^2} dx^1 dx^2 = \int_a^b \left[\int_{f(x^1)}^{g(x^1)} \frac{\partial F_1}{\partial x^2} dx^2 \right] dx^1 = \int_a^b F_1\left(x^1; x^2\right) \Big|_{f(x^1)}^{g(x^1)}$$

$$dx^1 = -\int_a^b F_1\left[x^1; f\left(x^1\right)\right] dx^1 - \int_a^b F_1\left[x^1; g\left(x^1\right)\right] dx^1$$

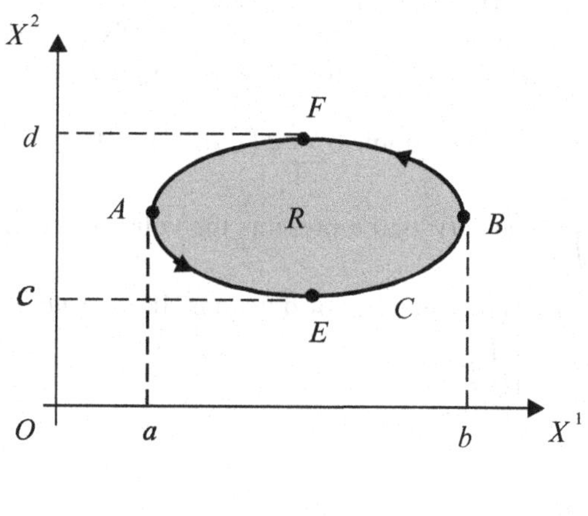

Fig. 3.5 Simply connected region

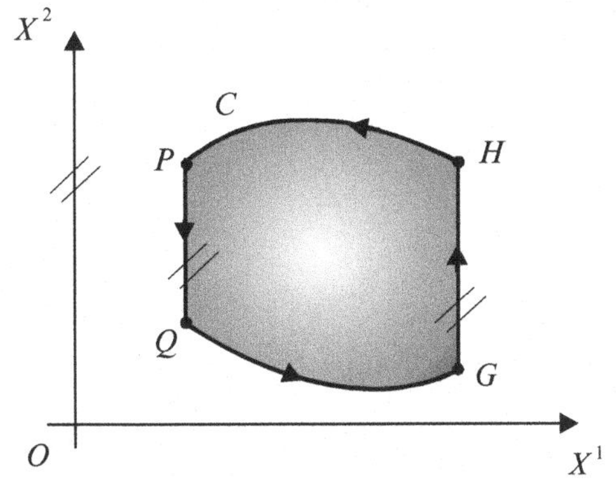

Fig. 3.6 Region simply connects with segments parallel to one of the coordinate axes

The two right members are the line integrals, then

$$\iint\limits_R \frac{\partial F_1}{\partial x^2} dx^1 dx^2 = -\int\limits_{BFC} F_1\left(x^1; x^2\right) dx^1 - \int\limits_{AEB} F_1\left(x^1; x^2\right) dx^1 = \oint\limits_C F_1\left(x^1; x^2\right) dx^1$$

If the segment of curve C is parallel to axis OX^2, the results of the integrals are not modified (Fig. 3.6). The integral $\int F_1 dx^1$ is cancelled in segment GH, for $x^1 = \text{constant}$ then $dx^1 = 0$. The same occurs for segment PQ.

With the segment QG given by $x^2 = f(x^1)$, and the segment HP given by $x^2 = g(x^1)$:

$$-\iint\limits_R \frac{\partial F_1}{\partial x^2} dx^1 dx^2 = \oint\limits_C F_1\left(x^1; x^2\right) dx^1 \tag{3.2.2}$$

and in the same way

$$-\iint\limits_R \frac{\partial F_2}{\partial x^1} dx^1 dx^2 = \oint\limits_C F_2\left(x^1; x^2\right) dx^2 \tag{3.2.3}$$

Adding expressions (3.2.2) and (3.2.3) results in

$$\iint\limits_R \left(\frac{\partial F_2}{\partial x^1} - \frac{\partial F_1}{\partial x^2}\right) dx^1 dx^2 = \oint\limits_C \left(F_1 dx^1 + F_2 dx^2\right) \qquad \text{Q.E.D.}$$

To prove the validity of this theorem for the more general cases being the region $R = R_1 \cup R_2$, in which the integrals are calculated for each subregion (Fig. 3.7).

Fig. 3.7 Division of the simply connected region into two simply connected regions

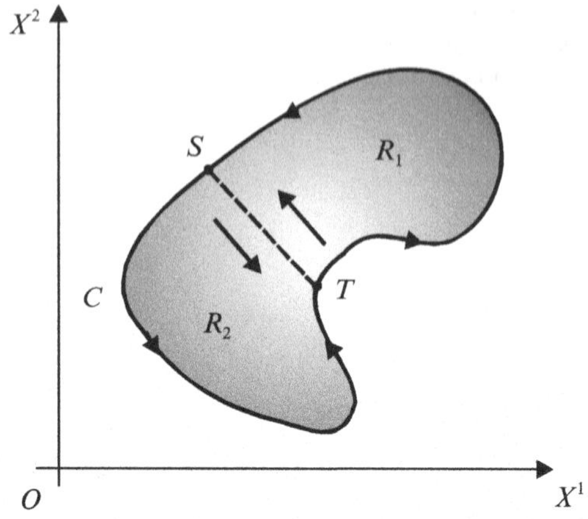

In the segment *ST* the line integrals are calculated twice, but as they are of different direction they cancel each other when they are added, hence

$$\oint_{TS} \left(F_1 dx^1 + F_2 dx^2\right) + \oint_{ST} \left(F_1 dx^1 + F_2 dx^2\right) = 0$$

Therefore, the expression of Green's theorem is valid for the subdivision of region R (Fig. 3.7). This ascertaining is generalized for a finite region $R = R_1 \cup R_2 \cdots R_N$ comprising N simple regions, with the outline curves $C_i, i = 1, 2, \ldots, N$, then

$$\iint_R \left(\frac{\partial F_2}{\partial x^1} - \frac{\partial F_1}{\partial x^2}\right) dx^1 dx^2 = \sum_{i=1}^{N} \oint_{C_i} \left(F_1 dx^1 + F_2 dx^2\right)$$

The consequence of this division of region R into parts is that this theorem can be applicable to multiply connected regions (Fig. 3.8). The region involved by the curve *TSBSTAT* is simply connected, so Green's theorem is valid for this region, hence

$$\iint_R \left(\frac{\partial F_2}{\partial x^1} - \frac{\partial F_1}{\partial x^2}\right) dx^1 dx^2 = \oint_{TSBSTAT} \left(F_1 dx^1 + F_2 dx^2\right)$$

To demonstrate the validity of Green's theorem for this kind of region, let the line integrals written in a symbolic way

$$\int_{TS} + \int_{C_2} + \int_{ST} + \int_{C_1} = \int_{C_2} + \int_{C_1} = \oint_{C}$$

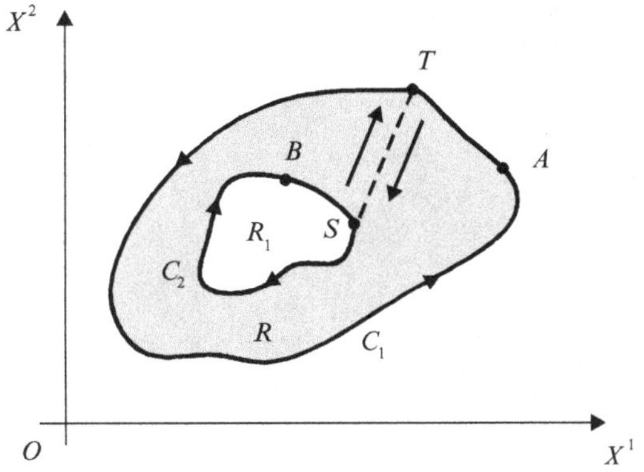

Fig. 3.8 Multiply connected regions

for

$$\int_{TS} = -\int_{ST}$$

therefore

$$\iint_R \left(\frac{\partial F_2}{\partial x^1} - \frac{\partial F_1}{\partial x^2}\right) dx^1 dx^2 = \oint_C \left(F_1 dx^1 + F_2 dx^2\right)$$

proves the previous statement.

With the condition $\frac{\partial F_2}{\partial x^1} = \frac{\partial F_1}{\partial x^2}$ in the region R it follows by Green's theorem

$$\oint_C \left(F_1 dx^1 + F_2 dx^2\right) = 0$$

thus the line integral is independent of the path on the closed curve C. To demonstrate that the admitted condition is necessary and sufficient being the segments C_1 and C_2 of the curve C shown in Fig. 3.9, for the line integral it follows that

$$\oint_{ADBEA} \left(F_1 dx^1 + F_2 dx^2\right) = 0$$

Writing the line integrals of the various segments of curve C under symbolic form

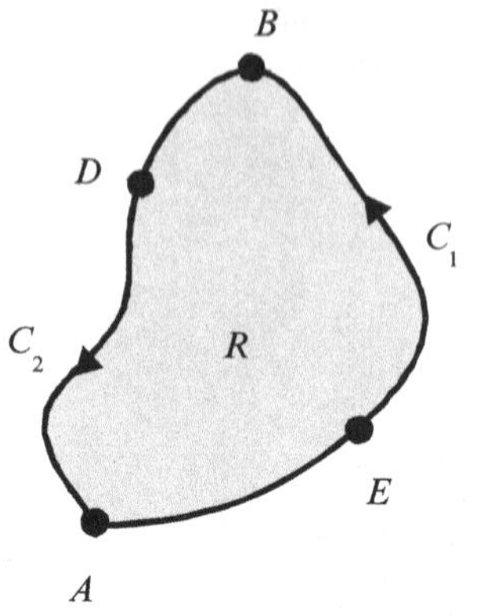

Fig. 3.9 Segments C_1 and C_2 of the closed curve C

$$\int_{ADB} + \int_{BEA} = 0$$

$$\int_{ADB} = -\int_{BEA} = \int_{AEB}$$

then

$$\oint_{C_1} \left(F_1 dx^1 + F_2 dx^2\right) = \oint_{C_2} \left(F_1 dx^1 + F_2 dx^2\right)$$

where by $\dfrac{\partial F_2}{\partial x^1} = \dfrac{\partial F_1}{\partial x^2}$ is the necessary and sufficient condition for this independence.

To admit that a parallel straight line of a coordinated axis intersects the region R in only two points is not essential, because R can be divided into a number of subregions which separately fulfill this property.

In vectorial notation with the function $\boldsymbol{F} = F_1\boldsymbol{i} + F_2\boldsymbol{j}$ and the position vector $\boldsymbol{r} = x^1\boldsymbol{i} + x^2\boldsymbol{j}$, and in differential form $d\boldsymbol{r} = dx^1\boldsymbol{i} + dx^2\boldsymbol{j}$, the line integral along the curve C is given by

$$\oint_C \left(F_1 dx^1 + F_2 dx^2\right) = \oint_C \boldsymbol{F} \cdot d\boldsymbol{r}$$

3.4 Stokes' Theorem

Consider the surface S with upward unit normal vector $\boldsymbol{n}$ involved by a closed smooth curve C with S to its left, which direction cosines are $n_i > 0$. Let the continuous real functions $F_1(x^1; x^2; x^3)$, $F_2(x^1; x^2; x^3)$, $F_3(x^1; x^2; x^3)$ with continuous partial derivatives in $S \cup C$. Then

$$\iint_S \left[\left(\frac{\partial F_3}{\partial x^2} - \frac{\partial F_2}{\partial x^3}\right) n_1 + \left(\frac{\partial F_1}{\partial x^3} - \frac{\partial F_3}{\partial x^1}\right) n_2 + \left(\frac{\partial F_2}{\partial x^1} - \frac{\partial F_1}{\partial x^2}\right) n_3\right] dS$$
$$= \oint_C \left(F_1 dx^1 + F_2 dx^2 + F_3 dx^3\right) \tag{3.3.1}$$

To demonstrate this theorem admit that a line parallel to axis OX^3 intersects S only in a point, then the projection of S on the plane OX^1X^2 will be the region S_{12} involved by the closed smooth curve C_{12} oriented positively (Fig. 3.10), then

$$dS_{12} = n_3 dS \tag{3.3.2}$$

and $n_3 > 0$.

The equation of surface S is given explicitly by $x^3 = \phi(x^1; x^2)$, which allows substituting the line integral along the curve C by the line integral along curve C_{12}:

$$\oint_C F_1\left(x^1; x^2; x^3\right) dx^1 = \oint_{C_{12}} F_1\left[x^1; x^2; \phi\left(x^1; x^2\right)\right] dx^1$$

In the term to the right the coordinate x^2 appears twice, in a direct way and in the function that represents the surface S. Applying Green's theorem it follows that

$$\oint_C F_1\left[x^1; x^2; \phi\left(x^1; x^2\right)\right] dx^1 = -\iint_{S_{12}} \left(\frac{\partial F_1}{\partial x^2} + \frac{\partial F_1}{\partial \phi}\frac{\partial \phi}{\partial x^2}\right) dx^1 dx^2$$

where

$$dS_{12} = dx^1 dx^2$$

and using expression (3.3.2)

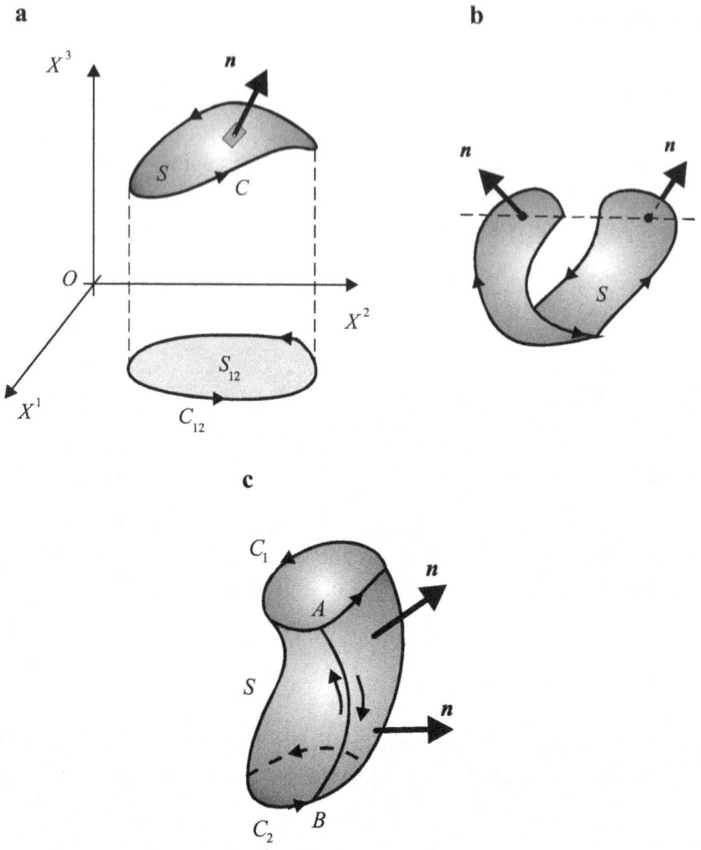

Fig. 3.10 Stokes theorem: (**a**) projection of the smooth surface S with upward unit normal vector $\boldsymbol{n}$ on the plane OX^1X^2; (**b**) surface delimited by the closed smooth curve C; and (**c**) surface with outline delimited by more than one curve

$$\oint_C F_1(x^1;x^2;x^3)\,dx^1 = -\iint_S \left\{ \frac{\partial F_1[x^1;x^2;\phi(x^1;x^2)]}{\partial x^2} + \frac{\partial F_1[x^1;x^2;\phi(x^1;x^2)]}{\partial \phi}\,\frac{\partial \phi(x^1;x^2)}{\partial x^2} \right\} n_3 dS \tag{3.3.3}$$

In Integral Calculus of Multiple Variables when studying the surface integrals of $x^3 = \phi(x^1;x^2)$ the following expressions are deducted for the direction cosines of its upward unit normal vector $\boldsymbol{n}$:

$$n_1 = \frac{\frac{\partial \phi}{\partial x^1}}{\pm\sqrt{1 + \left(\frac{\partial \phi}{\partial x^1}\right)^2 + \left(\frac{\partial \phi}{\partial x^2}\right)^2}} \tag{3.3.4}$$

$$n_2 = \frac{\frac{\partial \phi}{\partial x^2}}{\pm\sqrt{1 + \left(\frac{\partial \phi}{\partial x^1}\right)^2 + \left(\frac{\partial \phi}{\partial x^2}\right)^2}} \tag{3.3.5}$$

$$n_3 = \frac{1}{\mp\sqrt{1 + \left(\frac{\partial \phi}{\partial x^1}\right)^2 + \left(\frac{\partial \phi}{\partial x^2}\right)^2}} \tag{3.3.6}$$

As the direction cosines are positive, expression (3.3.5) provides

$$n_3 \frac{\partial \phi(x^1; x^2)}{\partial x_2} = -n_3$$

and substituting this expression in expression (3.3.3)

$$\oint_C F_1\, dx^1 = -\iint_S \left(n_2 \frac{\partial F_1}{\partial x^3} - n_3 \frac{\partial F_1}{\partial x^2} \right) dS$$

In an analogous way for the projections of S on the planes OX^2X^3 and OX^3X^1 it follows that

$$\oint_C F_2\, dx^2 = -\iint_S \left(n_3 \frac{\partial F_2}{\partial x^1} - n_1 \frac{\partial F_2}{\partial x^3} \right) dS$$

$$\oint_C F_3\, dx^3 = -\iint_S \left(n_1 \frac{\partial F_3}{\partial x^2} - n_2 \frac{\partial F_3}{\partial x^1} \right) dS$$

Adding these three expressions results in

$$\iint_S \left[\left(\frac{\partial F_3}{\partial x^2} - \frac{\partial F_2}{\partial x^3} \right) n_1 + \left(\frac{\partial F_1}{\partial x^3} - \frac{\partial F_3}{\partial x^1} \right) n_2 + \left(\frac{\partial F_2}{\partial x^1} - \frac{\partial F_1}{\partial x^2} \right) n_3 \right]$$
$$dS = \oint_C \left(F_1 dx^1 + F_2 dx^2 + F_3 dx^3 \right) \qquad \text{Q.E.D.}$$

Admit that a line parallel to one of the coordinate axis cuts the surface S only in a point is not an essential premise. Figure 3.10b, c shows two kinds of surface that do not fulfill this condition. In this case the surfaces must be divided into a finite number of subsurfaces, which separately fulfills this hypothesis, allowing Stokes' theorem to be applied to these subsurfaces, and add the partial results obtained. Then the line integrals referent to the outlines common to the projections of these surfaces on a plane of the coordinate system cancel each other, for they are integrated twice, but with the signs changed.

For a surface formed by several closed curves it is also possible to apply Stokes' theorem. Figure 3.10c shows a surface S limited by the closed and smooth curves C_1 and C_2. The section S along the curve AB generates a new surface, which outlines are the curves C_1, C_2 and AB, considered in opposite directions. Then the line integral referent to curve AB is calculated twice, but with opposite signs, whereby it cancels itself, leaving only the results referent to the line integrals of the curves C_1 and C_2.

The Stokes theorem is a generalization of Green's theorem for the tridimensional space. In vectorial notation with the function $\boldsymbol{F} = F_1\boldsymbol{i} + F_2\boldsymbol{j} + F_3\boldsymbol{k}$ and the vector $\boldsymbol{r} = x^1\boldsymbol{i} + x^2\boldsymbol{j} + x^3\boldsymbol{k}$, which differential is $d\boldsymbol{r} = dx^1\boldsymbol{i} + dx^2\boldsymbol{j} + dx^3\boldsymbol{k}$, the line integral along the curve C is given by

$$\oint_C \left(F_1 dx^1 + F_2 dx^2 + F_3 dx^3\right) = \oint_C \boldsymbol{F} \cdot d\boldsymbol{r} \tag{3.3.7}$$

The surface integrals that are present in Stokes' theorem also have a vectorial interpretation (item 4.4).

3.5 Gauß–Ostrogradsky Theorem

Consider the volume V with upward unit normal vector $\boldsymbol{n}$ *involved by a closed and smooth surface S, which direction cosines are* $n_i > 0$. *Let the continuous real functions* $F_1(x^1;x^2;x^3)$, $F_2(x^1;x^2;x^3)$, $F_3(x^1;x^2;x^3)$ *with continuous partial derivatives in* $V \cup S$. *Then*

$$\iiint_V \left(\frac{\partial F_1}{\partial x^1} + \frac{\partial F_2}{\partial x^2} + \frac{\partial F_3}{\partial x^3}\right) dx^1 dx^2 dx^3 = \iint_S (F_1 n_1 + F_2 n_2 + F_3 n_3)\, dS \tag{3.4.1}$$

Consider a line parallel to axis OX^2 that intersects the surface S in a maximum of two points P and P', with upward unit normal vector $\boldsymbol{n}(P)$ and $\boldsymbol{n}(P')$, respectively (Fig. 3.11). Then the projection of S on OX^3X^1 will be S_{31}, it follows that

$$\iiint_V \frac{\partial F_2}{\partial x^2} dx^1 dx^2 dx^3 = \iint_{S_{31}} \left(\int \frac{\partial F_2}{\partial x^2} dx^2\right) dS_{31} = \iint_{S_{31}} \left[F_2\left(P'\right) - F_2(P)\right] dS_{31}$$

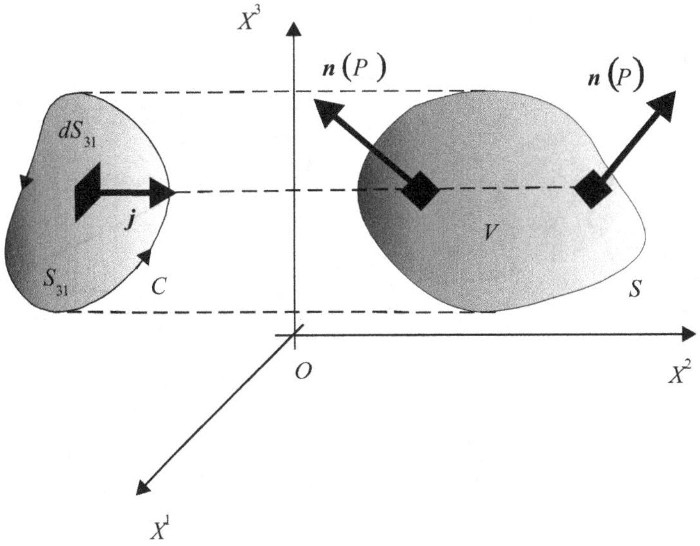

Fig. 3.11 Volume V with upward unit normal vector $\boldsymbol{n}$, which outline is a closed and smooth surface S

For the area element in this plane and with the direction cosines of the upward normal $\boldsymbol{n}(P)$ and $\boldsymbol{n}(P')$:

$$dS_{31} = dS(P)\,n_2(P) = -dS\left(P'\right)n_2\left(P'\right)$$

Substituting it results for the point P on S:

$$\iiint\limits_V \frac{\partial F_2}{\partial x^2} dx^1 dx^2 dx^3 = \iint\limits_S F_2(P)\,n_2(P)\,dS$$

In an analogous way, for the projections of S on the planes OX^1X^2 and OX^2X^3:

$$\iiint\limits_V \frac{\partial F_3}{\partial x^3} dx^1 dx^2 dx^3 = \iint\limits_S F_3(P)\,n_3(P)\,dS$$

$$\iiint\limits_V \frac{\partial F_1}{\partial x^1} dx^1 dx^2 dx^3 = \iint\limits_S F_1(P)\,n_1(P)\,dS$$

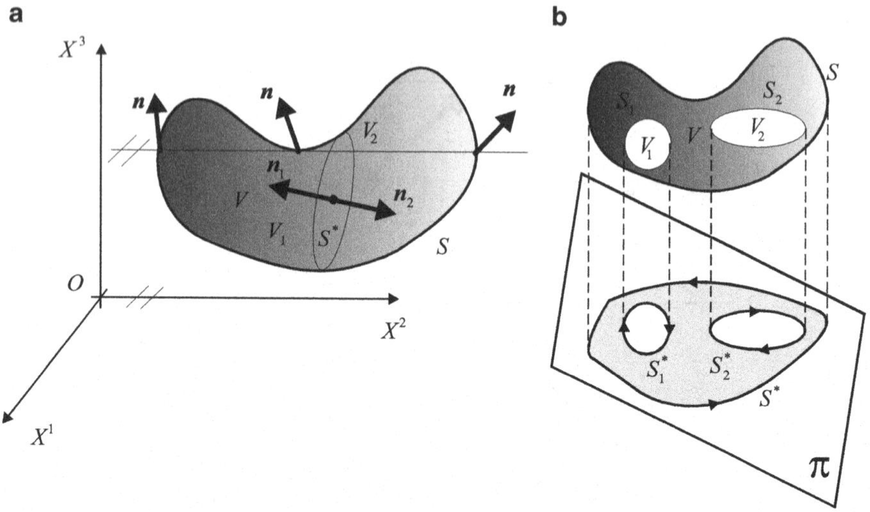

Fig. 3.12 Gauß–Ostrogradsky theorem: (**a**) volume cut in more than two points by a straight line parallel to a coordinated axis and (**b**) volume V with voids V_1 and V_2

The addition of these three expressions results

$$\iiint_V \left(\frac{\partial F_1}{\partial x^1} + \frac{\partial F_2}{\partial x^2} + \frac{\partial F_3}{\partial x^3}\right) dx^1 dx^2 dx^3 = \iint_S (F_1 n_1 + F_2 n_2 + F_3 n_3)\, dS \quad \text{Q.E.D.}$$

One of the premises adopted in the proof of the theorem of Carl Friedrich Gauß and Mikhail Vasilievich Ostrogradsky (1801–1861) is that the surface S has two sides, with a single upward and inward normal in each point. To admit that a straight line parallel to a coordinate axis intersects the volume V in only two points is not an essential hypothesis, for V can be divided into a number of subvolumes that separately fulfill the property admitted initially, allowing the Gauß–Ostrogradsky theorem to be applied to these subvolumes and adding the partial results obtained.

Figure 3.12a shows the volume V cut in more than two points by a straight line parallel to axis OX^2. The division of V into two volumes V_1 and V_2, separated by surface S^*, with opposite unit normal vector $\boldsymbol{n}_1$ and $\boldsymbol{n}_2$, being V_1 involved by $S_1 \cup S^*$, and V_2 by $S_2 \cup S^*$. Then the surface integrals referent to this part common to the two volumes cancel each other, remaining the integrals on the surfaces S_1 and S_2. This makes the applying of this theorem valid to volume V.

If the closed surface S that involves volume V is not smooth, it can be divided into a finite number of smooth surfaces represented by the functions $\phi(x^i)$, which have continuous partial derivatives, each one involving a subvolume. This procedure allows applying the Gauß–Ostrogradsky theorem to these subvolumes and adding the results obtained.

Figure 3.12b shows the volume V involved by the closed surface S with empty volumes V_1 and V_2, with which are involved, respectively, by the smooth closed surfaces S_1 and S_2. In this case it is necessary to cut the total volume and the volumes of the voids by a plane π and the surfaces of their outlines to project in this plane, originating the surfaces S^*, S_1^* and S_2^*, and then apply the Gauß–Ostrogradsky theorem considering these surfaces.

In vectorial notation with $\boldsymbol{F} = F_1\boldsymbol{i} + F_2\boldsymbol{j} + F_3\boldsymbol{k}$ results

$$\iiint_V \left(\frac{\partial F_1}{\partial x^1} + \frac{\partial F_2}{\partial x^2} + \frac{\partial F_3}{\partial x^3} \right) dx^1 dx^2 dx^3 = \iint_S \boldsymbol{F} \cdot \boldsymbol{n}\, dS \tag{3.4.2}$$

The volume integral that is present in Gauß–Ostrogradsky theorem also has a vectorial interpretation (item 4.3).

Chapter 4
Differential Operators

4.1 Scalar, Vectorial, and Tensorial Fields

4.1.1 Initial Notes

The study of the scalar, vectorial, and tensorial fields is strictly related with the differential operators which are applied to the analytic functions that represent these fields.

In this chapter the differential operators gradient, divergence, and curl will be defined, and their physical interpretations, as well as various fundamental relations with these operators, will be presented. These expressions form the mathematical backbone for the practical applications of the Field Theory. The conception of fields is of fundamental importance to the formulation of Tensor Calculus, and allows defining various concepts and deducing several expressions which form the framework for the study of the tensors contained in the tensorial space that defines the field.

The scalar, vectorial, and tensorial fields are formulations carried out on a point $x^i \in D$, the domain $D \subset E_N$ being an open subset and embedded in the ordinary geometric space. In these three kinds of fields the formulations are the functions smooth, continuous, and derivable.

By defining an arbitrary origin in the space E_N a biunivocal correspondence is determined for each domain with a variety, scalar, vector, or tensor that defines the kind of field.

The scalar and vectorial fields are particular cases of the tensorial fields. The behavior of a tensorial field is measured by the variation rate of the tensor in the points contained in the field. In the literature it is usual to call this variation rate as tensor derivative, which is incorrect, for what exists is the variation rate of the field defined by this variety, so the proper denomination is variation of the tensorial field. However, on account of being customary by use, the denomination **tensor derivative** will be used in this text to express this variation.

© Springer International Publishing Switzerland 2016

E. de Souza Sánchez Filho, *Tensor Calculus for Engineers and Physicists*,
DOI 10.1007/978-3-319-31520-1_4

4.1.2 Scalar Field

Let a scalar be associated to a point in the Euclidian space E_3 given by a function of the coordinates x^i, which is defined as $\phi = \phi(x^i;\ t), i = 1, 2, 3$, where t is the time in the instant in which the scalar is measured.

A scalar field is defined by the function of field $\phi(x^i;\ t)$, and if the time variable t is constant, the level surface of the field $\phi(x^i) = C$ is defined, where C is a constant. For several values of C there is a family of level surfaces, which characterize the field geometrically. These surfaces do not intersect, for if they did the function $\phi(x^i)$ would have to assume various values, which is impossible, for this function has only one value for each x^i.

As an example of scalar field there is a point in the interior of a reservoir containing liquid, where each particle of this fluid is submitted to a pressure proportional to the distance of this particle up to the top of the free surface. Another example is the field of temperatures due to a heat source, where the isotherms are spherical surfaces, with the temperature decreasing to the points farthest from this source.

4.1.3 Pseudoscalar Field

If the field function defines a pseudoscalar then the field is pseudoscalar. The specific mass $\rho(x^i)$ of the points of a solid body is an example of this sort of field.

4.1.4 Vectorial Field

If the vector $\boldsymbol{u}(x^i, t)$ is associated with the point $P(x^i)$ of the space E_N, then a vectorial field is defined, and if $t =$ constant the field is homogeneous. For the space E_3 which points are referenced to a Cartesian coordinate system there are three scalar functions of these points, $f_1(x^i), f_2(x^i), f_3(x^i), i = 1, 2, 3$, which express the field vector

$$\boldsymbol{u}(x^i) = f_1(x^i)\boldsymbol{i} + f_2(x^i)\boldsymbol{j} + f_3(x^i)\boldsymbol{k}$$

Field lines are defined for a vectorial field determined by the vectorial function $\boldsymbol{u}(x^i)$, in which for each point $P(x^i)$ the field vectors are collinear with the vectors tangents to these lines (Fig. 4.1). The condition of collinearity between the vector $\boldsymbol{u}(x^i)$ and the tangent vector $\boldsymbol{t}(x^i)$ is given by the nullity of cross product

$$\varepsilon_{ijk} u_j dx_k = 0$$

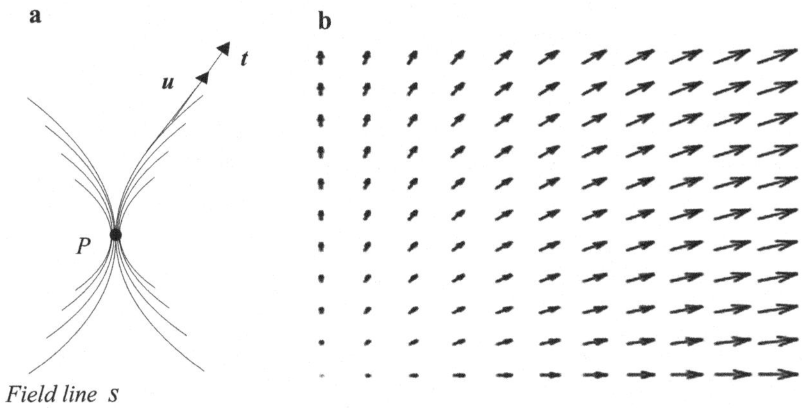

Fig. 4.1 Vectorial field: (**a**) field lines and field vector and (**b**) field vectors

Developing provides:

- $i = 1$

$$\varepsilon_{1jk}u_j dx_k = 0 \Rightarrow \varepsilon_{12k}u_2 dx_k = 0 \Rightarrow \varepsilon_{123}u_2 dx_3 = 0 \Rightarrow u_2 dx_3 = 0$$
$$\varepsilon_{13k}u_3 dx_2 = 0 \Rightarrow \varepsilon_{132}u_3 dx_2 = 0 \Rightarrow -u_3 dx_2 = 0$$

- $i = 2$

$$\varepsilon_{2jk}u_j dx_k = 0 \Rightarrow \varepsilon_{213}u_1 dx_3 = 0 \Rightarrow -u_1 dx_3 = 0$$
$$\varepsilon_{2jk}u_j dx_k = 0 \Rightarrow \varepsilon_{231}u_3 dx_1 = 0 \Rightarrow u_3 dx_1 = 0$$

- $i = 3$

$$\varepsilon_{3jk}u_j dx_k = 0 \Rightarrow \varepsilon_{312}u_1 dx_2 = 0 \Rightarrow u_1 dx_2 = 0$$
$$\varepsilon_{3jk}u_j dx_k = 0 \Rightarrow \varepsilon_{321}u_2 dx_1 = 0 \Rightarrow -u_2 dx_1 = 0$$

Thus the following system results

$$\begin{cases} u_3 dx_2 = u_2 dx_3 \\ u_1 dx_2 = u_2 dx_1 \\ u_3 dx_1 = u_1 dx_3 \end{cases} \Rightarrow \begin{cases} \dfrac{u_3}{u_2} = \dfrac{dx_3}{dx_2} \\ \dfrac{u_1}{u_2} = \dfrac{dx_1}{dx_2} \\ \dfrac{u_3}{u_1} = \dfrac{dx_3}{dx_1} \end{cases}$$

For a flat vectorial field the condition of collinearity between the field vector and the vector tangent to the field lines is given by

$$u_2 dx_1 - u_1 dx_2 = 0$$

The gravitational, the electric, and the magnetic are examples of vectorial fields.

4.1.5 Tensorial Field

The fundamental problem of Tensor Calculus is associated to the concept of tensorial field. If the tensorial field is fixed the tensor $\boldsymbol{T}(x^i)$ is a function of the coordinates of a point $P(x^i)$ situated in the tensorial space E_N. When this tensor is function of x^i and other parameters then the tensorial field is variable.

For tensor $\boldsymbol{T}(\overline{x}^i)$ which components are defined with respect to a curvilinear coordinates $\overline{X}^i$ which origin is the point $P(x^i)$, a few difficulties arise in the calculation of its derivatives, because the local coordinate system varies as a function of the point. The study of the tensorial fields in a tensorial space E_N, considering curvilinear local coordinate systems, is associated to the basis of this space.

Exercise 4.1 Calculate the parametric equation of the lines of the vectorial field $\boldsymbol{u} = -x_2 \boldsymbol{i} + x_1 \boldsymbol{j} + m\boldsymbol{k}$ that contains the point of coordinates $(1; 0; 0)$ where m is a scalar.

The differential equations of the field lines are

$$\frac{dx_1}{-x_2} = \frac{dx_2}{x_1} = \frac{dx_3}{m}$$

Following with the first two differential relations

$$x_1 dx_1 + x_2 dx_2 = 0 \Rightarrow \int x_1 dx_1 + \int x_2 dx_2 = C_0 \Rightarrow (x_1)^2 + (x_2)^2 = C_1; \quad C_1 > 0$$

and introducing a parameter t

$$x_1 = \sqrt{C_1} \cos t \quad x_2 = \sqrt{C_1} \sin t$$

so

$$dx_2 = \left(\sqrt{C_1} \cos t\right) dt$$

and with the differential relations

$$\frac{dx_2}{x_1} = \frac{dx_3}{m}$$

it follows that

$$\frac{(\sqrt{C_1}\cos t)dt}{\sqrt{C_1}\cos t} = \frac{dx_3}{m} \Rightarrow dx_3 = mdt \Rightarrow x_3 = mt + C_2$$

As the field line contains the point of coordinates (1; 0; 0), then

$$1 = \sqrt{C_1}\cos t \Rightarrow t = 2k\pi; \quad k = 0, \pm 1, \ldots$$

- $k = 0 \to C_1 = 1$;
- $k = 0 \to t = 0$ so $C_2 = 0$;
- *verifying that* $0 = mt + C_2$ *for* $t = 0$.

The parametric equations of the field lines represent a circular helix given by

$$x_1 = cos\ t;\ x_2 = \sin t;\ x_3 = mt$$

4.1.6 Circulation

Consider the field defined by the vectorial function $\boldsymbol{u}$ and the point $P(x^i)$ located on an open curve C, continuous by parts, smooth, and derivable, which is the hodograph of the position vector $\boldsymbol{r}(s)$, where s is the curvilinear abscissa, and admitting that this point varies in the interval $a \leq x^i \leq b$, then the line integral of this curve is given by

$$I = \int_a^b \boldsymbol{u} \cdot d\boldsymbol{r} \tag{4.1.1}$$

where line integral defines the circulation of the vectorial function $\boldsymbol{u}$ on the curve C.

Let $\boldsymbol{u} \cdot d\boldsymbol{r}$ be the differential total of the function $\phi(x^i)$, thus

$$I = \int_a^b d\phi(x^i) = \phi(x^i)\big|_a^b = \phi(b) - \phi(a) \tag{4.1.2}$$

The value of this integral depends only on the extreme points of the interval for which the function $\phi(x^i)$ is defined, regardless of the integration path. Expression (4.1.2) is a generalization of the fundamental theorem of the Integral Calculus.

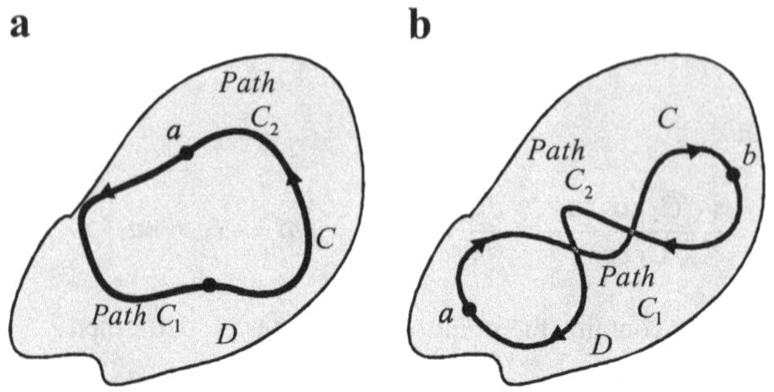

Fig. 4.2 Closed curve paths: (**a**) with no self-intersection and (**b**) with a finite number of self-intersections

For a closed curve the extreme points of this interval are coincident, which allows concluding that

$$\oint_C \boldsymbol{u} \cdot d\boldsymbol{r} = 0 \tag{4.1.3}$$

This expression defines the circulation of vector $\boldsymbol{u}$ along the closed curve C. The line integral of an open curve C defined in a certain interval will be independent of the path adopted in this calculation, and will be null if the curve C is closed.

Figure 4.2 shows two types of closed paths of curves defined in domain D—the single closed path in which there are no self-intersection points and the closed path with self-intersection points.

For the closed spatial curves defined in the domain D self-intersecting in a finite number of points, the line integral is calculated dividing the path in a finite number of single closed paths. For an infinite number of intersections, a reasonable approximation is obtained with the integrals on paths which are polygonal segments, using a limit process to achieve a finite number of intersections.

4.2 Gradient

In item 2.2 the gradient of a scalar field was defined, by a function $\phi(x^i)$ which differential is given by

$$d\phi = \frac{\partial \phi}{\partial x^i} dx^i \tag{4.2.1}$$

that is called differential parameter of the first order of Beltrami.

Expression (4.2.1) shows that there is no difference between the total differential $d\phi$ and the absolute differential, which allows adopting the notation $\phi_{,i}$ for the partial derivatives of this function. It was also shown that the gradient is a vector obtained by means of applying a vectorial operator to the scalar function $\phi(x^i)$, that with respect to a coordinate system X^i is given by

$$\nabla \cdots = \mathbf{e}^i \frac{\partial \cdots}{\partial x^i}$$

For a curvilinear coordinate system $\overline{X}^j$ by the chain rule

$$\frac{\partial \cdots}{\partial x^i} = \frac{\partial \overline{x}^j}{\partial x^i} \frac{\partial \cdots}{\partial \overline{x}^j}$$

and with the transformation law of unit vectors

$$\mathbf{e}^i = \mathbf{g}^k \frac{\partial x^i}{\partial \overline{x}^k}$$

it follows that

$$\nabla \cdots = \mathbf{e}^i \frac{\partial \cdots}{\partial x^i} = \mathbf{g}^k \frac{\partial x^i}{\partial \overline{x}^k} \frac{\partial \cdots}{\partial x^i} = \mathbf{g}^k \frac{\partial x^i}{\partial \overline{x}^k} \frac{\partial \overline{x}^j}{\partial x^i} \frac{\partial \cdots}{\partial \overline{x}^j} = \mathbf{g}^k \delta^j_k \frac{\partial \cdots}{\partial \overline{x}^j} = \mathbf{g}^k \frac{\partial \cdots}{\partial \overline{x}^k}$$

The several notations for the gradient vector are

$$\text{grad}\,\phi(x^i) = G(\phi) = \nabla(\phi) = \mathbf{g}^k \frac{\partial \phi(x^i)}{\partial x^k} = \mathbf{g}^k \phi_{,k} \tag{4.2.2}$$

This comma notation will hereafter be used in some special case for derivatives with respect to coordinates. The classic notation for the operator that defines the gradient of a scalar function is grad $\phi(x^i)$, and was introduced by Maxwell, Riemann, and Weber. The other notation is an inverted delta, called *nabla* operator (in Greek $\nu\alpha'\beta\lambda\alpha = harp$), *del*, *atled* (inverted delta), expressed as $\nabla \cdots = \mathbf{g}^k \frac{\partial \cdots}{\partial x^k}$. This notation was designed by Hamilton in 1837, initially was not used to represent the gradient of a function, but was written with the rotated delta $\lhd$, and represented symbolically the Laplace operator $\left(\frac{d}{dx}\right)^2 + \left(\frac{d}{dy}\right)^2 + \left(\frac{d}{dz}\right)^2$ that was already well used at the time, thereby the denomination Hamilton operator, or Hamiltonian operator. Another interpretation for the name *nabla* is due to Maxwell, who remarks that the rotated delta calls to cuneiform writing, which name in Hebrew would be this one.

The use of the *nabla* operator has many advantages with respect to the spelling *grad*, especially in the development of expressions, for it reinforces the tensorial characteristics of the gradient. This symbolic vector enables making the spelling for the differential operators uniform, and complies with the Vectorial Algebra rules.

Figure 4.3a shows schematically a scalar field defined by a function $\phi(x^i)$, where in a field line contained in the level surface the $\phi(x^i) = C$, being $C =$ constant, a point P is defined, and with an arbitrary origin O for the coordinate system X^i,

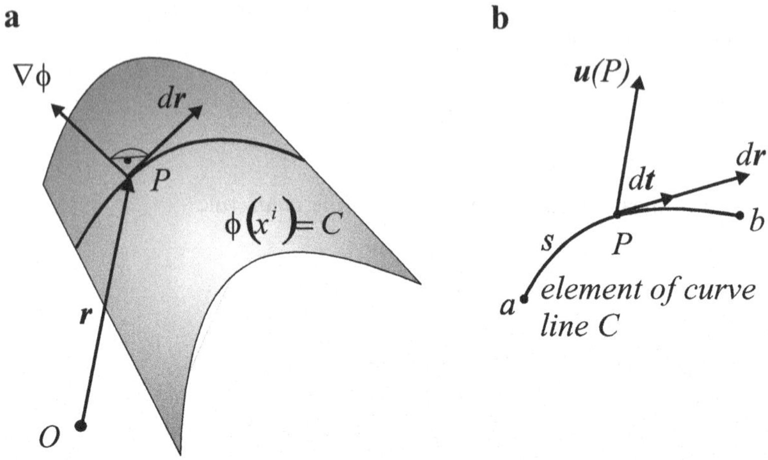

Fig. 4.3 Scalar field: (**a**) gradient and (**b**) line element

results in the position vector $\boldsymbol{r}$, which derivative is the vector $d\boldsymbol{r} = dx^k \boldsymbol{g}_k$ tangent to the field line, and denotes the line element. The differential of the scalar function that represents this field is given by the dot product

$$d\phi = d\boldsymbol{r} \cdot \nabla\phi = dx^k \boldsymbol{g}_k \cdot \boldsymbol{g}^\ell \frac{\partial \phi}{\partial x^\ell} = \delta_k^\ell \frac{\partial \phi}{\partial x^\ell} dx^k = \frac{\partial \phi}{\partial x^k} dx^k$$

The field represented by the gradient of a function is conservative, thus this function is called potential, or field gradient.

As the operator *nabla* is a vector, it is invariant for a change in the coordinate system, which can be proven admitting $\overline{\nabla}$ for a coordinate system $\overline{X}^j$, and ∇ for a coordinate system X^i, so by means of the vectors transformation law

$$\frac{\partial \cdots}{\partial \overline{x}^k} = \left(\frac{\partial \cdots}{\partial \overline{x}^\ell}\right) \frac{\partial \overline{x}^\ell}{\partial \overline{x}^k}$$

$$\overline{\nabla} \cdots = \overline{\boldsymbol{g}}_k \frac{\partial \cdots}{\partial \overline{x}^k} = \frac{\partial \overline{x}^k}{\partial x^m} \frac{\partial x^\ell}{\partial \overline{x}^k} \boldsymbol{g}^m \frac{\partial \cdots}{\partial x^\ell} = \delta_m^\ell \boldsymbol{g}^m \frac{\partial \cdots}{\partial x^\ell} = \boldsymbol{g}^\ell \frac{\partial \cdots}{\partial x^\ell} = \nabla \cdots$$

therefore the operator ∇ is a vector.

The gradient for the function $\phi = x^k$, where x^k represents a coordinate of the referential system, is given by $\nabla\phi = \frac{\partial x^k}{\partial x^i} \boldsymbol{g}^i$, then the gradient is the unit vector for the coordinate axis.

For the product of two scalar functions $\phi(x^i)$ and $\psi(x^i)$ the result is

$$\operatorname{grad}(\phi\psi) = \nabla(\phi\psi) = \boldsymbol{g}^k \frac{\partial(\phi\psi)}{\partial x^k} = \boldsymbol{g}^k \frac{\partial\phi}{\partial x^k}\psi + \phi\boldsymbol{g}^k \frac{\partial\psi}{\partial x^k} = (\nabla\phi)\psi + \phi(\nabla\psi)$$

In this demonstration it is observed that the *nabla* operator acts on each parcel of the expression in a distinct way, maintaining a parcel variable and the other constant. If it comes before the parcel it acts with a variable, if it comes after, it acts as a constant.

The gradient can be defined by means of the Gauß-Ostrogradsky theorem. Let the field be determined by the vectorial function $\boldsymbol{u} = \boldsymbol{v}\phi(x^i)$, $\boldsymbol{v}$ being a constant vector, then

$$\frac{\partial u^1}{\partial x^i} = \boldsymbol{v} \cdot \frac{\partial \phi(x^i)}{\partial x^i}$$

$$\frac{\partial u^1}{\partial x^1} + \frac{\partial u^2}{\partial x^2} + \frac{\partial u^3}{\partial x^3} = \boldsymbol{v} \cdot \left(\frac{\partial\phi}{\partial x^1} + \frac{\partial\phi}{\partial x^2} + \frac{\partial\phi}{\partial x^3} \right) = \boldsymbol{v} \cdot \nabla\phi$$

it follows that

$$\iiint_V \left(\frac{\partial u^1}{\partial x^1} + \frac{\partial u^2}{\partial x^2} + \frac{\partial u^3}{\partial x^3} \right) dV = \iint_S \boldsymbol{u} \cdot \boldsymbol{n}\, dS$$

$$\boldsymbol{v} \cdot \iiint_V \nabla\phi\, dV = \boldsymbol{v} \cdot \iint_S \boldsymbol{u} \cdot \boldsymbol{n}\, dS$$

$$\boldsymbol{v} \cdot \left(\iiint_V \nabla\phi\, dV - \iint_S \boldsymbol{u} \cdot \boldsymbol{n}\, dS \right) = 0$$

For the point P in the scalar field $\phi(x^i)$ contained in an elementary volume, and with the component $\frac{\partial\phi}{\partial x^1}$ of $\nabla\phi$ by the mean value theorem of the Integral Calculus

$$\iiint_V \frac{\partial\phi}{\partial x^1} dV = \left(\frac{\partial\phi}{\partial x^1} \right)_{P^*} V$$

where P^* is the midpoint of volume V.

Applying the Gauß-Ostrogradsky theorem

$$\left(\frac{\partial\phi}{\partial x^1} \right)_{P^*} = \frac{1}{V} \iint_S \phi n_1 dS$$

where n_1 is the direction cosine of the angle between the upward normal unit vector $\boldsymbol{n}$ and the coordinate axis OX^1.

When the point P approaches the point P^*, the volume V and the surface S also come close to P, and with continuous function $\phi(x^i)$ and its partial derivatives it results in

$$\left(\frac{\partial \phi}{\partial x^1}\right)_P = \lim_{V \to 0} \frac{1}{V} \iint_S \phi n_1 dS$$

For the coordinates x^2 and x^3 the result with analogous formulations is

$$\left(\frac{\partial \phi}{\partial x^2}\right)_P = \lim_{V \to 0} \frac{1}{V} \iint_S \phi n_2 dS \quad \left(\frac{\partial \phi}{\partial x^3}\right)_P = \lim_{V \to 0} \frac{1}{V} \iint_S \phi n_3 dS$$

If these limits exist, the gradient of the scalar function $\phi(x^i)$ in point P is determined by

$$\nabla \phi(x^i) = \lim_{V \to 0} \frac{1}{V} \iint_S \phi(x^i) \, \boldsymbol{n} dS \tag{4.2.3}$$

that is valid for any coordinate system, which shows that the gradient is independent of the coordinate system.

4.2.1 Norm of the Gradient

The norm of the gradient is given by

$$\|\nabla \phi\| = \sqrt{\|\nabla \phi \cdot \nabla \phi\|} = \sqrt{g_{ij} \frac{\partial \phi}{\partial x^i} \frac{\partial \phi}{\partial x^j}} \tag{4.2.4}$$

A few authors use the spelling $\Delta_1 \phi = g_{ij} \frac{\partial \phi}{\partial x^i} \frac{\partial \phi}{\partial x^j}$ to designate the first differential parameter of Beltrami.

For the orthogonal coordinate systems $g_{ij} = \delta_{ij}$:

$$\|\nabla \phi\| = \sqrt{\left(\frac{\partial \phi}{\partial x^i}\right)^2} \tag{4.2.5}$$

4.2.2 Orthogonal Coordinate Systems

Consider the point $P(x^i)$ be coincident with the origin of the curvilinear orthogonal coordinate $\overline{X}^j$, and $\boldsymbol{r}$ its position vector with respect to the Cartesian coordinate X^j. Rewriting expression (2.3.4) the result for this vector's differential is

$$d\boldsymbol{r} = \frac{\partial \boldsymbol{r}}{\partial \overline{\boldsymbol{x}}^i} d\overline{x}^i$$

with

$$\frac{\partial \boldsymbol{r}}{\partial \overline{\boldsymbol{x}}^i} = h_i \boldsymbol{g}_i$$

then

$$h_i = \left\| \frac{\partial \boldsymbol{r}}{\partial \overline{\boldsymbol{x}}^i} \right\|$$

where $\boldsymbol{g}_i$ are the unit vectors of the coordinate system $\overline{X}^j$, and $h_i = \sqrt{g_{(ii)}}$ are the metric tensor coefficients that represent scale factors of the magnitudes of the vectors tangents to the curves of this coordinate system, where the indexes in parenthesis do not indicate summation. They are called Lamé coefficients for orthogonal coordinate systems.

The differential of the scalar function $\phi(x^i)$ is given by

$$d\phi = \frac{\partial \phi}{\partial \overline{\boldsymbol{x}}^i} d\overline{x}^i$$

but

$$d\phi = \nabla \phi \cdot d\boldsymbol{r}$$

$$d\boldsymbol{r} = h_i d\overline{x}^i \boldsymbol{g}_i$$

then

$$d\phi = \left(\frac{\partial \phi}{\partial \overline{\boldsymbol{x}}^i} d\overline{x}^i \right) \boldsymbol{g}^i \cdot d\boldsymbol{r}$$

whereby

$$\nabla \phi = \left(\frac{1}{h_i} \frac{\partial \phi}{\partial x^i} \right) \boldsymbol{g}^i \tag{4.2.6}$$

that provides the components of the gradient vector in a curvilinear orthogonal coordinate system. The physical components of the vector $\nabla\phi$ are given by

$$(\nabla\phi)^* = \frac{1}{h_i}\frac{\partial\phi}{\partial x^i} \tag{4.2.7}$$

4.2.3 Directional Derivative of the Gradient

Figure 4.3b shows the differential element of the line contained in the level surface $\phi(x^i) = C$ and its tangent unit vector $\boldsymbol{t}$, collinear with vector $d\boldsymbol{r}$, which allows writing for the line element $d\boldsymbol{r} = \boldsymbol{t}\,ds$.

The geometric interpretation of the gradient of a scalar field is given by the dot product

$$\frac{d\phi}{ds} = \boldsymbol{t}\cdot\nabla\phi \tag{4.2.8}$$

that defines the field directional derivative. The symbol $\nabla\phi$ characterizes the field, and unit vector $\boldsymbol{t}$ being independent of the function $\phi(x^i)$, this indicates the direction in which the derivative is calculated. If $\nabla\phi(x^i)$ exists in the point P, defined by the field $\phi(x^i)$, it will be possible to calculate the directional derivative of this function in all the directions of the field. Then the field $\phi(x^i)$ is non-homogeneous.

Let α be the angle between the two vectors from expression (4.2.8), the dot product provides

$$\frac{d\phi}{ds} = \|\boldsymbol{t}\|\,\|\nabla\phi\|\cos\alpha = \|\nabla\phi\|\cos\alpha$$

As $\phi =$ constant, it results in $d\phi = 0$, so $d\boldsymbol{r}\cdot\nabla\phi = 0$, then the vector $\nabla\phi$ is perpendicular to the vector $d\boldsymbol{r}$.

The variation rate of the field defined by the function ϕ is maximum in the direction of $\nabla\phi$, for $\alpha = 0$ results in $\cos\alpha = 1$, then

$$\left(\frac{d\phi}{dn}\right)_{\max} = \|\nabla\phi\| > 0$$

The directional derivative is calculated in the direction of the unit normal vector $\boldsymbol{n}$ to the level surface $\phi(x^i) = C$ (Fig. 4.4), thus

$$\nabla\phi = \frac{d\phi}{dn}\boldsymbol{n} \tag{4.2.9}$$

Fig. 4.4 Interpretation of the gradient as a vector normal to surface

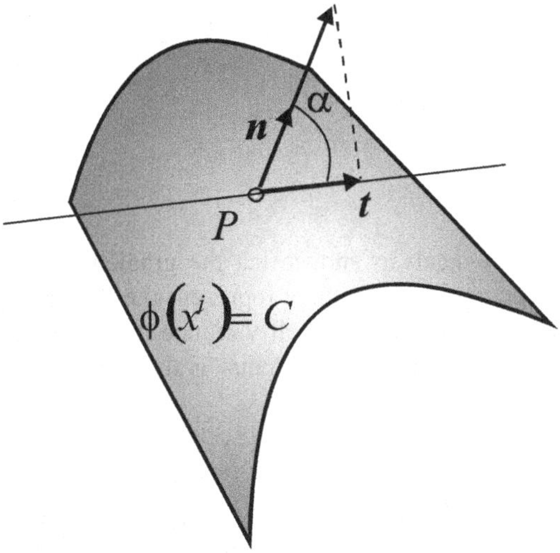

4.2.4 Dyadic Product

The *nabla* operator applied to a vectorial function $\boldsymbol{u} = u^k \boldsymbol{g}_k$ results in the dyadic product

$$\nabla \otimes \boldsymbol{u} = \nabla \boldsymbol{u} = \boldsymbol{T}$$

$$\boldsymbol{T} = \boldsymbol{g}^i \frac{\partial \boldsymbol{u}}{\partial \boldsymbol{x}^i} = \boldsymbol{g}^i \frac{\partial \left(u^k \boldsymbol{g}_k \right)}{\partial x^i} = \boldsymbol{g}^i \left(\frac{\partial u^k}{\partial x^i} \boldsymbol{g}_k + u^k \frac{\partial \boldsymbol{g}_k}{\partial x^i} \right)$$

and with expression (2.3.10)

$$\frac{\partial \boldsymbol{g}_k}{\partial x^i} = \Gamma^m_{ki} \boldsymbol{g}_m$$

it follows that

$$\boldsymbol{T} = \boldsymbol{g}^i \left(\frac{\partial u^k}{\partial x^i} \boldsymbol{g}_k + u^k \Gamma^m_{ki} \boldsymbol{g}_m \right)$$

Interchanging the indexes $m \leftrightarrow k$ in the second member to the right

$$\boldsymbol{T} = \left(\frac{\partial u^k}{\partial x^i} + u^m \Gamma^k_{mi} \right) \boldsymbol{g}^i \otimes \boldsymbol{g}_k$$

The covariant derivative of a contravariant vector results in a variance tensor $(1, 1)$

$$T_i^k = \partial_i u^k = \frac{\partial u^k}{\partial x^i} + u^m \Gamma_{mi}^k$$

then

$$\boldsymbol{T} = T_i^k \boldsymbol{g}^i \otimes \boldsymbol{g}_k \tag{4.2.10}$$

This analysis shows that the gradient and the covariant derivative represent a same concept, i.e., they represent the derivative of a scalar, vectorial, or tensorial function, increasing their variance from one unit.

Formulating an analogous analysis for a covariant vector

$$\boldsymbol{T} = \nabla \otimes \boldsymbol{u} = \nabla \boldsymbol{u} = \boldsymbol{g}^i \frac{\partial \boldsymbol{u}}{\partial \boldsymbol{x}^i}$$

$$\boldsymbol{T} = \boldsymbol{g}^i \frac{\partial \left(u_k \boldsymbol{g}^k\right)}{\partial x^i} = \boldsymbol{g}^i \left(\frac{\partial u_k}{\partial x^i} \boldsymbol{g}^k + u_k \frac{\partial \boldsymbol{g}^k}{\partial x^i}\right) = \boldsymbol{g}^i \left(\frac{\partial u_k}{\partial x^i} \boldsymbol{g}^k - u_m \Gamma_{ki}^m \boldsymbol{g}^m\right)$$

$$= \boldsymbol{g}^i \left(\frac{\partial u_k}{\partial x^i} - u_k \Gamma_{mi}^k\right) \boldsymbol{g}^i \otimes \boldsymbol{g}^k$$

then

$$\nabla \otimes \boldsymbol{u} = \nabla \boldsymbol{u} = (\partial_i u_k) \boldsymbol{g}^i \otimes \boldsymbol{g}^k = T_{ik} \boldsymbol{g}^i \otimes \boldsymbol{g}^k \tag{4.2.11}$$

The differential of a vectorial field is a vector, for the differential of vector $\boldsymbol{u}$:

$$d\boldsymbol{u} = d\boldsymbol{r} \cdot \nabla \otimes \boldsymbol{u} = \left(dx^i \boldsymbol{g}_i\right)\left(\partial_i u_k \boldsymbol{g}^i \otimes \boldsymbol{g}^k\right) = dx^i \partial_i u_k \boldsymbol{g}_i \cdot \boldsymbol{g}^i \otimes \boldsymbol{g}^k$$

whereby

$$d\boldsymbol{u} = dx^i \partial_i u_k \boldsymbol{g}^k \tag{4.2.12}$$

For the fields vectorial there is the directional derivative

$$\frac{d\boldsymbol{u}}{ds} = \boldsymbol{t} \cdot \nabla \otimes \boldsymbol{u} \tag{4.2.13}$$

The same considerations formulated for the directional derivative of a scalar field $\phi(x^i)$ are applicable to the vectorial fields. The physical components for the gradient of vector $\boldsymbol{u}^*$ are obtained considering the physical components of the second-order tensor.

4.2.5 Gradient of a Second-Order Tensor

The generalization of the concept of gradient for an arbitrary tensorial field is immediate. For a coordinate system X^i with unit vector $\boldsymbol{g}^\ell$, and with the tensor $\boldsymbol{T}$ defining the tensorial field

$$\nabla \otimes \boldsymbol{T} = \operatorname{grad} \boldsymbol{T} = \boldsymbol{g}^\ell \frac{\partial \boldsymbol{T}}{\partial x^\ell} \tag{4.2.14}$$

Then the gradient of a tensor $\boldsymbol{T}$ is calculated by *nabla* operator $\nabla = \boldsymbol{g}^\ell \frac{\partial \ldots}{\partial x^\ell}$ applying to this tensor. This operator is defined for a contravariant base.

The tensorial product of the *nabla* operator by the second-order tensor $\boldsymbol{T}$ is given by

$$\nabla \otimes \boldsymbol{T} = \nabla \boldsymbol{T} = \boldsymbol{g}^m \frac{\partial \left(T^{ki} \boldsymbol{g}_k \otimes \boldsymbol{g}_i\right)}{\partial x^m} = \boldsymbol{g}^m \left(\frac{\partial T^{ki}}{\partial x^m} \boldsymbol{g}_k \otimes \boldsymbol{g}_i + T^{ki} \frac{\partial \boldsymbol{g}_k}{\partial x^m} \otimes \boldsymbol{g}_i + T^{ki} \boldsymbol{g}_k \frac{\partial \boldsymbol{g}_i}{\partial x^m}\right)$$

and with the expressions

$$\frac{\partial \boldsymbol{g}_k}{\partial x^m} = \Gamma^p_{km} \boldsymbol{g}_p \qquad \frac{\partial \boldsymbol{g}_i}{\partial x^m} = \Gamma^p_{im} \boldsymbol{g}_p$$

it follows that

$$\nabla \otimes \boldsymbol{T} = \boldsymbol{g}^m \left(\frac{\partial T^{ki}}{\partial x^m} \boldsymbol{g}_k \otimes \boldsymbol{g}_i + T^{ki} \Gamma^p_{km} \boldsymbol{g}_p \otimes \boldsymbol{g}_i + T^{ki} \Gamma^p_{im} \boldsymbol{g}_k \otimes \boldsymbol{g}_p\right)$$

Interchanging the indexes $p \leftrightarrow k$ in the second term to the right, and the indexes $p \leftrightarrow i$ in the third term

$$\nabla \otimes \boldsymbol{T} = \left(\frac{\partial T^{ki}}{\partial x^m} + T^{ki} \Gamma^p_{km} + T^{ki} \Gamma^p_{im}\right) \boldsymbol{g}^m \otimes \boldsymbol{g}_k \otimes \boldsymbol{g}_i$$

this expression becomes

$$\nabla \otimes \boldsymbol{T} = \partial_m T^{ki} \boldsymbol{g}^m \otimes \boldsymbol{g}_k \otimes \boldsymbol{g}_p \tag{4.2.15}$$

and shows that the gradient of a second-order tensor is a variance tensor $(1, 2)$. The other components of the gradient of tensor $\boldsymbol{T}$ are given by expressions (2.5.18) and (2.5.21).

The generalization of the definition of the gradient of a third-order tensor $\boldsymbol{T}$ is immediate. The components of the fourth-order tensor that result from applying this operator to tensor $\boldsymbol{T}$ being given by their covariant derivatives, for instance, for tensor T_{ijk}:

$$\nabla \otimes \boldsymbol{T} = \left(\frac{\partial T_{ijk}}{\partial x^{\ell}} - T_{mjk}\Gamma_{i\ell}^{m} - T_{imk}\Gamma_{j\ell}^{m} - T_{ijm}\Gamma_{k\ell}^{m} \right) \boldsymbol{g}^{\ell} \otimes \boldsymbol{g}^{i} \otimes \boldsymbol{g}^{j} \otimes \boldsymbol{g}^{k}$$

For a tensor $\boldsymbol{T}$ of order p the variety $\nabla \otimes \boldsymbol{T}$ is a tensor of order $(p+1)$.

The differential of a tensorial field is a tensor, is gives by then the differential of the second-order tensor $\boldsymbol{T}$ thus:

$$\begin{aligned} d\boldsymbol{T} &= d\boldsymbol{r} \cdot \nabla \otimes \boldsymbol{T} = dx^j \boldsymbol{g}_j \cdot \partial_m T^{ki} \boldsymbol{g}^m \otimes \boldsymbol{g}_k \otimes \boldsymbol{g}_p = dx^j \partial_m T^{ki} \boldsymbol{g}_j \cdot \boldsymbol{g}^m \otimes \boldsymbol{g}_k \otimes \boldsymbol{g}_p \\ &= dx^j \partial_m T^{ki} \delta_j^m \boldsymbol{g}_k \otimes \boldsymbol{g}_p \end{aligned}$$

whereby

$$d\boldsymbol{T} = dx^j \partial_j T^{kj} \boldsymbol{g}_k \otimes \boldsymbol{g}_p \tag{4.2.16}$$

The physical components for the gradient of tensor $\boldsymbol{T}^*$ are obtained considering the physical components of the tensor of the third order.

The same considerations formulated for the directional derivative of a scalar field $\phi(x^i)$ and of a vectorial field are applicable to the tensorial fields, where

$$\frac{d\boldsymbol{T}}{ds} = \boldsymbol{t} \cdot \nabla \otimes \boldsymbol{T} \tag{4.2.17}$$

4.2.6 Gradient Properties

The ascertaining achieved in the previous paragraphs allow establishing the conditions so that a vector is gradient of a scalar function, for if the vector $\boldsymbol{u}(\boldsymbol{x}^i)$ defined in a single or multiply connected region, and if the line integral $\int_C \boldsymbol{u} \cdot d\boldsymbol{r}$ is independent of the path, then a scalar function $\phi(x^i)$ exists and fulfills the condition $\boldsymbol{u} = \nabla\phi(x^i)$ in all of this region of the space.

The gradient operator applied to the addition of two tensors provides

$$\nabla \otimes (\boldsymbol{T} + \boldsymbol{A}) = \boldsymbol{g}^{\ell} \frac{\partial(\boldsymbol{T} + \boldsymbol{A})}{\partial x^{\ell}} = \boldsymbol{g}^{\ell} \frac{\partial \boldsymbol{T}}{\partial x^{\ell}} + \boldsymbol{g}^{\ell} \frac{\partial \boldsymbol{A}}{\partial x^{\ell}} = \nabla \otimes \boldsymbol{T} + \nabla \otimes \boldsymbol{A}$$

The applying of this operator to the multiplication of the scalar m by the tensor $\boldsymbol{T}$ provides

$$\nabla \otimes (m\boldsymbol{T}) = \boldsymbol{g}^{\ell} \frac{\partial(m\boldsymbol{T})}{\partial x^{\ell}} = m\boldsymbol{g}^{\ell} \frac{\partial \boldsymbol{T}}{\partial x^{\ell}} = m\nabla \otimes \boldsymbol{T}$$

These two demonstrations prove that the gradient is a linear operator, which is already implicit, because it is a vector.

Exercise 4.2 Calculate: (a) $\boldsymbol{v} \cdot \nabla \boldsymbol{u}$; (b) $\nabla(\boldsymbol{u} \cdot \boldsymbol{v})$.

(a) The gradient for the field defined by a vectorial function is given by

$$\nabla \boldsymbol{u} = \partial_i u_k \boldsymbol{g}^i \otimes \boldsymbol{g}^k$$

With

$$\boldsymbol{v} = v^\ell \boldsymbol{g}_\ell$$

it follows that

$$\boldsymbol{v} \cdot \nabla \boldsymbol{u} = v^\ell \boldsymbol{g}_\ell \cdot \partial_i u_k \boldsymbol{g}^i \otimes \boldsymbol{g}^k = v^\ell \partial_i u_k \boldsymbol{g}_\ell \cdot \boldsymbol{g}^i \otimes \boldsymbol{g}^k = v^\ell \partial_i u_k \delta^i_\ell \boldsymbol{g}^k = v^i (\partial_i u_k) \boldsymbol{g}^k$$

Thus for the Cartesian coordinates

$$\boldsymbol{v} \cdot \nabla \boldsymbol{u} = v_i \frac{(\partial u_k)}{\partial x^i} \boldsymbol{g}_k$$

(b) The gradient of the scalar field represented by the dot product of the vectorial functions $\boldsymbol{u}$ and $\boldsymbol{v}$ is given by

$$\nabla(\boldsymbol{u} \cdot \boldsymbol{v}) = \boldsymbol{g}^k \frac{\partial (u^i v_i)}{\partial x^k} = \left[(\partial_k u^i) \, v_i + u^i (\partial_k v_i) \right] \boldsymbol{g}^k$$

and with the expressions

$$\partial_k u^i = \frac{\partial u^i}{\partial x^k} + u^m \Gamma^i_{mk} \quad \partial_k v_i = \frac{\partial v_i}{\partial x^k} - v_m \Gamma^m_{ik}$$

it follows that

$$\nabla(\boldsymbol{u} \cdot \boldsymbol{v}) = \left(\frac{\partial u^i}{\partial x^k} v_i + u^m \Gamma^i_{mk} v_i + u^i \frac{\partial v_i}{\partial x^k} - u^i v_m \Gamma^m_{ik} \right) \boldsymbol{g}^k$$

In the last term in parenthesis interchanging the indexes $i \leftrightarrow m$:

$$\nabla(\boldsymbol{u} \cdot \boldsymbol{v}) = \left(\frac{\partial u^i}{\partial x^k} v_i + u^m \Gamma^i_{mk} v_i + u^i \frac{\partial v_i}{\partial x^k} - u^m v_i \Gamma^i_{mk} \right) \boldsymbol{g}^k$$

then

$$\nabla(\boldsymbol{u} \cdot \boldsymbol{v}) = \left(\frac{\partial u^i}{\partial x^k} v_i + u^i \frac{\partial v_i}{\partial x^k} \right) \boldsymbol{g}^k$$

For the Cartesian coordinates

$$\nabla(\boldsymbol{u}\cdot\boldsymbol{v}) = \left(\frac{\partial u_i}{\partial x^k}v_i + u_i\frac{\partial v_i}{\partial x^k}\right)\boldsymbol{g}_k$$

Another way of expressing $\nabla(\boldsymbol{u}\cdot\boldsymbol{v})$ is to use the expression between the covariant derivative of a covariant vector and the covariant derivative of a contravariant vector, which is given by

$$\partial_k u_m = g_{im}\partial_k u^i$$

The multiplying of this expression by g^{in} provides

$$g^{in}\partial_k u_m = g^{in}g_{im}\partial_k u^i = \delta^n_m\partial_k u^i \Rightarrow \partial_k u^i = g^{im}\partial_k u_m{}^m$$

whereby

$$v_i\partial_k u^i = v_i g^{im}\partial_k u_m = v^m\partial_k u_m$$

Replacing the dummy index $m \to i$:

$$v_i\partial_k u^i = v^i\partial_k u_i$$

and by substitution

$$\nabla(\boldsymbol{u}\cdot\boldsymbol{v}) = \left(v^i\partial_k u_i + u^i\partial_k v_i\right)\boldsymbol{g}^k$$

Adding and subtracting the terms $v^i\partial_i u_k$ and $u^i\partial_i v_k$

$$\nabla(\boldsymbol{u}\cdot\boldsymbol{v}) = \left[v^i(\partial_k u_i - \partial_i u_k) + v^i\partial_i u_k + u^i(\partial_k v_i - \partial_i v_k) + u^i\partial_i v_k\right]\boldsymbol{g}^k$$

and with the expressions

$$\boldsymbol{v}\times(\nabla\times\boldsymbol{u}) = v^i(\partial_k u_i - \partial_i u_k)\boldsymbol{g}^k \quad \boldsymbol{u}\times(\nabla\times\boldsymbol{v}) = u^i(\partial_k v_i - \partial_i v_k)\boldsymbol{g}^k$$

$$\boldsymbol{v}\cdot\nabla\boldsymbol{u} = v^i(\partial_i u_k)\boldsymbol{g}^k \qquad \boldsymbol{u}\cdot\nabla\boldsymbol{v} = u^i(\partial_i v_k)\boldsymbol{g}^k$$

it results in

$$\nabla(\boldsymbol{u}\cdot\boldsymbol{v}) = \boldsymbol{v}\cdot\nabla\boldsymbol{u} + \boldsymbol{v}\times(\nabla\times\boldsymbol{u}) + \boldsymbol{u}\cdot\nabla\boldsymbol{v} + \boldsymbol{u}\times(\nabla\times\boldsymbol{v})$$

For the particular case in which $\boldsymbol{u} = \boldsymbol{v}$:

$$\boldsymbol{v}\cdot\nabla\boldsymbol{v} = \frac{1}{2}\nabla v^2 - \boldsymbol{v}\times(\nabla\times\boldsymbol{v})$$

Exercise 4.3 Calculate the gradient of the scalar function $\phi(x^i)$ expressed in cylindrical coordinates.

For the cylindrical coordinates $\sqrt{g_{11}} = \sqrt{g_{33}} = 1, \sqrt{g_{22}} = r$, then

$$\nabla\phi = \frac{1}{\sqrt{g_{ii}}}\frac{\partial\phi}{\partial x^i}g_k$$

it follows that

$$\nabla\phi = \frac{\partial\phi}{\partial r}g_r + \frac{1}{r}\frac{\partial\phi}{\partial\theta}g_\theta + \frac{\partial\phi}{\partial z}g_z$$

Exercise 4.4 Calculate the gradient of the scalar function $\phi(x^1)$ expressed in spherical coordinates.

For the spherical coordinates $\sqrt{g_{11}} = 1, \sqrt{g_{22}} = r, \sqrt{g_{33}} = r\sin\phi$, then

$$\nabla\phi = \frac{\partial\phi}{\partial r}g_r + \frac{1}{r}\frac{\partial\phi}{\partial\phi}g_\phi + \frac{1}{r\,\sin\,\phi}\frac{\partial\phi}{\partial x\theta}g_\theta$$

Exercise 4.5 Show that $\dfrac{\partial^2\phi(x^i)}{\partial x^i\partial x^j}$ is a second-order tensor, where $\phi(x^i)$ is a scalar function.

Putting

$$T_{ij} = \frac{\partial^2\phi}{\partial x^i\partial x^j} = \phi_{,ij}$$

for the coordinate system $\overline{X}^i$

$$\frac{\partial^2\phi}{\partial\overline{x}^i\partial\overline{x}^j} = \frac{\partial\phi}{\partial\overline{x}^i}\left(\frac{\partial\phi}{\partial x^k}\frac{\partial x^k}{\partial\overline{x}^k}\right) = \frac{\partial\phi}{\partial x^m}\left[\frac{\partial\phi}{\partial x^k}\left(\frac{\partial x^k}{\partial\overline{x}^j}\right)\right]\frac{\partial x^m}{\partial\overline{x}^i} = \left(\frac{\partial x^k}{\partial\overline{x}^j}\frac{\partial x^m}{\partial\overline{x}^i}\right)\frac{\partial^2\phi}{\partial\overline{x}^m\partial\overline{x}^k}$$

This transformation law proves that $\dfrac{\partial^2\phi(x^i)}{\partial x^i\partial x^j}$ is a second-order tensor.

Exercise 4.6 Calculate the directional derivative of the function $\phi(x, y) = x^2 + y^2 - 3xy^3$, at the point P(1; 2) in the direction of vector $\boldsymbol{u} = \frac{1}{2}\boldsymbol{e_1} + \frac{\sqrt{3}}{2}\boldsymbol{e_2}$, being $\boldsymbol{e_1}(1; 0), \boldsymbol{e_2}(0; 1)$.

The gradient of the scalar field is given by

$$\nabla\phi = \left(2x - 3y^3\right)\boldsymbol{e_1} + \left(2y - 9xy^2\right)\boldsymbol{e_2}$$

and in point P(1; 2)

$$\nabla\phi = -22\boldsymbol{e_1} - 32\boldsymbol{e_2}$$

The vector $\boldsymbol{u}$ is a unit vector, for $\frac{\boldsymbol{u}}{\|\boldsymbol{u}\|} = \frac{1}{2}\boldsymbol{e}_1 + \frac{\sqrt{3}}{2}\boldsymbol{e}_2$, whereby it follows that for the directional derivative

$$\nabla\phi\cdot\boldsymbol{u} = -22\times\frac{1}{2} - 32\times\frac{\sqrt{3}}{2} = -11 - 16\sqrt{3}$$

4.3 Divergence

The analysis of the flow magnitude for the vectorial field $\boldsymbol{u}$ that passes through the volume V involved by the closed surface S with respect to this volume leads to the conception of a differential operator (Fig. 4.5a). In volume V let the elementary parallelepiped with sides dx^1, dx^2, dx^3, and the vectorial function $\boldsymbol{u}$ continuous with continuous partial derivatives (Fig. 4.5b).

The study of the flow of $\boldsymbol{u}$ that passes through the volume V with respect to this volume is carried out considering the point $P(x^1; x^2; x^3)$ at the center of an elementary parallelepiped (Fig. 4.5). For the face dS_1, with upward normal unit vector $\boldsymbol{n}$ (1; 0; 0), the component of $\boldsymbol{u}$ in the direction of axis OX^1 is given by

$$\boldsymbol{u}\cdot\boldsymbol{n} = u^1$$

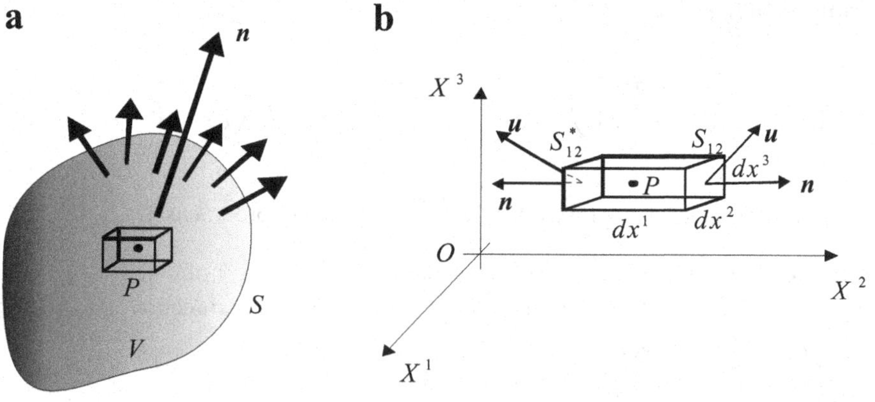

Fig. 4.5 Flow of the vectorial function $\boldsymbol{u}$: (**a**) that passes through the volume V and (**b**) in the elementary parallelepiped

The center of the elementary area $dS_1 = dx^1 dx^2$ has coordinates $\left(x^1 + \frac{dx^1}{2};\right.$ $\left.dx^2;\, dx^3\right)$, whereby it follows that for the surface integral in this face of the elementary parallelepiped dV

$$\iint_{dS_1} \boldsymbol{u} \cdot \boldsymbol{n} dS \cong u^1 \left(x^1 + \frac{dx^1}{2}; dx^2; dx^3\right) dx^2 dx^3$$

for the area considered is elementary, which allows calculating approximately the surface integral as the dot product

$$\boldsymbol{u}.\boldsymbol{n} dS = u^1 dx^2 dx^3$$

For the elementary face $dS_1^* = dx^2 dx^3$ with upward normal unit vector $\boldsymbol{n}(-1;\, 0; 0)$ centered in the midpoint $\left(x^1 - \frac{dx^1}{2};\, dx^2;\, dx^3\right)$ it follows that in an analogous way

$$\iint_{dS_1^*} \boldsymbol{u}.\boldsymbol{n}\, dS \cong -u^1 \left(x^1 - \frac{dx^1}{2};\, dx^2;\, dx^3\right) dx^2 dx^3$$

Adding these contributions

$$\begin{aligned}\iint_{dS_1 + dS_1^*} \boldsymbol{u}.\boldsymbol{n}\, dS &\cong \left[u^1 \left(x^1 + \frac{dx^1}{2};\, dx^2;\, dx^3\right) - u^1 \left(x^1 - \frac{dx^1}{2};\, dx^2;\, dx^3\right)\right] dx^2 dx^3 \\ &\cong u^1 \left(x^1\right) dx^2 dx^3\end{aligned}$$

The component u^1 in the point of coordinate dx^1 varies according to the rate

$$du^1 = \frac{\partial u^1}{\partial x^1} dx^1$$

then

$$\iint_{dS_1 + dS_1^*} \boldsymbol{u} \cdot \boldsymbol{n}\, dS \cong \frac{\partial u^1}{\partial x^1} dx^1 dx^2 dx^3 \cong \frac{\partial u^1}{\partial x^1} dV$$

and the same way for the components u^2 and u^3 in the other faces of the parallelepiped the result is, respectively

$$\iint\limits_{dS_2+dS_2^*} \boldsymbol{u}\cdot\boldsymbol{n}\,dS \cong \frac{\partial u^2}{\partial x^2}dV \qquad \iint\limits_{dS_3+dS_3^*} \boldsymbol{u}\cdot\boldsymbol{n}\,dS \cong \frac{\partial u^3}{\partial x^3}dV$$

Adding these three expressions results for the six faces of the elementary parallelepiped

$$\iint\limits_{S} \boldsymbol{u}\cdot\boldsymbol{n}\,dS = \iiint\limits_{V}\left(\frac{\partial u^1}{\partial x^1}+\frac{\partial u^2}{\partial x^2}+\frac{\partial u^3}{\partial x^3}\right)dV$$

Putting

$$\operatorname{div}\boldsymbol{u} = \frac{\partial u^1}{\partial x^1}+\frac{\partial u^2}{\partial x^2}+\frac{\partial u^3}{\partial x^3} \tag{4.3.1}$$

this analysis leads to the following definition for the divergence of the vectorial function $\boldsymbol{u}$ at point $P(x^i)$

$$\operatorname{div}\boldsymbol{u} = \lim_{V\to 0}\frac{1}{V}\iint\limits_{S}\boldsymbol{u}\cdot\boldsymbol{n}\,dS \tag{4.3.2}$$

that can be interpreted as the dot product between the *nabla* operator and the vectorial function $\boldsymbol{u}$, thus

$$\operatorname{div}\boldsymbol{u} = \nabla\cdot\boldsymbol{u} = \left(\frac{\partial\cdots}{\partial x^j}\right)\boldsymbol{g}^j\cdot u^i\boldsymbol{g}_i$$

To demonstrate that expression (4.3.2) represents the divergence of the vectorial function $\boldsymbol{u}$, consider the sphere of radius $R > 0$, of surface $S(R)$ and volume $V(R)$, centered at point P located in the vectorial space E_3. For the vectorial field $\boldsymbol{u}$ acting in the space

$$\operatorname{div}\boldsymbol{u}(P) = \lim_{R\to 0}\frac{1}{V(R)}\iint\limits_{S(R)}\boldsymbol{u}\cdot\boldsymbol{n}\,dS \tag{4.3.3}$$

Let $g(P) = \operatorname{div}\boldsymbol{u}$, and admitting that $g(x^i)$ is a continuous function that can be written as

$$g\left(x^i\right) = g(P) + h\left(x^i\right)$$

where

$$h\left(x^i\right)_{x^i\to P} = 0$$

Applying the divergence theorem to the vectorial field

$$\frac{1}{V(R)}\iint\limits_{S(R)} \boldsymbol{u}.\boldsymbol{n}\,dS = \frac{1}{V(R)}\iiint\limits_{V(R)} h(x^i)\,dV = \frac{1}{V(R)}\iiint\limits_{V(R)} h(P)\,dV + \frac{1}{V(R)}\iiint\limits_{V(R)} h(x^i)\,dV$$

As $g(P) = \operatorname{div}\boldsymbol{u}$:

$$\frac{1}{V(R)}\iiint\limits_{V(R)} g(P)\,dV = \frac{1}{V(R)}g(P)\iiint\limits_{V(R)} dV = \frac{1}{V(R)}g(P)V(R) = g(P)$$

For the function $h(x^i)$ the result when $R \to 0$ is

$$\left\| \frac{1}{V(R)}\iiint\limits_{V(R)} h(x^i)\,dV \right\| = \underset{\|x^i - P\| \leq R}{Max} \|h(x^i)\| \frac{1}{V(R)}\iiint\limits_{V(R)} dV \leq \underset{\|x^i - P\| \leq R}{Max} \|h(x^i)\|$$

The maximum value of this function fulfills the condition $\|h(x^i)\| \to 0$ when $\|x^i - P\| \to 0$, then the expression (4.3.2) represents the divergence of the vectorial function $\boldsymbol{u}$ at the point P.

This expression is valid for any kind of coordinate system, which shows that the divergence is independent of the referential system. This analysis was formulated for a Cartesian coordinate system for a question of simplicity, being that for the case of curvilinear coordinate systems it was enough to adopt the local trihedron with unit vectors $(\boldsymbol{g}_1; \boldsymbol{g}_2; \boldsymbol{g}_3)$, and one elementary parallelepiped of volume dV.

The scalar field generated by the applying of the divergence to the vectorial field defined by the vectorial function $\boldsymbol{u}$ is called solenoidal or vorticular field, when $\operatorname{div}\boldsymbol{u} = 0$, where $\boldsymbol{u}$ is a solenoidal vector, and this field is called field without source.

4.3.1 Divergence Theorem

The divergence allows writing the Gauß-Ostrogradsky theorem as

$$\iint\limits_{S} \boldsymbol{u} \cdot \boldsymbol{n}\,dS = \iiint\limits_{V} \nabla \cdot \boldsymbol{u}\,dV \tag{4.3.4}$$

which is called the divergence theorem. The symbology adopted in expression (4.3.4) does not change the characteristics and properties shown in item 3.4.

Let a solenoidal field acting in a region R be located between the two closed surfaces S_1 and S_2 (Fig. 4.6), then

Fig. 4.6 Solenoidal field in a region R between two volumes

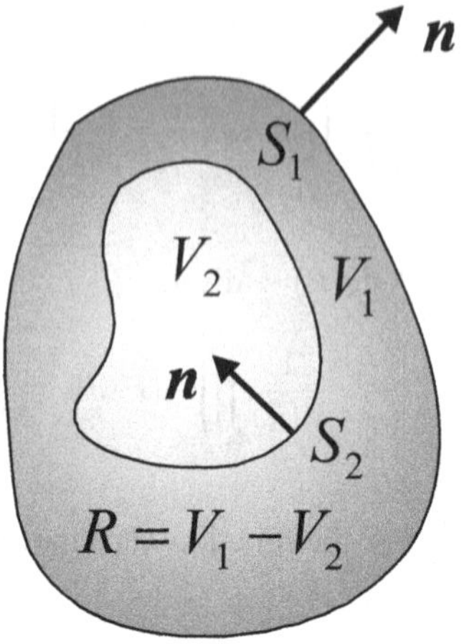

$$\iint_{S_1} \boldsymbol{u}\cdot\boldsymbol{n}\,dS = \iint_{S_2} \boldsymbol{u}\cdot\boldsymbol{n}\,dS$$

To demonstrate this equality consider the closed surface S_1 with upward normal unit vector $\boldsymbol{n}$, with R to the left of the outline of this surface that involves the volume V_1, and the closed surface S_2 with unit downward unit normal vector $\boldsymbol{n}$, involving the volume V_2. Applying the divergence theorem

$$\operatorname{div}\boldsymbol{u} = \iint_{S_1} \boldsymbol{u}\cdot\boldsymbol{n}\,dS - \iint_{S_2} \boldsymbol{u}\cdot\boldsymbol{n}\,dS = 0 \Rightarrow \iint_{S_1} \boldsymbol{u}\cdot\boldsymbol{n}\,dS = \iint_{S_2} \boldsymbol{u}\cdot\boldsymbol{n}\,dS$$

then it is enough to calculate only the integral of a surface.

For a field represented by the vectorial function $\boldsymbol{u} = \phi\nabla\psi$, where ϕ and ψ are scalar functions

$$\nabla\cdot\boldsymbol{u} = \nabla(\phi\nabla\psi) = \phi\nabla\cdot\nabla\psi + \nabla\phi\cdot\nabla\psi = \phi\nabla^2\psi + \nabla\phi\cdot\nabla\psi$$

The component of $\boldsymbol{u}$ in the direction of the normal unit vector $\boldsymbol{n}$ is given by

$$\boldsymbol{u}\cdot\boldsymbol{n} = \phi\boldsymbol{n}\cdot\nabla\psi = \phi\frac{\partial\psi}{\partial n}$$

and applying the divergence theorem

$$\iint_{S} \boldsymbol{u}\cdot\boldsymbol{n}\,dS = \iiint_{V} \nabla\cdot\boldsymbol{u}\,dV$$

results in

$$\iint\limits_S \phi \frac{\partial \psi}{\partial n}\, dS = \iiint\limits_V \left(\phi \nabla^2 \psi + \nabla \phi \cdot \nabla \psi \right) dV$$

that is called Green's first formula.

If the vectorial function is given by $\boldsymbol{u} = \phi \nabla \psi + \psi \nabla \phi$, in an analogous way

$$\nabla \cdot \boldsymbol{u} = \phi\, \nabla^2 \psi + \psi\, \nabla^2 \phi$$

$$\boldsymbol{u} \cdot \boldsymbol{n} = \phi \frac{\partial \psi}{\partial n} - \psi \frac{\partial \phi}{\partial n}$$

whereby

$$\iint\limits_S \phi \frac{\partial \psi}{\partial n} - \psi \frac{\partial \phi}{\partial n}\, dS = \iiint\limits_V \left(\phi\, \nabla^2 \psi + \psi\, \nabla^2 \phi \right) dV$$

that is called Green's second formula.

4.3.2 Contravariant and Covariant Components

The vectorial function $\boldsymbol{u}$ can be expressed by means of their contravariant or covariant components, so it is necessary to calculate this function's divergence for these components. For the vector's contravariant coordinates

$$\operatorname{div} \boldsymbol{u} = \nabla \cdot \boldsymbol{u} = \boldsymbol{g}^j \cdot \left(\frac{\partial u^i \boldsymbol{g}_i}{\partial x^j} \right) \tag{4.3.5}$$

and for its covariant coordinates

$$\operatorname{div} \boldsymbol{u} = \nabla \cdot \boldsymbol{u} = \boldsymbol{g}_j \cdot \left(\frac{\partial u_i \boldsymbol{g}^i}{\partial x^j} \right) \tag{4.3.6}$$

The terms in parenthesis in these expressions indicate that this definition can be amplified considering the vector's covariant derivatives, expressed in their contravariant and covariant coordinates.

Let the covariant derivative of the contravariant vector u^i be:

$$\partial_k u^i = \frac{\partial u^i}{\partial x^k} + u^j \Gamma^i_{jk}$$

that generates a tensor which contraction for $i = k$ provides

$$\partial_i u^i = \frac{\partial u^i}{\partial x^i} + u^j \Gamma^i_{ji}$$

and rewriting the expression (2.4.23)

$$\Gamma^i_{ji} = \frac{\partial\left(\ell n\sqrt{g}\right)}{\partial x^j} = \frac{1}{\sqrt{g}}\frac{\partial\left(\sqrt{g}\right)}{\partial x^j}$$

The use of this expression is more adequate, for it abbreviates the calculation of the Christoffel symbol. Substituting this expression in the previous expression

$$\partial_i u^i = \frac{\partial u^i}{\partial x^i} + u^j \frac{1}{\sqrt{g}}\frac{\partial\left(\sqrt{g}\right)}{\partial x^j} = \frac{1}{\sqrt{g}}\left[\sqrt{g}\frac{\partial u^i}{\partial x^i} + \frac{u^j}{\sqrt{g}}\frac{\partial\left(\frac{1}{2}\ell n\, g\right)}{\partial x^j}\right]$$

and replacing the indexes $i \rightarrow j$ of the first term to the right

$$\partial_i u^i = \frac{1}{\sqrt{g}}\left[\sqrt{g}\frac{\partial u^j}{\partial x^j} + \frac{u^j}{2\sqrt{g}}\frac{\partial\left(\frac{1}{2}\ell n\, g\right)}{\partial x^j}\right]$$

or in a compact form

$$\partial_i u^i = \frac{1}{\sqrt{g}}\frac{\partial\left(\sqrt{g}u^j\right)}{\partial x^j} \tag{4.3.7}$$

It is verified that expression (4.3.7), deducted by means of the contravariant vector u^i, represents a scalar, for it was obtained by means of contraction of the second-order tensor. The other way of formulating this analysis is by means of the covariant derivative of their covariant components.

Let the covariant derivative of the covariant vector u_i be:

$$\partial_i u_i = \partial_i\left(g^{ij} u_j\right)$$

that developed leads to the following expression

$$\partial_i u_i = \partial_i\left(g^{ij}\right) u_j + g^{ij}\partial_i\left(u_j\right)$$

Ricci's lemma shows that $\partial_i(g^{ij}) = 0$ whereby

$$\partial_i u_i = g^{ij}\partial_i\left(u_j\right) = \partial_i u^j$$

and the contraction of this tensor for $i = j$ provides

$$\partial_i u_i = \partial_i u^i = \text{div} u^i = \frac{1}{\sqrt{g}} \frac{\partial \left(\sqrt{g} u^i \right)}{\partial x^i} \tag{4.3.8}$$

Expressions (4.3.7) and (4.3.8) provide the same result, i.e., $\partial_i u_i = \partial_i u^i$. Then the covariant derivative of a vector is independent of the type of the component.

The divergence defined by expressions (4.3.5) and (4.3.6) is the dot product of the *nabla* operator by the vector to which it is applied. The development of the derivatives indicated in these expressions leads to the same results of expressions (4.3.7) and (4.3.8), whereby these last expressions represent the divergence of a vectorial function.

For the Cartesian coordinates

$$\nabla \cdot \boldsymbol{u} = \frac{\partial u_i}{\partial x^i} \tag{4.3.9}$$

4.3.3 Orthogonal Coordinate Systems

Consider the elementary parallelepiped with sides ds_1, ds_2, ds_3, defined in the curvilinear orthogonal coordinates OX^j, by means of which the flow of the field is represented by the vectorial function $\boldsymbol{u}$ (Fig. 4.7).

The divergence of this field is given by

$$\nabla \cdot \boldsymbol{u} = \lim_{V \to 0} \frac{1}{V} \iint_S \boldsymbol{u} \cdot \boldsymbol{n} dS$$

Let

$$ds_i = h_i d\overline{x}^i$$

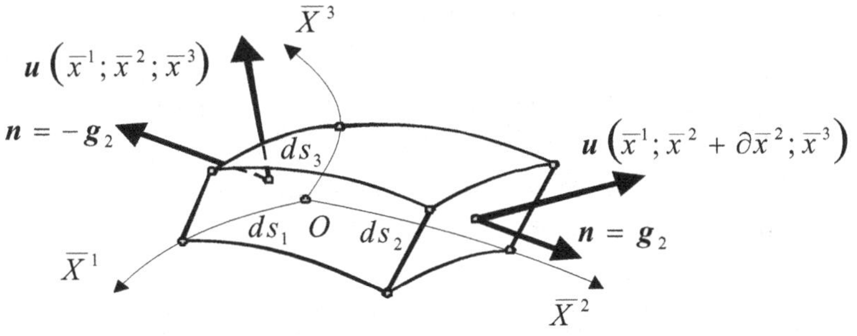

Fig. 4.7 Divergence of the vectorial function $\boldsymbol{u}$ in the curvilinear orthogonal coordinates

$$dV = ds_1 ds_2 ds_3 = h_1 h_2 h_3 d\bar{x}^1 d\bar{x}^2 d\bar{x}^3$$

there is, respectively, for the face with upward normal unit vector $\boldsymbol{n} = -\boldsymbol{g}_2$ and $\boldsymbol{n} = \boldsymbol{g}_2$

$$-\boldsymbol{u} \cdot \boldsymbol{g}_2 ds_1 ds_3 = -u^2 h_1 h_3 d\bar{x}^1 d\bar{x}^3 = -u^2 \left[h_1 h_3 + \frac{\partial (u^2 h_1 h_3)}{\partial \bar{x}^2} d\bar{x}^2 \right] d\bar{x}^1 d\bar{x}^3$$

In an analogous way, for the other faces

$$-\boldsymbol{u} \cdot \boldsymbol{g}_1 ds_2 ds_3 = -u^1 h_2 h_3 d\bar{x}^2 d\bar{x}^3 = -u^1 \left[h_2 h_3 + \frac{\partial (u^1 h_2 h_3)}{\partial \bar{x}^1} d\bar{x}^2 \right] d\bar{x}^2 d\bar{x}^3$$

$$-\boldsymbol{u} \cdot \boldsymbol{g}_2 ds_1 ds_3 = -u^2 h_1 h_3 d\bar{x}^1 d\bar{x}^3 = -u^2 \left[h_1 h_3 + \frac{\partial (u^2 h_1 h_3)}{\partial \bar{x}^2} d\bar{x}^1 \right] d\bar{x}^1 d\bar{x}^3$$

$$-\boldsymbol{u} \cdot \boldsymbol{g}_3 ds_1 ds_2 = -u^3 h_1 h_2 d\bar{x}^1 d\bar{x}^2 = -u^3 \left[h_1 h_2 + \frac{\partial (u^3 h_1 h_2)}{\partial \bar{x}^3} d\bar{x}^3 \right] d\bar{x}^1 d\bar{x}^2$$

Adding the expressions relative to the six faces of the parallelepiped

$$\iint_S \boldsymbol{u} \cdot \boldsymbol{n} dS = \left[\frac{\partial (u^1 h_2 h_3)}{\partial \bar{x}^1} + \frac{\partial (u^2 h_1 h_3)}{\partial \bar{x}^2} + \frac{\partial (u^3 h_1 h_2)}{\partial \bar{x}^3} \right] d\bar{x}^1 d\bar{x}^2 d\bar{x}^3$$

but

$$d\bar{x}^1 d\bar{x}^2 d\bar{x} = \frac{dV}{h_1 h_2 h_3}$$

then

$$\nabla \cdot \boldsymbol{u} = \lim_{V \to 0} \iint_S \boldsymbol{u} \cdot \boldsymbol{n} dS = \frac{1}{h_1 h_2 h_3} \left[\frac{\partial (u^1 h_2 h_3)}{\partial \bar{x}^1} + \frac{\partial (u^2 h_1 h_3)}{\partial \bar{x}^2} + \frac{\partial (u^3 h_1 h_2)}{\partial \bar{x}^3} \right]$$

The result for the orthogonal coordinate system is

$$\nabla \cdot \boldsymbol{u} = div\, \boldsymbol{u} = \frac{1}{h_1 h_2 h_3} \frac{\partial}{\partial x^i} \left(\frac{h_1 h_2 h_3 u^i}{h_i} \right) \tag{4.3.10}$$

where $h_i = \sqrt{g_{(ii)}}$ are the components of the metric tensor, and the indexes in parenthesis do not indicate summation.

4.3.4 Physical Components

With expression (4.3.6) the physical components of the divergence of a vector takes the form

$$\nabla \cdot \boldsymbol{u}^{*} = div\,\boldsymbol{u}^{*} = \frac{1}{h_1 h_2 h_3} \frac{\partial}{\partial x^i} \left(\frac{h_1 h_2 h_3 u^{*i}}{h_i} \right) \qquad (4.3.11)$$

where $u^{*\,i}$ are the vector's physical components.

4.3.5 Properties

As the divergence is the dot product of the *nabla* operator for a vectorial function, the distributive property of the dot product is valid. For the sum of two vectorial functions $\boldsymbol{u}$ and $\boldsymbol{v}$:

$$div\,(\boldsymbol{u} + \boldsymbol{v}) = \nabla \cdot (\boldsymbol{u} + \boldsymbol{v}) = \nabla \cdot \boldsymbol{u} + \nabla \cdot \boldsymbol{v} \qquad (4.3.12)$$

and in terms of the covariant derivative

$$\nabla \cdot (\boldsymbol{u} + \boldsymbol{v}) = \partial_{;i} u^i + \partial_{;i} v^i \qquad (4.3.13)$$

and for the Cartesian coordinates the result is

$$\nabla \cdot (\boldsymbol{u} + \boldsymbol{v}) = \frac{\partial u^i}{\partial x^i} + \frac{\partial v^i}{\partial x^i} \qquad (4.3.14)$$

Considering the vectorial function $m\boldsymbol{u}$, where m is a scalar, the result of the dot product of vectors is

$$div\,(m\boldsymbol{u}) = \nabla \cdot (m\boldsymbol{u}) = m\nabla \cdot \boldsymbol{u} \qquad (4.3.15)$$

These two demonstrations prove that the divergence, for these cases, is a linear operator. In general, the divergence is not a linear operator, as it will be shown in Exercise 4.7.

4.3.6 Divergence of a Second-Order Tensor

The generalization of the divergence theorem for tensorial fields is immediate. Consider, for example, the field represented by the tensorial function of the second order $\boldsymbol{T}(\boldsymbol{r})$ in space E_3, which components depend on the position vector, i.e., $T_{ij} = T_{ij}(\boldsymbol{r})$. For the surface S smooth and continuous by parts in its two faces, with normal unit vector $\boldsymbol{n}(n_1; n_2; n_3)$ varying on each point of the surface, the flow of this tensorial function through S is given by the vector $\boldsymbol{v}$ of components

$$v_i = \iint_S T_{ij} n_j dS; \quad i, j = 1, 2, 3$$

or

$$v_i = \iint_S T_{ji} n_j dS; \quad i, j = 1, 2, 3$$

In absolute notation for the flow $\boldsymbol{v}$ the result is

$$\boldsymbol{v} = \iint_S \boldsymbol{T} \otimes \boldsymbol{n} dS \tag{4.3.16}$$

The flow of the unit tensor δ_{ij} through the closed surface S is given by the components of vector $\boldsymbol{n}$:

$$v_i = \iint_S \delta_{ij} n_j dS = \iint_S n_i dS$$

or in absolute notation

$$v_i = \iint_S \boldsymbol{n} dS$$

The comparison with expression (4.2.3)

$$\nabla \phi(x^i) = \lim_{V \to 0} \frac{1}{V} \iint_S \phi(x^i)\, \boldsymbol{n} dS$$

shows that $\phi(x^i) = 1$, i.e., $\phi(x^i) =$ constant, so $\nabla \phi(x^i) = 0$, which indicates that $\boldsymbol{v} = 0$. Concluding that for a unitary tensorial field the flow through the closed surface S is null.

The concept of a field's divergence can be extended to the tensorial fields, for it is enough that the tensors be contravariant. In the case of covariant tensors their indexes must be raised by means of the metric tensor, next they must be derived and contracted.

There are distinct divergences, depending on the index to be contracted. For T^{ij} there are two divergences: $\partial_i T^{ij}$ and $\partial_j T^{ij}$. If the tensor is symmetrical $T^{ij} = T^{ji}$ then $\partial_i T^{ij} = \partial_j T^{ij}$, i.e., the divergence is unique.

The divergence components of a contravariant second-order tensor are given by

$$\operatorname{div} T^{ij} = \partial_j T^{ij} \tag{4.3.17}$$

and the covariant derivative of the components for this tensor is

$$\partial_k T^{ij} = \left(\frac{\partial T^{ij}}{\partial x^k} + T^{mk} \Gamma^{\,i}_{mk} + T^{im} \Gamma^{\,j}_{mk} \right)$$

With $k = j$:

$$\partial_j T^{ij} = \left(\frac{\partial T^{ij}}{\partial x^j} + T^{mj} \Gamma^{\,i}_{mj} + T^{im} \Gamma^{\,j}_{mj} \right)$$

being

$$\Gamma^{\,j}_{mj} = \frac{\partial \left(\ell n \sqrt{g} \right)}{\partial x^m}$$

it follows that

$$\partial_j T^{ij} = \left[\frac{\partial T^{ij}}{\partial x^j} + T^{mj} \Gamma^{\,i}_{mj} + T^{im} \frac{\partial \left(\ell n \sqrt{g} \right)}{\partial x^m} \right]$$

The change of the indexes $m \rightarrow j$ in the last term in brackets provides

$$\partial_k T^{ij} = \left[\frac{\partial T^{ij}}{\partial x^j} + T^{mj} \Gamma^{\,i}_{mj} + T^{ij} \frac{\partial \left(\ell n \sqrt{g} \right)}{\partial x^j} \right]$$

or

$$\partial_k T^{ij} = T^{mj} \Gamma^{\,i}_{mj} + \frac{1}{\sqrt{g}} \left[\sqrt{g} \frac{\partial T^{ij}}{\partial x^j} + T^{ij} \frac{\partial \left(\sqrt{g} \right)}{\partial x^j} \right]$$

then

$$\operatorname{div} T^{ij} = T^{mj} \Gamma^{\,i}_{mj} + \frac{1}{\sqrt{g}} \frac{\partial \left(\sqrt{g} T^{ij} \right)}{\partial x^k} \tag{4.3.18}$$

that shows that the divergence of a second-order tensor is a vector.

For a mixed second-order tensor the result is

$$\operatorname{div} T^i_j = \partial_i T^i_j$$

and rewriting expression (2.5.21)

$$\partial_k T^i_j = \frac{\partial T^i_j}{\partial x^k} + T^m_j \Gamma^i_{mk} - T^i_m \Gamma^m_{jk}$$

Assuming $i = k$:

$$\partial_i T^i_j = \frac{\partial T^i_j}{\partial x^k} + T^m_j \Gamma^i_{mi} - T^i_m \Gamma^m_{ji}$$

and with

$$\Gamma^i_{mi} = \frac{\partial \left(\ell n \sqrt{g}\right)}{\partial x^m}$$

then

$$\partial_i T^i_j = \left[\frac{\partial T^i_j}{\partial x^k} + T^m_j \frac{\partial \left(\ell n \sqrt{g}\right)}{\partial x^m} - T^i_m \Gamma^m_{ji}\right]$$

The change of indexes $m \to i$ in the second term in brackets provides

$$\partial_i T^i_j = \left[\frac{\partial T^i_j}{\partial x^k} + T^i_j \frac{\partial \left(\ell n \sqrt{g}\right)}{\partial x^i} - T^i_m \Gamma^m_{ji}\right]$$

or

$$\partial_i T^i_j = \sqrt{g} \frac{\partial T^i_j}{\partial x^k} + \frac{1}{\sqrt{g}} T^i_j \frac{\partial \left(\sqrt{g}\right)}{\partial x^i} - T^i_m \Gamma^m_{ji}$$

then

$$\partial_i T^i_j = \frac{1}{\sqrt{g}} \frac{\partial \left(T^i_j \sqrt{g}\right)}{\partial x^i} - T^i_m \Gamma^m_{ji} \tag{4.3.19}$$

The generalization of the definition of the divergence for a third-order tensor is immediate

$$\nabla \cdot \boldsymbol{T} = \boldsymbol{g}^j \cdot \partial_j \boldsymbol{T} = \boldsymbol{g}^j \cdot \partial_j T^{kim} \boldsymbol{g}_k \otimes \boldsymbol{g}_i \otimes \boldsymbol{g}_m = \left(\partial_j T^{kim}\right) \boldsymbol{g}^j \cdot \boldsymbol{g}_k \otimes \boldsymbol{g}_i \otimes \boldsymbol{g}_m$$
$$= \left(\partial_j T^{kim}\right) \delta_k^j \boldsymbol{g}_i \otimes \boldsymbol{g}_m$$

thus

$$\nabla \cdot \boldsymbol{T} = \partial_j T^{jim} \boldsymbol{g}_i \otimes \boldsymbol{g}_m \tag{4.3.20}$$

This expression shows that $\nabla \cdot \boldsymbol{T}$ is a second-order tensor.

For a tensor $\boldsymbol{T}$ of order p then $\nabla \cdot \boldsymbol{T}$ is a tensor of order $(p-1)$. In absolute or invariant notation the result for the divergence of tensor $\boldsymbol{T}$ is

$$\nabla \cdot \boldsymbol{T} = (\nabla \otimes \boldsymbol{T}) \overline{\overline{\otimes}} \boldsymbol{G} \tag{4.3.21}$$

where $\boldsymbol{G}$ is the metric tensor.

In the particular case in which $\operatorname{div} \boldsymbol{T} = 0$ the tensor $\boldsymbol{T}$ defines a tensorial solenoidal field.

The divergence theorem also applies to a tensorial field. Let the field be defined by $\boldsymbol{u} = \boldsymbol{T}\boldsymbol{v}$, which in terms of the components of vectors and tensor is given by $u_i = T_{ik} v_k$, being $\boldsymbol{v}$ an arbitrary and constant vector. Applying the divergence theorem to this field

$$\iint_S \boldsymbol{u} \cdot \boldsymbol{n} dS = \iiint_V \nabla \cdot \boldsymbol{u} dV$$

where

$$\nabla \cdot \boldsymbol{u} = \nabla(\boldsymbol{T}\boldsymbol{v}) = \nabla \boldsymbol{T} \cdot \boldsymbol{v}$$

This vector has components $\dfrac{\partial T_{ik}}{\partial x^k} v_i$, and the component of vector $\boldsymbol{u}$ in the direction of the normal unit vector $\boldsymbol{n}$ is given by the dot product

$$\boldsymbol{u} \cdot \boldsymbol{n} = (T_{ik} v_i) n_k$$

then

$$\iint_S (T_{ik} v_i) n_k \, dS = \iiint_V \frac{\partial T_{ik}}{\partial x^k} v_i dV$$

whereby in terms of the tensor components the result is

$$\iint_S T_{ik} n_k \, dS = \iiint_V \frac{\partial T_{ik}}{\partial x^k} dV \tag{4.3.22}$$

and in absolute notation this expression becomes

$$\iint_S \boldsymbol{T} \otimes \boldsymbol{n} dS = \iiint_V \nabla \cdot \boldsymbol{T} dV \tag{4.3.23}$$

Exercise 4.7 Calculate: (a) $\nabla \cdot (\phi \boldsymbol{u})$; (b) $\nabla \cdot (\boldsymbol{u} \times \boldsymbol{v})$.

(a) The field divergence defined by the product of a scalar function $\phi(x^i)$ by a vector $\boldsymbol{u}$ is given by

$$\operatorname{div}(\phi \boldsymbol{u}) = \nabla \cdot (\phi \boldsymbol{u}) = \boldsymbol{g}^m \frac{\partial (\phi u^k \boldsymbol{g}_k)}{\partial x^m} = \boldsymbol{g}^m \left(\frac{\partial \phi}{\partial x^m} u^k \boldsymbol{g}_k + \phi \frac{\partial u^k}{\partial x^m} \boldsymbol{g}_k + \phi u^k \frac{\partial \boldsymbol{g}_k}{\partial x^m} \right)$$

and substituting (2.3.10)

$$\frac{\partial \boldsymbol{g}_k}{\partial x^m} = \Gamma^p_{km} \boldsymbol{g}_p$$

in the previous expression

$$\operatorname{div}(\phi \boldsymbol{u}) = \boldsymbol{g}^m \left(\frac{\partial \phi}{\partial x^m} u^k \boldsymbol{g}_k + \phi \frac{\partial u^k}{\partial x^m} \boldsymbol{g}_k + \phi u^k \Gamma^p_{km} \boldsymbol{g}_p \right)$$

The permutation of the indexes $p \leftrightarrow k$ in the third member in parenthesis provides

$$\operatorname{div}(\phi \boldsymbol{u}) = \left(\frac{\partial \phi}{\partial x^m} u^k + \phi \frac{\partial u^k}{\partial x^m} + \phi u^p \Gamma^k_{pm} \right) \boldsymbol{g}^m \cdot \boldsymbol{g}_k = \frac{\partial \phi}{\partial x^m} u^k + \phi \frac{\partial u^k}{\partial x^m} + \phi u^p \Gamma^k_{pm}$$

and with

$$\partial_m u^k = \frac{\partial \phi}{\partial x^m} u^k + \phi \frac{\partial u^k}{\partial x^m} \Rightarrow \operatorname{div}(\phi \boldsymbol{u}) = \frac{\partial \phi}{\partial x^m} u^k + \phi \partial_m u^k$$

Putting

$$\partial_m u^k = \nabla u^k \quad \frac{\partial \phi}{\partial x^m} = \nabla \phi$$

thus

$$\nabla \cdot (\phi \boldsymbol{u}) = (\nabla \phi) \cdot \boldsymbol{u} + \phi (\nabla \boldsymbol{u})$$

or

$$\operatorname{div}(\phi \boldsymbol{u}) = \operatorname{grad}\phi \cdot \boldsymbol{u} + \phi \operatorname{grad}\boldsymbol{u}$$

In this case the divergence is not a linear operator, but for $\phi = m$, where m is a constant, the result is expression (4.3.15), verifying the linearity of this operator.

(b) The field represented by the vectorial function generated by the cross product $\boldsymbol{w} = \boldsymbol{u} \times \boldsymbol{v}$ is given by

$$\boldsymbol{w} = w^p \boldsymbol{g}_p = \varepsilon^{pqr} u_q v_r \boldsymbol{g}_p$$

The divergence of this function is given by

$$\nabla \cdot (\boldsymbol{u} \times \boldsymbol{v}) = \boldsymbol{g}^i \cdot \frac{\partial \left(w^p \boldsymbol{g}_p \right)}{\partial x^i} = \left(\frac{\partial w^p}{\partial x^i} \boldsymbol{g}_p + w^p \frac{\partial \boldsymbol{g}_p}{\partial x^i} \right) \cdot \boldsymbol{g}^i$$

and the expression

$$\frac{\partial \boldsymbol{g}_p}{\partial x^i} = \Gamma^j_{pi} \boldsymbol{g}_j$$

substituted in the previous expression provides

$$\nabla \cdot (\boldsymbol{u} \times \boldsymbol{v}) = \left(\frac{\partial w^p}{\partial x^i} \boldsymbol{g}_p + w^p \Gamma^j_{pi} \boldsymbol{g}_j \right) \cdot \boldsymbol{g}^i$$

Interchanging indexes $p \leftrightarrow j$ in the second term in parenthesis it follows that

$$\nabla \cdot (\boldsymbol{u} \times \boldsymbol{v}) = \left(\frac{\partial w^p}{\partial x^i} + w^j \Gamma^p_{ji} \right) \boldsymbol{g}^i \cdot \boldsymbol{g}_p = \delta^i_p \partial_i w^p$$

or

$$\partial_p w^p = \partial_p \left(\varepsilon^{pqr} u_q v_r \right) = \left(\partial_p \varepsilon^{pqr} \right) u_q v_r + \varepsilon^{pqr} \partial_p \left(u_q v_r \right)$$

and with

$$\partial_p w^p = \varepsilon^{pqr} \partial_p \left(u_q v_r \right)$$

thus

$$\nabla \cdot (\boldsymbol{u} \times \boldsymbol{v}) = \varepsilon^{pqr} \partial_p \left(u_q v_r \right)$$

whereby

$$\nabla \cdot (\boldsymbol{u} \times \boldsymbol{v}) = \varepsilon^{pqr}\left[\left(\partial_p u_q\right) v_r + u_q \partial_p v_r\right]$$

With the $\varepsilon^{pqr} = \varepsilon^{qpr}$ and $\varepsilon^{pqr} = -\varepsilon^{rpq}$ the results for the terms to the right are

$$\varepsilon^{pqr}\left(\partial_p u_q\right) v_r = \varepsilon^{rpq}\left(\partial_p u_q\right) v_r = \boldsymbol{v} \cdot \nabla \boldsymbol{u}$$
$$\varepsilon^{pqr} u_q\left(\partial_p v_r\right) = -\varepsilon^{qpr} u_q\left(\partial_p v_r\right) = -\boldsymbol{u} \cdot \nabla \boldsymbol{v}$$

whereby

$$\nabla \cdot (\boldsymbol{u} \times \boldsymbol{v}) = \boldsymbol{v} \cdot \nabla \times \boldsymbol{u} - \boldsymbol{u} \cdot \nabla \times \boldsymbol{v} \quad \Rightarrow \quad div\,(\boldsymbol{u} \times \boldsymbol{v}) = \boldsymbol{v} \cdot rot\,\boldsymbol{u} - \boldsymbol{u} \cdot rot\,\boldsymbol{v}$$

For the Cartesian coordinates

$$\nabla \cdot (\boldsymbol{u} \times \boldsymbol{v}) = \varepsilon_{ijk} \frac{\partial\left(u_j v_k\right)}{\partial x^i}$$

Exercise 4.8 Let T^{ij} and T^i_j be associated tensors, write $\operatorname{div} T^i_j$ in terms of the symmetrical tensor T^{ij}.

The divergence of a second-order tensor is given by

$$\operatorname{div} T^i_j = \partial_i T^i_j = \frac{1}{\sqrt{g}} \frac{\partial\left(T^i_j \sqrt{g}\right)}{\partial x^i} - T^i_m \Gamma^m_{ji}$$

and with

$$\Gamma^m_{ij} = g^{mk} \Gamma_{ij,k}$$

thus

$$\operatorname{div} T^i_j = \partial_i T^i_j = \frac{1}{\sqrt{g}} \frac{\partial\left(T^i_j \sqrt{g}\right)}{\partial x^i} - T^i_m g^{mk} \Gamma_{ij,k}$$

Let

$$\Gamma_{ij,k} = \frac{1}{2}\left(\frac{\partial g_{jk}}{\partial x^i} + \frac{\partial g_{ik}}{\partial x^j} + \frac{\partial g_{ij}}{\partial x^k}\right)$$
$$T^i_m g^{mk} = T^{ik}$$

then

$$\partial_i T^i_j = \frac{1}{\sqrt{g}} \frac{\partial\left(T^i_j \sqrt{g}\right)}{\partial x^i} - T^{ik} \frac{1}{2}\left(\frac{\partial g_{jk}}{\partial x^i} + \frac{\partial g_{ik}}{\partial x^j} + \frac{\partial g_{ij}}{\partial x^k}\right)$$

or

$$\partial_i T^i_j = \frac{1}{\sqrt{g}} \frac{\partial\left(T^i_j \sqrt{g}\right)}{\partial x^i} - \frac{1}{2} T^{ik} \frac{\partial g_{jk}}{\partial x^i} - \frac{1}{2} T^{ik} \frac{\partial g_{ik}}{\partial x^j} - \frac{1}{2} T^{ik} \frac{\partial g_{ij}}{\partial x^k}$$

Interchanging the indexes $i \leftrightarrow j$ in the last term to the right

$$\partial_i T^i_j = \frac{1}{\sqrt{g}} \frac{\partial\left(T^i_j \sqrt{g}\right)}{\partial x^i} - \frac{1}{2} T^{ik} \frac{\partial g_{jk}}{\partial x^i} - \frac{1}{2} T^{ik} \frac{\partial g_{ik}}{\partial x^j} - \frac{1}{2} T^{ki} \frac{\partial g_{kj}}{\partial x^k}$$

As

$$g_{jk} = g_{kj} \quad T^{ik} = T^{ki}$$

thus

$$\partial_i T^i_j = \frac{1}{\sqrt{g}} \frac{\partial\left(T^i_j \sqrt{g}\right)}{\partial x^i} - \frac{1}{2} T^{ik} \frac{\partial g_{ik}}{\partial x^j}$$

Exercise 4.9 Calculate the divergence of vector u^i expressed in cylindrical coordinates.

For the cylindrical coordinates $\sqrt{g} = r$, and with the contravariant components of vector (u^r, u^θ, u^z) it follows that

$$\begin{aligned} \operatorname{div} u^i &= \frac{1}{\sqrt{g}} \frac{\partial\left(\sqrt{g} u^i\right)}{\partial x^i} = \frac{1}{r}\left[\frac{\partial (r u_i)}{\partial x^i}\right] = \frac{1}{r}\left[\frac{\partial (r u^r)}{\partial r} + \frac{\partial\left(r u^\theta\right)}{\partial \theta} + \frac{\partial (r u^z)}{\partial z}\right] \\ &= \frac{\partial u^r}{\partial r} + \frac{\partial u^\theta}{\partial \theta} + \frac{\partial u^z}{\partial z} + u^r \end{aligned}$$

In an analogous way in terms of the vector's covariant components

$$\operatorname{div} u_j = \frac{\partial u_r}{\partial r} + \frac{1}{r^2} \frac{\partial u_\theta}{\partial \theta} + \frac{\partial u_z}{\partial z} + u_r$$

Exercise 4.10 Calculate the divergence of vector u^i expressed in spherical coordinates.

For the spherical coordinates $\sqrt{g} = r^2 \sin \phi$, and with the contravariant components of vector (u^r, u^ϕ, u^θ) it follows that

$$\operatorname{div} u^i = \frac{1}{\sqrt{g}}\frac{\partial\left(\sqrt{g}u^i\right)}{\partial x^i} = \frac{1}{r^2 \sin\phi}\left[\frac{\partial(r^2 \sin\phi u^i)}{\partial x^i}\right]$$
$$= \frac{1}{r^2 \sin\phi}\left[\frac{\partial(r^2 \sin\phi u^r)}{\partial r} + \frac{\partial(r^2 \sin\phi u^\phi)}{\partial\phi} + \frac{\partial(r^2 \sin\phi u^\theta)}{\partial\theta}\right]$$
$$= \frac{\partial u^r}{\partial r} + \frac{\partial u^\phi}{\partial\phi} + \frac{\partial u^\theta}{\partial\theta} + \frac{2u^r}{r} + (\cot\phi)u^\phi$$

For the vector's covariant components the result is

$$\operatorname{div} u_j = \frac{\partial u_r}{\partial r} + \frac{1}{r^2}\frac{\partial u_\phi}{\partial\phi} + \frac{1}{r^2 \sin^2\phi}\frac{\partial u_\theta}{\partial\theta} + \frac{2u_r}{r} + \frac{(\cot\phi)}{r^2}u_\phi$$

Exercise 4.11 Let $\boldsymbol{r}$ be the position vector of the points in the space E_3, show that: (a) $\operatorname{div}\boldsymbol{r} = 3$; (b) $\operatorname{div}(r^n\boldsymbol{r}) = (n+3)\boldsymbol{r}^n$; (c) $\operatorname{div}\left(\frac{\boldsymbol{r}}{r^3}\right) = 0$; (d) $\operatorname{div}\left(\frac{\boldsymbol{r}}{r}\right) = \frac{2}{r}$.

(a) With the definition of divergence

$$\nabla\cdot\boldsymbol{r} = \left(\boldsymbol{g}_i\frac{\partial}{\partial x^i}\right)\cdot\boldsymbol{r} = \boldsymbol{g}_i\cdot\frac{\partial\boldsymbol{r}}{\partial x^i}$$

but

$$\frac{\partial\boldsymbol{r}}{\partial x^i} = \boldsymbol{g}_i$$

then

$$\nabla\cdot\boldsymbol{r} = \boldsymbol{g}_i\cdot\boldsymbol{g}_i$$

For $i = 1, 2, 3$ the result is

$$\operatorname{div}\boldsymbol{r} = 3 \quad \text{Q.E.D.}$$

(b) Let

$$\operatorname{div}(\phi\boldsymbol{u}) = \phi\operatorname{div}\boldsymbol{u} + \boldsymbol{u}\operatorname{grad}\phi$$

and putting

$$\boldsymbol{u} = r \quad \phi = r^n$$

thus

$$\operatorname{div}(r^n\boldsymbol{r}) = r^n\operatorname{div}\boldsymbol{r} + \boldsymbol{r}\operatorname{grad}r^n = 3r^n + \boldsymbol{r}\cdot\left(nr^{n-1}\frac{1}{r}\boldsymbol{r}\right) = 3r^n + \left(nr^{n-2}r^2\right)$$

then

$$\operatorname{div}(r^n \boldsymbol{r}) = (n+3)\boldsymbol{r}^n \quad \text{Q.E.D.}$$

(c) Putting

$$\operatorname{div}\left(\frac{\boldsymbol{r}}{r^3}\right) = \operatorname{div}\left(r^{-3}\boldsymbol{r}\right)$$

it follows that

$$\begin{aligned}\operatorname{div}(r^{-3}\boldsymbol{r}) &= r^{-3}\operatorname{div}\boldsymbol{r} + \boldsymbol{r}\cdot\operatorname{grad} r^{-3} = 3r^{-3} + \boldsymbol{r}\cdot(-3r^{-4}\operatorname{grad} r) = 3r^{-3} + \boldsymbol{r}\cdot\left(-3r^{-4}\frac{\boldsymbol{r}}{r}\right)\\ &= 3r^{-3} + r^2\left(-3r^{-4}\frac{1}{r}\right)\end{aligned}$$

whereby

$$\operatorname{div}\left(r^{-3}\boldsymbol{r}\right) = 0 \qquad \text{Q.E.D.}$$

This conclusion shows that $r^{-3}\boldsymbol{r}$ is a solenoidal vectorial function.

(d) Putting

$$\operatorname{div}\left(\frac{\boldsymbol{r}}{r}\right) = \operatorname{div}\left(\frac{1}{r}\boldsymbol{r}\right)$$

$$\boldsymbol{r} = x\boldsymbol{i} + y\boldsymbol{j} + z\boldsymbol{k}$$

it follows that

$$\begin{aligned}\operatorname{div}\left(\frac{1}{r}\boldsymbol{r}\right) &= \operatorname{div}\left(\frac{x}{r}\boldsymbol{i} + \frac{y}{r}\boldsymbol{j} + \frac{z}{r}\boldsymbol{k}\right) = \frac{\partial}{\partial x}\left(\frac{x}{r}\right) + \frac{\partial}{\partial y}\left(\frac{y}{r}\right) + \frac{\partial}{\partial z}\left(\frac{z}{r}\right)\\ &= \left(\frac{1}{r} - \frac{x}{r^2}\frac{\partial r}{\partial x}\right) + \left(\frac{1}{r} - \frac{y}{r^2}\frac{\partial r}{\partial y}\right) + \left(\frac{1}{r} - \frac{z}{r^2}\frac{\partial r}{\partial z}\right)\end{aligned}$$

and with

$$r^2 = x^2 + y^2 + z^2$$

$$\frac{\partial r}{\partial x} = \frac{x}{r} \quad \frac{\partial r}{\partial y} = \frac{y}{r} \quad \frac{\partial r}{\partial z} = \frac{z}{r}$$

thus

$$\operatorname{div}\left(\frac{1}{r}\boldsymbol{r}\right) = \frac{3}{r} - \left(\frac{x\,x}{r^2 r} + \frac{y\,y}{r^2 r} + \frac{z\,z}{r^2 r}\right)$$

whereby

$$\operatorname{div}\left(\frac{\boldsymbol{r}}{r}\right) = \frac{2}{r} \quad \text{Q.E.D.}$$

4.4 Curl

The vector product of the *nabla* operator by a vector generates a differential operator linked to the direction of rotation of the coordinate system defining the curl, also called rotation or whirl. In absolute notation this operator is written as

$$\nabla \times \boldsymbol{u} = rot\,\boldsymbol{u} = \boldsymbol{v} \tag{4.4.1}$$

In English literature the notation curl $\boldsymbol{u}$ is used to designate rotational of vector $\boldsymbol{u}$, which was adopted firstly by Maxwell. The term *curl* literally means ring, and it designates the pseudovector $\nabla \times \boldsymbol{u}$. With

$$\boldsymbol{v} = \boldsymbol{g}^i \times \frac{\partial\left(u_j \boldsymbol{g}^j\right)}{\partial x^i} = \boldsymbol{g}^i \times \frac{\partial u_j}{\partial x^i}\boldsymbol{g}^j + \boldsymbol{g}^i \times u_j \frac{\partial \boldsymbol{g}^j}{\partial x^i}$$

and rewriting expression (2.4.4)

$$\frac{\partial \boldsymbol{g}^j}{\partial x^i} = -\Gamma^j_{ki}\boldsymbol{g}^m$$

it follows for the second term of the member to the right of expression (4.4.1)

$$u_j \boldsymbol{g}^i \times \frac{\partial \boldsymbol{g}^m}{\partial x^i} = -u_j \Gamma^j_{ki} \boldsymbol{g}^i \times \boldsymbol{g}^m$$

The cross product of these vectors is given by

$$\boldsymbol{g}^i \times \boldsymbol{g}^m = \frac{e^{imk}}{\sqrt{g}} g_k = \begin{cases} +1 & \text{for an even number of} \\ & \text{permutations of the indexes} \\ -1 & \text{for an odd number of} \\ & \text{permutations of the indexes} \\ 0 & \text{when there are repeated indexes} \end{cases}$$

and substituting results in

$$u_j \boldsymbol{g}^i \times \frac{\partial \boldsymbol{g}^m}{\partial x^i} = -\frac{u_j}{\sqrt{g}}\left(\Gamma^j_{ki} - \Gamma^j_{ik}\right) g_k = 0$$

whereby

$$\nabla \times \boldsymbol{u} = \boldsymbol{v} = \frac{\partial u_j}{\partial x^i} \boldsymbol{g}^i \times \boldsymbol{g}^m$$

As

$$\boldsymbol{g}^i \times \boldsymbol{g}^j = \frac{e^{ijk}}{\sqrt{g}} \boldsymbol{g}_k$$

in tensorial terms

$$\nabla \times \boldsymbol{u} = \frac{\mathrm{e}^{ijk}}{\sqrt{g}} \frac{\partial u_j}{\partial x^i} \boldsymbol{g}_k \tag{4.4.2}$$

As a function of Ricci's pseudotensor

$$\nabla \times \boldsymbol{u} = \varepsilon^{ijk} \frac{\partial u_j}{\partial x^i} \boldsymbol{g}_k \tag{4.4.3}$$

and with

$$\varepsilon^{ijk} = \frac{\mathrm{e}^{ijk}}{\sqrt{g}}$$

the expression (4.4.3) after a cyclic permutation of the indexes $i, j, k = 1, 2, 3$ takes the form

$$\nabla \times \boldsymbol{u} = \frac{1}{\sqrt{g}} \left(\frac{\partial u_j}{\partial x^i} - \frac{\partial u_i}{\partial x^j} \right) \boldsymbol{g}_k \tag{4.4.4}$$

In a space provided with metric, the curl of a vectorial function can also be defined by means of its contravariant components, for these relate with its covariant components by means of the metric tensor.

In an analogous way, the results for the contravariant coordinates are

$$\nabla \times \boldsymbol{u} = \boldsymbol{g}^\ell \times \frac{\partial \left(u^k \boldsymbol{g}_k \right)}{\partial x^\ell} = \boldsymbol{g}^\ell \times \left(\frac{\partial u^k}{\partial x^\ell} \boldsymbol{g}_k + u^k \frac{\partial \boldsymbol{g}_k}{\partial x^\ell} \right)$$

$$\frac{\partial \boldsymbol{g}_k}{\partial x^\ell} = \Gamma^m_{k\ell} \boldsymbol{g}_m$$

$$\nabla \times \boldsymbol{u} = \boldsymbol{g}^\ell \times \left(\frac{\partial u^k}{\partial x^\ell} \boldsymbol{g}_k + u^k \Gamma^m_{k\ell} \boldsymbol{g}_m \right) = \left(\frac{\partial u^k}{\partial x^\ell} + u^m \Gamma^k_{m\ell} \right) \boldsymbol{g}^\ell \times \boldsymbol{g}_k$$

$$\begin{aligned}
\nabla \times \boldsymbol{u} &= \partial_\ell u^k \boldsymbol{g}^\ell \times \boldsymbol{g}_k \\
\boldsymbol{g}^\ell &= g^{\ell j} \boldsymbol{g}_j \\
\nabla \times \boldsymbol{u} &= \partial_\ell u^k g^{\ell j} \boldsymbol{g}_j \times \boldsymbol{g}_k \\
\boldsymbol{g}_j \times \boldsymbol{g}_k &= \varepsilon_{ijk} \boldsymbol{g}^i \\
\nabla \times \boldsymbol{u} &= \varepsilon_{ijk} \left(\partial_\ell u^k \right) g^{\ell j} \boldsymbol{g}^i
\end{aligned} \tag{4.4.5}$$

A vectorial field is called an irrotational field when $\nabla \times \boldsymbol{u} = 0$, then

$$\frac{\partial u_j}{\partial x^i} = \frac{\partial u_i}{\partial x^j}$$

In space E_3 the curl $\nabla \times \boldsymbol{u}$ is an axial vector (vectorial density), so it is associated to an antisymmetric second-order tensor, which components are

$$A^{ij} = \begin{bmatrix} 0 & \dfrac{\partial u^2}{\partial x^1} - \dfrac{\partial u^1}{\partial x^2} & \dfrac{\partial u^3}{\partial x^1} - \dfrac{\partial u^1}{\partial x^3} \\ \dfrac{\partial u^1}{\partial x^2} - \dfrac{\partial u^2}{\partial x^1} & 0 & \dfrac{\partial u^3}{\partial x^2} - \dfrac{\partial u^2}{\partial x^3} \\ \dfrac{\partial u^1}{\partial x^3} - \dfrac{\partial u^3}{\partial x^1} & \dfrac{\partial u^2}{\partial x^3} - \dfrac{\partial u^3}{\partial x^2} & 0 \end{bmatrix} \tag{4.4.6}$$

For the space E_N the curl $\nabla \times \boldsymbol{u}$ has $\frac{1}{2}N(N-1)$ independent components. In space E_2 the curl is a pseudoscalar. For the Cartesian coordinates

$$\nabla \times \boldsymbol{u} = e_{ijk} \frac{\partial u_j}{\partial x^i} \boldsymbol{g}_k \tag{4.4.7}$$

or in a determinant form

$$\nabla \times \boldsymbol{u} = \begin{vmatrix} \boldsymbol{i} & \boldsymbol{j} & \boldsymbol{k} \\ \dfrac{\partial}{\partial x^1} & \dfrac{\partial}{\partial x^2} & \dfrac{\partial}{\partial x^3} \\ u_1 & u_2 & u_3 \end{vmatrix} \tag{4.4.8}$$

4.4.1 Stokes Theorem

In expression (3.3.7) with $\boldsymbol{F} = \boldsymbol{u}$, and with the expression (4.4.8) Stokes theorem in vectorial notation is given by

$$\iint_S \boldsymbol{n} \cdot \nabla \times \boldsymbol{u} \, dS = \oint_C \boldsymbol{u} \cdot d\boldsymbol{r} \tag{4.4.9}$$

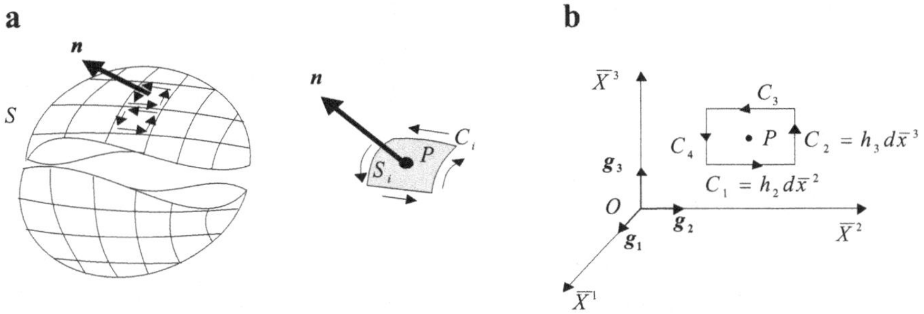

Fig. 4.8 Concept of curl: (**a**) circulation in a closed surface and (**b**) elementary rectangle

A more consistent definition of the curl can be formulated analyzing the circulation of the vectorial field $\boldsymbol{u}$ in a closed surface S with upward unit normal vector $\boldsymbol{n}$ (Fig. 4.8a).

Consider the elementary rectangle dS determined in the orthogonal curvilinear coordinate system $\overline{X}^j$, with sides $h_1 d\overline{x}^1$ and $h_2 d\overline{x}^2$ located in the plane $O\overline{X}^2\overline{X}^3$, with the point $P\left(\overline{x}^1;\overline{x}^2;\overline{x}^3\right)$ located in its center. Locally, the coordinate system $\overline{X}^j$ is considered as a Cartesian orthogonal system (Fig. 4.8b), with the scale factors h_i, $i = 1, 2, 3$.

The line integral $\oint\limits_C \boldsymbol{u}\cdot d\boldsymbol{r}$ along the perimeter of this rectangle is carried out dividing this perimeter into segments C_1, C_2, C_3, C_4. The center of segment C_1 of perimeter of the rectangle is given by the coordinates $\left(\overline{x}^1;\overline{x}^2;\overline{x}^3 - \frac{d\overline{x}^3}{2}\right)$ then $\boldsymbol{u}\cdot d\boldsymbol{r} = u^2 d\overline{x}^2$. As this length is elementary its contribution to the line integral is given by

$$\oint\limits_{C_1} \boldsymbol{u}\cdot d\boldsymbol{r} \cong u^2\left(\overline{x}^1;\overline{x}^2;\overline{x}^3 - \frac{d\overline{x}^3}{2}\right) h_2 d\overline{x}^2$$

For segment C_3 with center $\left(\overline{x}^1;\overline{x}^2;\overline{x}^3 + \frac{d\overline{x}^3}{2}\right)$:

$$\oint\limits_{C_3} \boldsymbol{u}\cdot d\boldsymbol{r} \cong -u^2\left(\overline{x}^1;\overline{x}^2;\overline{x}^3 + \frac{d\overline{x}^3}{2}\right) h_2 dx^2$$

where the negative sign indicates that the direction of the path is contrary to the coordinate axis.

Adding the contributions of these two segments

$$\oint_{C_1+C_3} \boldsymbol{u}\cdot d\boldsymbol{r} \cong \left[u^2\left(\overline{x}^1;\overline{x}^2;\overline{x}^3-\frac{d\overline{x}^3}{2}\right)-u^2\left(\overline{x}^1;\overline{x}^2;\overline{x}^3+\frac{d\overline{x}^3}{2}\right)\right]h_2 d\overline{x}^2$$

The component u^2 varies according to the rate

$$du^2 = -\frac{\partial u^2}{\partial \overline{x}^3} d\overline{x}^3$$

where the negative sign indicates that this variation decreases in the positive direction of axis $O\overline{X}^1$, it follows that

$$\oint_{C_1+C_3} \boldsymbol{u}\cdot d\boldsymbol{r} \cong -\frac{\partial u^2}{\partial \overline{x}^3} d\overline{x}^3 h_2 d\overline{x}^2$$

and dividing by $dS = h_2 h_3 d\overline{x}^2 d\overline{x}^3$

$$\frac{1}{dS}\oint_{C_1+C_3} \boldsymbol{u}\cdot d\boldsymbol{r} \cong -\frac{1}{h_2 h_3}\frac{\partial (u^2 h_2)}{\partial \overline{x}^3}$$

Adopting analogous formulations for the other segments

$$\oint_{C_2+C_4} \boldsymbol{u}\cdot d\boldsymbol{r} \cong \left[u^3\left(\overline{x}^1;\overline{x}^2;\overline{x}^3+\frac{d\overline{x}^3}{2}\right)-u^3\left(\overline{x}^1;\overline{x}^2;\overline{x}^3-\frac{d\overline{x}^3}{2}\right)\right]h_3 d\overline{x}^3 \cong \frac{\partial u^3}{\partial \overline{x}^2} d\overline{x}^2 h_3 d\overline{x}^3$$

$$\frac{1}{dS}\oint_{C_2+C_4} \boldsymbol{u}\cdot d\boldsymbol{r} \cong \frac{1}{h_2 h_3}\frac{\partial (u^3 h_3)}{\partial \overline{x}^2}$$

Adding these contributions the result when $dS \to 0$ is

$$\left[\frac{\partial (h_3 u^3)}{\partial \overline{x}^2} - \frac{\partial (h_2 u^2)}{\partial \overline{x}^3}\right] = \lim_{dS\to 0}\frac{1}{dS}\oint_C \boldsymbol{u}\cdot d\boldsymbol{r}$$

or

$$\boldsymbol{e}_1\cdot\nabla\times\boldsymbol{u} = \frac{\partial (h_3 u^3)}{\partial \overline{x}^2} - \frac{\partial (h_2 u^2)}{\partial \overline{x}^3}$$

For the components u^1 and u^2 the result is, respectively,

$$e_2 \cdot \nabla \times \boldsymbol{u} = \frac{\partial (h_1 u^1)}{\partial \overline{x}^3} - \frac{\partial (h_3 u^3)}{\partial \overline{x}^1}$$

$$e_3 \cdot \nabla \times \boldsymbol{u} = \frac{\partial (h_2 u^2)}{\partial \overline{x}^1} - \frac{\partial (h_1 u^1)}{\partial \overline{x}^2}$$

Concluding that the curl components of the vectorial field $\boldsymbol{u}$ in the direction of the upward unit normal vector to the closed surface S are given by

$$\boldsymbol{n} \cdot \nabla \times \boldsymbol{u} = \lim_{S \to 0} \frac{1}{S} \oint_C \boldsymbol{u} \cdot d\boldsymbol{r} \tag{4.4.10}$$

This expression is valid for any type of referential system, which shows that the curl is independent of the coordinate system.

For demonstrating that expression (4.4.9) represents the Stokes theorem, let the surface S which outline is curve C, and the field represented by the vectorial function $\boldsymbol{u}(\boldsymbol{r})$, continuous and with continuous partial derivatives in $S \cup C$. Dividing S in N cells S_i, $i = 1, 2, \ldots N$, which components of the upward normal unit vectors are n_i, with closed outline curves C_i (Fig. 4.8a), and with expression (4.4.10) the result for each cell of S is

$$\boldsymbol{n_i} \cdot \nabla \times \boldsymbol{u} = \lim_{S_i \to 0} \frac{1}{S_i} \oint_{C_i} \boldsymbol{u} \cdot d\boldsymbol{r}$$

Applying this expression to point P contained in cell S_i with boundary C_i, the result when the area of this cell is reduced approaching the outline P is

$$(\boldsymbol{n_i} \cdot \nabla \times \boldsymbol{u}) S_i = \oint_{C_i} \boldsymbol{u} \cdot d\boldsymbol{r} + h(\overline{x}^i) S_i$$

where $\|h(\overline{x}^i)\| > 0$ is a function with very small value, which decreases with the reduction of size of S_i.

With the division of the surface S into N parts the result is that $N[h_i(\overline{x}^i)] > 0$, then $\underset{1 \le i \le N}{\mathrm{Max}}\ h_i(\overline{x}^i) < h(\overline{x}^i)$. For $N \to \infty$ the result is $h(\overline{x}^i) \to 0$, so

$$\left\| (\boldsymbol{n_i} \cdot \nabla \times \boldsymbol{u}) S_i - \sum_{i=1}^{N} \oint_{C_i} \boldsymbol{u} \cdot d\boldsymbol{r} \right\| < h(\overline{x}^i) \sum_{i=1}^{N} S_i = h(\overline{x}^i) S$$

and with

$$\sum_{i=1}^{N} \oint_{C_i} \boldsymbol{u} \cdot d\boldsymbol{r} = \oint_{C} \boldsymbol{u} \cdot d\boldsymbol{r}$$

for the outlines of the cells S_i are calculated twice, but in opposite directions, whereby these parcels cancel each other, leaving only the parcel of boundary C of S. Then

$$\left\| (\boldsymbol{n}_i \cdot \nabla \times \boldsymbol{u}) S_i - \oint_{C} \boldsymbol{u} \cdot d\boldsymbol{r} \right\| < h(\overline{x}^i)\, S$$

As $N \to \infty$ the result is

$$\lim_{N \to \infty} (\boldsymbol{n}_i \cdot \nabla \times \boldsymbol{u}) S_i = \oint_{C} \boldsymbol{u} \cdot d\boldsymbol{r}$$

whereby the result of the expression of Stokes theorem is

$$\iint_{S} \boldsymbol{n} \cdot \nabla \times \boldsymbol{u}\, dS = \oint_{C} \boldsymbol{u} \cdot d\boldsymbol{r}$$

This theorem is a particular case of the divergence theorem. To demonstrate this assertion let the vectorial function $\boldsymbol{u} = \boldsymbol{v} \times \boldsymbol{w}$, and an arbitrary and constant vector, then

$$\nabla \cdot (\boldsymbol{v} \times \boldsymbol{w}) = \boldsymbol{w} \cdot \nabla \times \boldsymbol{v}$$

and

$$\boldsymbol{n} \cdot (\boldsymbol{v} \times \boldsymbol{w}) = \boldsymbol{w} \cdot \boldsymbol{n} \times \boldsymbol{v}$$

Applying the divergence theorem to the function $\boldsymbol{u}$ it is written as

$$\iint_{S} \boldsymbol{u}.\boldsymbol{n}\, dS = \iiint_{V} \nabla \cdot \boldsymbol{u}\, dV$$

The substitution of the previous expressions in this expression shows that

$$\iiint_{V} \nabla \cdot \boldsymbol{u}\, dV = \iint_{S} (\boldsymbol{v} \times \boldsymbol{w}) \cdot \boldsymbol{n}\, dS = \iint_{S} \boldsymbol{w} \cdot (\boldsymbol{n} \times \boldsymbol{v})\, dS$$

whereby

$$\boldsymbol{w}\cdot\iiint\limits_V \nabla\times\boldsymbol{v} = \boldsymbol{w}\cdot\iint\limits_S \boldsymbol{n}\times\boldsymbol{v}\,dS$$

As $\boldsymbol{w}$ is arbitrary it results in

$$\iiint\limits_V \nabla\times\boldsymbol{v} = \iint\limits_S \boldsymbol{n}\times\boldsymbol{v}\,dS$$

The concept of curl of a vector $\boldsymbol{u}$ can be generalized for a space E_N, in which the vector is associated to an antisymmetric tensor $\boldsymbol{A}$, and its order depends on the dimension of the space. This tensor is generated by means of the dot product between the Ricci pseudotensor and the vector's covariant derivative

$$A^{i_1\,i_2\cdots i_{p-2}} = \varepsilon^{i_1\,i_2\cdots i_{p-2}\,j\,k}\partial_j u_k \tag{4.4.11}$$

4.4.2 Orthogonal Curvilinear Coordinate Systems

With the expressions used in the previous item to demonstrate expression (4.4.10), there is in index notation for the curl coordinates of vector $\boldsymbol{u}$ in a curvilinear orthogonal coordinate system

$$\nabla\times\boldsymbol{u} = \frac{h_k}{h_1h_2h_3}\left(\frac{\partial h_j u^j}{\partial\overline{x}^i} - \frac{\partial h_i u^i}{\partial\overline{x}^j}\right)\boldsymbol{g}_k \tag{4.4.12}$$

where $h_i = \sqrt{g_{(ii)}}, h_j = \sqrt{g_{(jj)}}, h_k = \sqrt{g_{(kk)}}$ are the components of the metric tensor, and the indexes in parenthesis indicate no summation.

In a determinant form the result is

$$\nabla\times\boldsymbol{u} = \frac{1}{h_1h_2h_3}\begin{vmatrix} h_1\mathbf{g}_1 & h_2\mathbf{g}_2 & h_3\mathbf{g}_3 \\ \dfrac{\partial}{\partial\overline{x}^1} & \dfrac{\partial}{\partial\overline{x}^2} & \dfrac{\partial}{\partial\overline{x}^3} \\ h_1u^1 & h_2u^2 & h_3u^3 \end{vmatrix} \tag{4.4.13}$$

and with the physical components of vector u^{*i} it follows that

$$\nabla\times\boldsymbol{u}^* = \frac{h_k}{h_1h_2h_3}\left(\frac{\partial h_j u^{*j}}{\partial\overline{x}^i} - \frac{\partial h_i u^{*i}}{\partial\overline{x}^j}\right)\boldsymbol{g}_k \tag{4.4.14}$$

4.4.3 Properties

As the curl is the cross product of the *nabla* operator by a vectorial function, the properties of this vector product are valid.

For the sum of two vectorial functions $\boldsymbol{u}$ and $\boldsymbol{v}$:

$$\nabla \times (\boldsymbol{u} + \boldsymbol{u}) = \nabla \times \boldsymbol{u} + \nabla \times \boldsymbol{v} \tag{4.4.15}$$

and the successive applying of the curl to this sum provides

$$\nabla \times \nabla \times (\boldsymbol{u} + \boldsymbol{u}) = \nabla \times \nabla \times \boldsymbol{u} + \nabla \times \nabla \times \boldsymbol{v} \tag{4.4.16}$$

Considering the vectorial function $m\boldsymbol{u}$, where m is a scalar, the result of the cross product

$$\nabla \times (m\boldsymbol{u}) = m\nabla \times \boldsymbol{u} \tag{4.4.17}$$

Expressions (4.4.16) and (4.4.17) show that the curl, for these cases, is a linear operator, which is valid for the general case as will be shown in item 4.5.

4.4.4 Curl of a Tensor

The concept of curl of a vector in space E_N is developed in an analogous way. For instance, for the second-order tensor $T_{k_1 k_2}$ in space E_N exists the curl of order $(p-3)$, given by the cross product between the Ricci pseudotensor of order p and the tensor's covariant derivative, then

$$A^{i_1\, i_2 \cdots i_{p-3}} = \varepsilon^{i_1\, i_2 \cdots i_{p-3}\, j\, k_1 k_2} \partial_j T_{k_1 k_2} \tag{4.4.18}$$

Expression (4.4.18) shows that the Ricci pseudotensor is the generator of the antisymmetric tensor that represents the rotational of the tensor.

In absolute notation the result is

$$\nabla \times \boldsymbol{T} = (\nabla \otimes \boldsymbol{T}) \overline{\overline{\overline{\otimes}}} \boldsymbol{E} \tag{4.4.19}$$

where $\boldsymbol{E}$ is the Ricci pseudotensor.

In the particular case of the space E_4 the curl for the second-order tensor $T_{k\ell}$ is given by the components of vector A^i:

$$A^i = \varepsilon^{ijk\ell} \partial_j T_{k\ell}$$

Assuming that the second-order tensor is decomposed into two tensors, one symmetric and the other antisymmetric

$$\boldsymbol{T} = \boldsymbol{S} + \boldsymbol{A}$$

then

$$\operatorname{rot} \boldsymbol{T} = \operatorname{rot} \boldsymbol{S} + \operatorname{rot} \boldsymbol{A}$$

The components of the curl of the symmetric tensor are given by $\varepsilon^{i_1 i_2 \cdots i_{p-3} j k_1 k_2} \partial_j S_{k_1 k_2}$, i.e., are obtained by means of the dot product of the Ricci pseudotensor (antisymmetric) by the symmetric tensor which is null, whereby $\operatorname{rot} \boldsymbol{S} = 0$. Concluding that the curl of a symmetric tensor is null, and that only the antisymmetric tensor $\boldsymbol{A}$ generates the rotational of tensor $\boldsymbol{T}$. In the particular case in which $\operatorname{rot} \boldsymbol{T} = 0$ the tensor $\boldsymbol{T}$ defines an irrotational tensorial field.

The definition of curl of a second-order tensor can be applied to a tensor of order $p > 2$, whereby

$$A^{i_1 i_2 \cdots i_{q-p-1}} = \varepsilon^{i_1 i_2 \cdots i_{q-p-1} j k_1 \cdots k_p} \partial_j T_{k_1 \cdots k_p} \tag{4.4.20}$$

being $(q-1)$ the order of the Ricci pseudotensor, and $(q-p-1)$ the order of the antisymmetric tensor that represents the curl of the tensor.

Exercise 4.12 Calculate: (a) $\nabla \times \phi \boldsymbol{u}$; (b) $\boldsymbol{u} \times (\nabla \times \boldsymbol{v})$; (c) $\nabla \times (\boldsymbol{u} \times \boldsymbol{v})$.

(a) The curl of the field defined by the product of a scalar function $\phi(x^i)$ for a vectorial function $\boldsymbol{u}$ is given by

$$\nabla \times \phi \boldsymbol{u} = \boldsymbol{g}^j \times \frac{\partial (\phi u_k \boldsymbol{g}^k)}{\partial x^j} = \boldsymbol{g}^j \times \left(\frac{\partial \phi}{\partial x^j} u_k \boldsymbol{g}^k + \phi \frac{\partial u_k}{\partial x^j} \boldsymbol{g}^k + \phi u_k \frac{\partial \boldsymbol{g}^k}{\partial x^j} \right)$$

Substituting expression (2.4.4)

$$\frac{\partial \boldsymbol{g}^k}{\partial x^j} = -\Gamma^k_{mj} \boldsymbol{g}^m$$

in this expression

$$\nabla \times \phi \boldsymbol{u} = \boldsymbol{g}^j \times \left(\frac{\partial \phi}{\partial x^j} u_k \boldsymbol{g}^k + \phi \frac{\partial u_k}{\partial x^j} \boldsymbol{g}^k - \phi u_k \Gamma^k_{mj} \boldsymbol{g}^m \right)$$

The permutation of the indexes $k \leftrightarrow m$ in the last term provides

$$\nabla \times \phi \boldsymbol{u} = \left(\frac{\partial \phi}{\partial x^j} u_k + \phi \frac{\partial u_k}{\partial x^j} - \phi u_m \Gamma^m_{kj} \right) \boldsymbol{g}^j \times \boldsymbol{g}^k$$

and with expressions

$$\boldsymbol{g}^j \times \boldsymbol{g}^k = \varepsilon^{ijk}\boldsymbol{g}_i \quad \partial_j u_k = \frac{\partial u_k}{\partial x^j} - u_m \Gamma^m_{kj}$$

it follows that

$$\nabla \times \phi \boldsymbol{u} = \left(\frac{\partial \phi}{\partial x^j} u_k + \phi \partial_j u_k \right) \varepsilon^{ijk} \boldsymbol{g}_i = \frac{\partial \phi}{\partial x^j} u_k \varepsilon^{ijk} \boldsymbol{g}_i + \phi \left(\partial_j u_k \right) \varepsilon^{ijk} \boldsymbol{g}_i$$

Putting $\partial_j u_k = \nabla u_k$

$$\nabla \times \phi \boldsymbol{u} = \nabla \phi \times \boldsymbol{u} + \phi \nabla \times \boldsymbol{u} \Rightarrow \operatorname{rot} \phi \boldsymbol{u} = \operatorname{grad} \phi \times \boldsymbol{u} + \phi \operatorname{rot} \boldsymbol{u}$$

For the Cartesian coordinates

$$\nabla \times \phi \boldsymbol{u} = \left(\varepsilon_{ijk} u_k \frac{\partial \phi}{\partial x^j} + \varepsilon_{ijk} \phi \frac{\partial u_k}{\partial x^j} \right) \boldsymbol{g}_i$$

It is verified for this case that the curl is not a linear operator. For $\phi = m$, where m is a constant, the result with expression (4.4.17) is that this operator's linearity is valid by this particular case.

(b) The curl $\nabla \times \boldsymbol{v}$ is given by

$$\nabla \times \boldsymbol{v} = \boldsymbol{w} = \varepsilon^{kmn} \partial_m v_n \boldsymbol{g}^k$$

then

$$\boldsymbol{u} \times \boldsymbol{w} = \varepsilon_{ijk} u^j w^k \boldsymbol{g}_i$$

whereby substituting

$$\boldsymbol{u} \times (\nabla \times \boldsymbol{v}) = \varepsilon_{ijk} u^j \varepsilon^{kmn} \partial_m v_n \boldsymbol{g}_i = \varepsilon_{ijk} \varepsilon^{kmn} u^j \partial_m v_n \boldsymbol{g}_i$$

and with

$$\varepsilon_{ijk} \varepsilon^{kmn} = \delta^{mn}_{ij} \; \delta^{mn}_{ij} = \delta^m_i \delta^n_j - \delta^m_j \delta^n_i$$

is it follows that

$$\boldsymbol{u} \times (\nabla \times \boldsymbol{v}) = \left[\delta^m_i \delta^n_j u^j \partial_m v_n - \delta^m_j \delta^n_i u^j \partial_m v_n \right] \quad \boldsymbol{g}_i = \left(u^j \partial_i v_j - u^j \partial_j v_i \right) \boldsymbol{g}_i$$

For the Cartesian coordinates the result is

$$\boldsymbol{u} \times (\nabla \times \boldsymbol{v}) = \left(u_j \frac{\partial v_j}{\partial x^i} - u_j \frac{\partial v_i}{\partial x^j} \right) \boldsymbol{g}_i$$

(c) The cross product $\boldsymbol{u} \times \boldsymbol{v} = \boldsymbol{w}$ is given by

$$\boldsymbol{u} \times \boldsymbol{v} = \boldsymbol{w} = w^\ell \boldsymbol{g}_\ell = \varepsilon_{\ell mn} u^m v^n \boldsymbol{g}_\ell$$

thereby

$$\begin{aligned} \nabla \times (\boldsymbol{u} \times \boldsymbol{v}) &= \varepsilon^{ij\ell} \partial_j w^\ell \boldsymbol{g}_i = \varepsilon^{ij\ell} \partial_j (\varepsilon_{\ell mn} u^m v^n) \boldsymbol{g}_i = \delta^{ij}_{mn} \partial_j (u^m v^n) \boldsymbol{g}_i \\ &= \left(\delta^i_m \delta^j_n - \delta^i_n \delta^j_m \right) \partial_j (u^m v^n) \boldsymbol{g}_i = \left[\partial_j (u^i v^j) - \partial_j (u^j v^i) \right] \boldsymbol{g}_i \end{aligned}$$

For the Cartesian coordinates

$$\nabla \times (\boldsymbol{u} \times \boldsymbol{v}) = \left(v^j \frac{\partial u^i}{\partial x^j} + u^i \frac{\partial v^j}{\partial x^j} - v^i \frac{\partial u^j}{\partial x^j} - u^j \frac{\partial v^i}{\partial x^j} \right) \boldsymbol{g}_i = \boldsymbol{v} \cdot \nabla \times \boldsymbol{u} - \boldsymbol{u} \cdot \nabla \times \boldsymbol{v}$$

Exercise 4.13 Calculate $\nabla \times \boldsymbol{u}$ for the vector u expressed in cylindrical coordinates.

For the cylindrical coordinates the result is $h_1 = 1, h_2 = r, h_3 = 1$ and (u_r, u_θ, u_z). The determinant is given by the expression (4.4.13)

$$\nabla \times \boldsymbol{u} = \frac{1}{r} \begin{vmatrix} \mathbf{g}_r & r\mathbf{g}_\theta & \mathbf{g}_z \\ \dfrac{\partial}{\partial x^r} & \dfrac{\partial}{\partial x^\theta} & \dfrac{\partial}{\partial x^z} \\ u_r & r u_\theta & u_z \end{vmatrix}$$

which development provides

$$\nabla \times \boldsymbol{u} = \left(\frac{1}{r} \frac{\partial u_z}{\partial \theta} - \frac{\partial u_\theta}{\partial z} \right) \boldsymbol{g}_r + \left(\frac{\partial u_r}{\partial z} - \frac{\partial u_z}{\partial r} \right) \boldsymbol{g}_\theta + \left(\frac{1}{r} \frac{\partial r u_\theta}{\partial r} - \frac{1}{r} \frac{\partial u_r}{\partial \theta} \right) \boldsymbol{g}_z$$

Exercise 4.14 Calculate $\nabla \times \boldsymbol{u}$ for the vector $\boldsymbol{u}$ expressed in spherical coordinates.

For the spherical coordinates the result is $h_1 = 1$, $h_2 = r$, $h_3 = r \sin \phi$ and (u_r, u_ϕ, u_θ). The determinant given by expression (4.4.13)

$$\nabla \times \boldsymbol{u} = \frac{1}{r^2 \sin\phi} \begin{vmatrix} \mathbf{g}_r & r\mathbf{g}_\kappa & r\sin\phi\,\mathbf{g}_\theta \\ \dfrac{\partial}{\partial x^r} & \dfrac{\partial}{\partial x^\phi} & \dfrac{\partial}{\partial x^\theta} \\ u_r & ru_\phi & r\sin\phi\, u_\theta \end{vmatrix}$$

which development provides

$$\nabla \times \boldsymbol{u} = \frac{1}{r\sin\phi}\left(\frac{\partial u_\theta \sin\phi}{\partial\phi} - \frac{\partial u_\phi}{\partial\theta}\right)\boldsymbol{g}_r + \left(\frac{1}{r\sin\phi}\frac{\partial u_\theta r\sin\phi}{\partial r} - \frac{1}{r}\frac{\partial u_r}{\partial\theta}\right)$$
$$\boldsymbol{g}_\phi + \frac{1}{r}\left(\frac{\partial r u_\phi}{\partial r} - \frac{\partial u_r}{\partial\phi}\right)\boldsymbol{g}_\theta$$

Exercise 4.15 Let $\boldsymbol{r}$ be the position vector of the point in space E_3, show that: (a) $\nabla \times \boldsymbol{r} = 0$; (b) $\phi(r)\boldsymbol{r}$ is irrotational.

(a) With the definition of curl

$$\nabla \times \boldsymbol{r} = \left(\boldsymbol{g}_i \frac{\partial}{\partial x^i}\right) \times \boldsymbol{r} = \boldsymbol{g}_i \times \frac{\partial \boldsymbol{r}}{\partial x^i}$$

but

$$\frac{\partial \boldsymbol{r}}{\partial x^i} = \boldsymbol{g}_i$$

then

$$\nabla \cdot \boldsymbol{r} = \boldsymbol{g}_i \times \boldsymbol{g}_i = 0 \quad \text{Q.E.D.}$$

(b) A condition that a vectorial function must fulfill so that the field that it represents is irrotational is

$$\nabla \times [\phi(r)\boldsymbol{r}] = 0$$

and putting

$$\phi(r) = \psi$$

it follows that

$$\nabla \times [\phi(r)\boldsymbol{r}] = \operatorname{grad}\phi \times \boldsymbol{r} + \phi\nabla \times \boldsymbol{r} = \left[\phi'(r)\operatorname{grad} r\right] \times \boldsymbol{r} + \phi(r) \cdot \nabla \times \boldsymbol{r}$$

but

$$\nabla \times \boldsymbol{r} = 0$$

then

$$\nabla \times [\phi(r)\boldsymbol{r}] = \left[\phi'(r)\frac{1}{r}\boldsymbol{r}\right] \times \boldsymbol{r}$$

as

$$\boldsymbol{r} \times \boldsymbol{r} = 0$$

thus

$$\nabla \times [\phi(r)\boldsymbol{r}] = 0 \quad \text{Q.E.D.}$$

4.5 Successive Applications of the Nabla Operator

The operator ∇ can be applied successively to a field. The number of combinations of two out of the three differential operators, the gradient, the divergence, and the curl, are $3^2 = 9$ types of double operators. The combinations $\nabla \cdot (\nabla \cdot \boldsymbol{u})$ and $\nabla \times (\nabla \cdot \boldsymbol{u})$ have no mathematical meaning.

4.5.1 Basic Relations

(1) $$\nabla \cdot (\nabla \times \boldsymbol{u})$$

The curl of a vectorial function is given by

$$\nabla \times \boldsymbol{u} = \varepsilon^{k\ell m} \partial_\ell u_m \boldsymbol{g}_k = \boldsymbol{w} = w^k \boldsymbol{g}_k \tag{4.5.1}$$

$$w^k = \varepsilon^{k\ell m} \partial_\ell u_m \tag{4.5.2}$$

then

$$\nabla \cdot (\nabla \times \boldsymbol{u}) = \boldsymbol{g}^i \cdot \frac{\partial (w^k \boldsymbol{g}_k)}{\partial \boldsymbol{x}^i} = \boldsymbol{g}^i \cdot \left(\frac{\partial w^k}{\partial \boldsymbol{x}^i} \boldsymbol{g}_k + w^k \frac{\partial \boldsymbol{g}_k}{\partial \boldsymbol{x}^i} \right)$$

and with expression

$$\frac{\partial \boldsymbol{g}_k}{\partial \boldsymbol{x}^i} = \Gamma^n_{ki} \boldsymbol{g}_n$$

it follows that

$$\nabla \cdot (\nabla \times \boldsymbol{u}) = \boldsymbol{g}^i \cdot \left(\frac{\partial w^k}{\partial \boldsymbol{x}^i} \boldsymbol{g}_k + w^k \Gamma^n_{ki} \boldsymbol{g}_n \right)$$

The permutation of indexes $n \leftrightarrow k$ in the second member in parenthesis provides

$$\nabla \cdot (\nabla \times \boldsymbol{u}) = \left(\frac{\partial w^k}{\partial \boldsymbol{x}^i} + w^n \Gamma^k_{ni} \right) \boldsymbol{g}^i \cdot \boldsymbol{g}_k$$

and with

$$\boldsymbol{g}^i \cdot \boldsymbol{g}_k = \delta^i_k$$

the result is

$$\nabla \cdot (\nabla \times \boldsymbol{u}) = \left(\frac{\partial w^k}{\partial \boldsymbol{x}^k} + w^n \Gamma^k_{nk} \right) = \partial_k w^k$$

Substituting expression (4.5.2)

$$\nabla \cdot (\nabla \times \boldsymbol{u}) = \partial_k \left(\varepsilon^{k\ell m} \partial_\ell u_m \right) = \varepsilon^{k\ell m} \partial_k (\partial_\ell u_m) = \frac{\mathrm{e}^{k\ell m}}{\sqrt{g}} \partial_k (\partial_\ell u_m)$$

and interchanging the indexes $i, j, k = 1, 2, 3$ cyclically

$$\nabla \cdot (\nabla \times \boldsymbol{u}) = \frac{1}{\sqrt{g}} (\partial_k \partial_\ell u_m - \partial_\ell \partial_k u_m) \tag{4.5.3}$$

whereby

$$\nabla \cdot (\nabla \times \boldsymbol{u}) = 0 \tag{4.5.4}$$

Vector $\nabla \times \boldsymbol{u}$ represents a vectorial field associated to the vectorial function $\boldsymbol{u}$. Expression (4.5.4) defines the condition of existence for this function. The property of the field defined by the curl of the vectorial function $\boldsymbol{u}$ shows that the divergence of this field is null, i.e., the field is solenoidal.

In a reciprocal way for a solenoidal field $\nabla \cdot (\nabla \times \boldsymbol{u}) = 0$ a solenoidal vector $\boldsymbol{v}$ can be determined, such that $\boldsymbol{v} = (\nabla \times \boldsymbol{u})$. In this case the vector $\boldsymbol{v}$ derives from the potential function $\boldsymbol{u}$, being linked to this function.

(2)

$$\nabla \times (\nabla \phi)$$

The gradient of a scalar function $\phi(x^i)$ is given by

$$(\nabla \phi) = \boldsymbol{u} = \frac{\partial \phi}{\partial x^k} \boldsymbol{g}^k = u_k \boldsymbol{g}^k \tag{4.5.5}$$

then

$$\nabla \times (\nabla \phi) = \nabla \boldsymbol{u} = \varepsilon^{ijk} \partial_j u_k \boldsymbol{g}_i$$

it follows that

$$\nabla \times (\nabla \phi) = \varepsilon^{ijk} \partial_j \left(\frac{\partial \phi}{\partial x^k} \right) \boldsymbol{g}_i = \varepsilon^{ijk} \left(\frac{\partial^2 \phi}{\partial x^j \partial x^k} \right) \boldsymbol{g}_i = \frac{\mathrm{e}^{ijk}}{\sqrt{g}} \left(\frac{\partial^2 \phi}{\partial x^j \partial x^k} \right) \boldsymbol{g}_i$$

Interchanging the indexes $i, j, k = 1, 2, 3$ cyclically

$$\nabla \times (\nabla \phi) = \frac{1}{\sqrt{g}} \left(\frac{\partial^2 \phi}{\partial x^j \partial x^k} - \frac{\partial^2 \phi}{\partial x^k \partial x^j} \right) \boldsymbol{g}_i$$

As

$$\frac{\partial^2 \phi}{\partial x^j \partial x^k} = \frac{\partial^2 \phi}{\partial x^k \partial x^j}$$

it results in

$$\nabla \times (\nabla \phi) = 0 \tag{4.5.6}$$

The field that fulfills the condition given by expression (4.5.6) is called a conservative field, i.e., every vectorial field with potential is an irrotational field.

Let $\nabla \phi = \boldsymbol{u}$ the result is $\nabla \times (\nabla \phi) = \nabla \times \boldsymbol{u} = 0$, then

$$\frac{\partial u_i}{\partial x^j} = \frac{\partial u_j}{\partial x^i}$$

As $u_i dx^i$ is an exact differential it follows that for a scalar function $\phi(x^i)$:

$$\phi(x^i) = \frac{\partial \phi}{\partial x^i} dx^i \Rightarrow \left(u_i - \frac{\partial \phi}{\partial x^i}\right) dx^i = 0$$

whereby

$$u_i = \frac{\partial \phi}{\partial x^i}$$

This analysis shows that the vector $\boldsymbol{u}$ can be considered as the gradient of a scalar function $\phi(x^i)$ as long as it fulfills the condition $\nabla \times \boldsymbol{u} = 0$.
Expression (4.5.6) can be demonstrated changing only the order of the operations, for $(\nabla \times \nabla)\phi = 0$, where the term in parenthesis indicates the cross product of a vector by itself, which results in the null vector.
The condition $\nabla \times \boldsymbol{u} = 0$ being $\boldsymbol{u} = \nabla \phi(x^i)$, where the scalar field defined by the function $\phi(x^i)$ is divided into families of level surfaces $\phi(x^i) = C$, which do not intersect, so they form level surface "*layers*," leads to the denomination of lamellar field.

(3)
$$\nabla \times (\nabla \times \boldsymbol{u})$$

For $\nabla \times (\nabla \times \boldsymbol{u})$ using the Grassmann identity

$$\boldsymbol{u} \times (\boldsymbol{v} \times \boldsymbol{w}) = (\boldsymbol{u} \cdot \boldsymbol{w})\,\boldsymbol{v} - (\boldsymbol{u} \cdot \boldsymbol{v})\,\boldsymbol{w}$$

whereby

$$\nabla \times (\nabla \times \boldsymbol{u}) = \nabla(\nabla \cdot \boldsymbol{u}) - \nabla \cdot (\nabla \boldsymbol{u}) \tag{4.5.7}$$

In terms of the vector coordinates it follows that

$$\nabla \times \boldsymbol{u} = \varepsilon_{tjk}\left(\partial_\ell u^k\right) g^{\ell j} \boldsymbol{g}^t = \boldsymbol{w} = w_t \boldsymbol{g}^t$$

$$w_t = \varepsilon_{tjk}\left(\partial_\ell u^k\right) g^{\ell j}$$

$$\nabla \times \boldsymbol{w} = \boldsymbol{g}^s \times \frac{\partial (w_t \boldsymbol{g}^t)}{\partial x^s} = \boldsymbol{g}^s \times \left(\frac{\partial w_t}{\partial x^s}\boldsymbol{g}^t + w_t \frac{\partial \boldsymbol{g}^t}{\partial x^s}\right)$$

$$\frac{\partial \boldsymbol{g}^t}{\partial x^s} = -\Gamma^t_{sn} \boldsymbol{g}^n$$

$$\nabla \times \boldsymbol{w} = \boldsymbol{g}^s \times \left(\frac{\partial w_t}{\partial x^s} \boldsymbol{g}^t - w_t \Gamma^t_{sn} \boldsymbol{g}^n \right) = \left(\frac{\partial w_t}{\partial x^s} - w_n \Gamma^n_{st} \right) \boldsymbol{g}^s \times \boldsymbol{g}^t$$

$$\boldsymbol{g}^s \times \boldsymbol{g}^t = \varepsilon^{rst} \boldsymbol{g}_r$$

$$\nabla \times \boldsymbol{w} = \varepsilon^{rst} \left(\frac{\partial w_t}{\partial x^s} - w_n \Gamma^n_{st} \right) \boldsymbol{g}_r$$

$$\partial_s \varepsilon^{rst} = 0$$

$$\nabla \times \boldsymbol{w} = \varepsilon^{rst} (\partial_s w_t) \boldsymbol{g}_r = \varepsilon^{rst} \varepsilon_{ijk} \left(\partial_s \partial_\ell u^k g^{\ell j} \right) \boldsymbol{g}_r$$

$$\partial_s g^{\ell j} = 0$$

$$\nabla \times \boldsymbol{w} = \varepsilon^{rst} \varepsilon_{tjk} \left(\partial_s \partial_\ell u^k \right) g^{\ell j} \boldsymbol{g}_r$$

and with

$$\varepsilon^{rst} \varepsilon_{tjk} = \delta^{rs}_{jk} \quad \delta^{rs}_{jk} = \delta^r_j \delta^s_k - \delta^s_j \delta^r_k$$

it follows that

$$\nabla \times (\nabla \times \boldsymbol{u}) = \left[\delta^r_j \delta^s_k \left(\partial_s \partial_\ell u^k \right) g^{\ell j} - \delta^s_j \delta^r_k \left(\partial_s \partial_\ell u^k \right) g^{\ell j} \right] \boldsymbol{g}_r$$

whereby

$$\nabla \times (\nabla \times \boldsymbol{u}) = \left[(\partial_k \partial_\ell u^k) g^{\ell r} - (\partial_j \partial_\ell u^r) g^{\ell j} \right] \boldsymbol{g}_r \tag{4.5.8}$$

$$\nabla \times (\nabla \times \boldsymbol{u}) = \left[(\partial_k \partial^r u^k) g^{\ell r} - \left(\partial_j \partial^j u^r \right) \right] \boldsymbol{g}_r \tag{4.5.9}$$

For the Cartesian coordinates the result is

$$\nabla \times (\nabla \times \boldsymbol{u}) = \left(\frac{\partial^2 u_k}{\partial x^k \partial x^r} - \frac{\partial^2 u_r}{\partial^2 x^j} \right) \boldsymbol{g}_r \tag{4.5.10}$$

(4) $$\nabla (\nabla \cdot \boldsymbol{u})$$

For the gradient of a vector

$$\nabla \cdot \boldsymbol{u} = \left(\frac{\partial u^i}{\partial x^i} + u^j \Gamma^i_{ji} \right)$$

then

$$\nabla(\nabla \cdot \boldsymbol{u}) = \boldsymbol{g}^m \frac{\partial}{\partial x^m}\left(\frac{\partial u^i}{\partial x^i} + u^j \Gamma^i_{ji}\right)$$

The development provides

$$\nabla(\nabla \cdot \boldsymbol{u}) = \boldsymbol{g}^m \left(\frac{\partial^2 u^i}{\partial x^m \partial x^i} + \frac{\partial u^j}{\partial x^m}\Gamma^i_{ji} + u^j \frac{\partial \Gamma^i_{ji}}{\partial x^m}\right)$$

or

$$\nabla(\nabla \cdot \boldsymbol{u}) = \partial_m\left(\partial_i u^i\right)\boldsymbol{g}^m \tag{4.5.11}$$

and with

$$u = \phi'(r)$$
$$\frac{du}{dr} = \phi'(r)$$

it follows that

$$\nabla(\nabla \cdot \boldsymbol{u}) = g^{mk}\partial_m\left(\partial_i u^i\right)\boldsymbol{g}_k$$

whereby

$$\nabla(\nabla \cdot \boldsymbol{u}) = \partial^k\left(\partial_i u^i\right)\boldsymbol{g}_k \tag{4.5.12}$$

Exercise 4.16 Let $\phi(x^i)$ and $\psi(x^i)$ be scalar functions, show that: (a) $\nabla \times (\psi\nabla\phi + \phi\nabla\psi) = 0$; (b) $\nabla \cdot (\nabla\phi \times \nabla\psi) = 0$; (c) $tr(\nabla \otimes \boldsymbol{u}) = \nabla \cdot \boldsymbol{u}$.

(a) Putting

$$\nabla\phi = \boldsymbol{u} \quad \nabla\psi = \boldsymbol{v}$$

then

$$\nabla \times (\psi\nabla\phi + \phi\nabla\psi) = \nabla \times (\psi\boldsymbol{u} + \phi\boldsymbol{v}) = \nabla \times \psi\boldsymbol{u} + \nabla \times \phi\boldsymbol{v}$$

and with the expression shown in Exercise 4.12 it follows that

$$\nabla \times \psi\boldsymbol{u} = \nabla\psi \times \boldsymbol{u} + \psi \times \nabla\boldsymbol{u} = \nabla\psi \times \nabla\phi + \psi\nabla \times \nabla\phi$$
$$\nabla \times \phi\boldsymbol{v} = \nabla\phi \times \boldsymbol{v} + \phi \times \nabla\boldsymbol{v} = \nabla\phi \times \nabla\psi + \phi\nabla \times \nabla\psi$$

and with expression (4.5.6)

$$\nabla \times \nabla\phi = \nabla \times \nabla\psi = 0$$

and

$$\nabla\phi \times \nabla\psi = -\nabla\psi \times \nabla\phi$$

then

$$\nabla \times (\psi\nabla\phi + \phi\nabla\psi) = 0 \qquad \text{Q.E.D.}$$

(b) Putting

$$\nabla\phi = \boldsymbol{u} \quad \nabla\psi = \boldsymbol{v} \quad \nabla \times (\nabla\phi \times \nabla\psi) = 0$$

then

$$\nabla \cdot (\nabla\phi \times \nabla\psi) = \nabla \cdot (\boldsymbol{u} \times \boldsymbol{v})$$

With expression deducted in Exercise 4.7b it follows that

$$\nabla \cdot (\boldsymbol{u} \times \boldsymbol{v}) = \boldsymbol{v} \cdot \nabla \times \boldsymbol{u} - \boldsymbol{u} \cdot \nabla \times \boldsymbol{v}$$

$$\nabla \times (\nabla\phi \times \nabla\psi) = 0$$

and with expression (4.5.6)

$$\nabla \times \nabla\phi = \nabla \times \nabla\psi = 0$$

then

$$\nabla \cdot (\nabla\phi \times \nabla\psi) = 0 \qquad \text{Q.E.D.}$$

(c) With expression (4.2.11)

$$\nabla \otimes \boldsymbol{u} = (\partial_i u_k)\boldsymbol{g}^i \otimes \boldsymbol{g}^k$$

the result for $i = k$ is

$$\text{tr}(\nabla \otimes \boldsymbol{u}) = \partial_i u_i \infty$$

and comparing this result with expression (4.3.6)

$$\nabla \cdot \boldsymbol{u} = \partial_i u_i$$

it is verified that

$$tr(\nabla \otimes \boldsymbol{u}) = \nabla \cdot \boldsymbol{u} \quad \text{Q.E.D.}$$

4.5.2 *Laplace Operator*

The combination of the divergence and the gradient, in this order, defines the Laplace operator or Laplacian

$$\nabla^2 = \nabla \cdot \nabla = \Delta = D_k \cdot D^k = \partial_k \partial^k = \operatorname{div}\operatorname{grad} = \operatorname{lap} \tag{4.5.13}$$

A few authors denominate this operator of differential parameter of the second order of Beltrami, and use the spelling Δ_2 to represent it.

With the expression the contravariant derivative

$$\partial^k = g^{kj}\partial_j$$

it follows that for the Laplacian of an arbitrary tensor

$$\partial_k \partial^k T^{\cdots}_{\cdots} = \partial_k\left(\partial^k T^{\cdots}_{\cdots}\right) = \partial_k\left(g^{kj}\partial_j T^{\cdots}_{\cdots}\right) = g^{kj}\partial_k\left(\partial_j T^{\cdots}_{\cdots}\right) = \partial^j \partial_j T^{\cdots}_{\cdots}$$

that shows that the Laplacian operator is commutative.

For Cartesian coordinates the covariant and contravariant derivatives are equal

$$\partial_k = \frac{\partial \cdots}{\partial x^k} \qquad \partial^k = \frac{\partial \cdots}{\partial x^k}$$

$$\partial_k = \partial^k$$

resulting for the Laplacian

$$\int d\phi(r) = \int \frac{m_1}{r^2}\, dr$$

4.5.2.1 Laplacian of a Scalar Function

The Laplacian of the scalar function $\phi(x^i)$ expresses in a curvilinear coordinate system, with covariant derivative given by

$$\phi(r) = \frac{m_1}{r} + m_2$$

thus

$$\boldsymbol{H}(\cdots) = \boldsymbol{g}^i \nabla \otimes \boldsymbol{g}^j \nabla (\cdots)$$

The development of the covariant derivative of the term in parenthesis provides

$$\phi\left(x^i\right)$$

The contracted Christoffel symbol

$$\Gamma^k_{mk} = \frac{1}{\sqrt{g}} \frac{\partial\left(\sqrt{g}\right)}{\partial x^m}$$

provides

$$\nabla^2 \phi = \frac{\partial}{\partial x^k}\left(g^{kj} \frac{\partial \phi}{\partial x^j}\right) + g^{mj} \frac{\partial \phi}{\partial x^j} \frac{1}{\sqrt{g}} \frac{\partial\left(\sqrt{g}\right)}{\partial x^m}$$

whereby

$$\boldsymbol{H}(\phi) = \left\{ \frac{\partial}{\partial x^1} \frac{\partial}{\partial x^2} \frac{\partial}{\partial x^3} \right\} \boldsymbol{g}^i \otimes \boldsymbol{g}^j \left\{ \frac{\partial}{\partial x^1} \frac{\partial}{\partial x^2} \frac{\partial}{\partial x^3} \right\} \phi$$

$$= \begin{bmatrix} \dfrac{\partial^2 \phi}{\partial x^1 \partial x^1} & \dfrac{\partial^2 \phi}{\partial x^1 \partial x^2} & \dfrac{\partial^2 \phi}{\partial x^1 \partial x^3} \\ \dfrac{\partial^2 \phi}{\partial x^2 \partial x^1} & \dfrac{\partial^2 \phi}{\partial x^2 \partial x^2} & \dfrac{\partial^2 \phi}{\partial x^2 \partial x^3} \\ \dfrac{\partial^2 \phi}{\partial x^3 \partial x^1} & \dfrac{\partial^2 \phi}{\partial x^3 \partial x^2} & \dfrac{\partial^2 \phi}{\partial x^3 \partial x^3} \end{bmatrix} \boldsymbol{g}^i \otimes \boldsymbol{g}^j$$

or

$$\nabla^2 \phi = g^{ik}\left(\partial_{jk} \phi\right) \tag{4.5.14}$$

it follows that

$$\nabla^2 \phi = g^{ik}\left(\frac{\partial^2 \phi}{\partial x^j \partial x^k} - \frac{\partial \phi}{\partial x^m} \Gamma^m_{jk}\right) \tag{4.5.15}$$

In vectorial notation

$$\nabla \cdot (\nabla \phi) = \nabla^2 \phi \tag{4.5.16}$$

or

$$\mathrm{div}(\mathrm{grad}\phi) = \nabla^2 \phi \tag{4.5.17}$$

In space E_3 and in orthogonal Cartesian coordinates the result is $g_{ij} = \delta_{ij}$, then the Laplacian of a scalar function is the sum of its derivatives of the second order

$$\nabla^2 \phi = \frac{\partial^2 \phi}{\partial x^j \partial x^j} \tag{4.5.18}$$

4.5.3 Properties

The Laplacian of the sum of two scalar functions $\phi(x^i)$ and $\psi(x^i)$ provides

$$\nabla^2(\phi + \psi) = \nabla \cdot \nabla(\phi + \psi) = \nabla \cdot (\nabla \phi + \nabla \psi) = \nabla \cdot \nabla \phi + \nabla \cdot \nabla \psi$$

whereby

$$\nabla^2(\phi + \psi) = \nabla^2 \phi + \nabla^2 \psi$$

For the function $m\phi(x^i)$, where m is a scalar, this operator provides

$$\nabla^2(m\phi) = \nabla \cdot \nabla(m\phi) = \nabla \cdot m\nabla(\phi) = m\nabla \cdot \nabla(\phi)$$

whereby

$$\nabla^2(m\phi) = m\nabla \cdot \nabla(\phi)$$

These two demonstrations prove that the Laplacian is a linear operator.

The gradient of the product of two scalar functions is given by

$$\nabla(\phi\psi) = \psi\nabla(\phi) + \phi\nabla(\psi)$$

then

$$\nabla \cdot \nabla(\phi\psi) = \nabla \cdot [\psi\nabla(\phi) + \phi\nabla(\psi)]$$

Putting

$$\nabla\phi = \boldsymbol{u} \quad \nabla\psi = \boldsymbol{v}$$

thus

$$\nabla \cdot [\psi\nabla(\phi) + \phi\nabla(\psi)] = \nabla \cdot (\psi\boldsymbol{u} + \phi\boldsymbol{v})$$

Applying the distributive property of the divergence to this expression, and using the expression deducted in Exercise 4.7a it follows that

$$\nabla \cdot (\psi\boldsymbol{u} + \phi\boldsymbol{v}) = \nabla \cdot \psi\boldsymbol{u} + \nabla \cdot \phi\boldsymbol{v}$$
$$\nabla \cdot \psi\boldsymbol{u} = \nabla\psi \cdot \boldsymbol{u} + \psi\nabla\boldsymbol{u}$$
$$\nabla \cdot \phi\boldsymbol{v} = \nabla\phi \cdot \boldsymbol{v} + \phi\nabla\boldsymbol{v}$$

then

$$\begin{aligned}\nabla^2(\phi\psi) &= (\nabla\psi \cdot \boldsymbol{u} + \psi\nabla\boldsymbol{u}) + (\nabla\phi \cdot \boldsymbol{v} + \phi\nabla\boldsymbol{v})\\ &= (\boldsymbol{v} \cdot \boldsymbol{u} + \psi\nabla\boldsymbol{u}) + (\boldsymbol{u} \cdot \boldsymbol{v} + \phi\nabla\boldsymbol{v})\\ &= \psi\nabla\boldsymbol{u} + \phi\nabla\boldsymbol{v} + 2(\boldsymbol{v} \cdot \boldsymbol{u})\end{aligned}$$

but

$$\nabla\boldsymbol{u} = \nabla\nabla\phi = \nabla^2\phi \quad \nabla\boldsymbol{v} = \nabla\nabla\psi = \nabla^2\psi$$

whereby substituting

$$\nabla^2(\phi\psi) = \psi\nabla^2(\phi) + \phi\nabla^2(\psi) + 2\nabla\phi\nabla\psi$$

An equation involving the Laplacian of a scalar function that appears in various problems of physics and engineering, the Laplace equation, is given by

$$\nabla^2\phi(x^i) = 0 \tag{4.5.19}$$

The function $\phi(x^i) = x^4z + 3xy^2 - zxy + 1$ that fulfills this equation is said to be harmonic. In addition to satisfying the Laplace equation it must be regular in the domain D, with partial derivatives of the first order continuous in the interior and in the boundary of D, and derivatives of the second order also continuous in D, which can be discontinuous in the boundary of this domain.

The successive applying of the Laplacian to a scalar function $(r; z\sin\theta; e^\theta \cos z)$ results in the bi-harmonic equation

$$\phi(x^i) = xy + yz + xz \tag{4.5.20}$$

For

$$\frac{\partial \cdots}{\partial t^2} \tag{4.5.21}$$

where $\psi(x^i)$ is a scalar function, this partial differential equation is called Poisson's equation.

As a consequence of the definition of the Laplacian the result is that $\nabla^2 m = 0$, where m is a scalar. The Laplacian of a scalar function $\phi(x^i)$ is a scalar, then its physical components are equal to its ordinary components.

4.5.4 Orthogonal Coordinate Systems

With the gradient of the scalar function $\phi(x^i)$:

$$\nabla \phi(x^i) = \boldsymbol{g}^i \frac{\partial \phi}{\partial x^i} = \boldsymbol{u}$$

and the orthogonal components of the vectorial function $\boldsymbol{u}$ given by

$$\nabla \cdot \boldsymbol{u} = \frac{1}{h_1 h_2 h_3} \frac{\partial}{\partial x^i} \left(\frac{u_i h_1 h_2 h_3}{h_i} \right)$$

results for the Laplacian of this function expressed in an orthogonal coordinate system

$$\nabla^2 \phi = \frac{1}{h_1 h_2 h_3} \frac{\partial}{\partial x^i} \left(\frac{h_1 h_2 h_3}{h_i} \frac{\partial \phi}{\partial x^i} \right) \tag{4.5.22}$$

where h_1, h_2, h_3 are the components of the metric tensor.

4.5.5 Laplacian of a Vector

With expression (4.5.7)

$$\nabla^2 \boldsymbol{u} = \nabla(\nabla \cdot \boldsymbol{u}) - \nabla \times (\nabla \times \boldsymbol{u})$$

and substituting expressions (4.5.12) and (4.5.9) this expression becomes

$$\nabla^2 \boldsymbol{u} = \boldsymbol{g}_k \partial^k \left(\partial_i u^i\right) - \left[\left(\partial_k \partial^r u^k\right) - \left(\partial_j \partial^j u^r\right)\right] \boldsymbol{g}_r \tag{4.5.23}$$

The change of the indexes $k \rightarrow r$ in the first term to the right and indexes $k \rightarrow i$ in the second term to the right provides

$$\nabla^2 \boldsymbol{u} = \left[\left(\partial^r \partial_i u^i\right) - \left(\partial_i \partial^r u^i\right) + \left(\partial_j \partial^j u^r\right)\right] \boldsymbol{g}_r$$

As

$$\partial^r \partial_i u^i = \partial_i \partial^r u^i$$

thus

$$\nabla^2 \boldsymbol{u} = \left(\partial_j \partial^j u^r\right) \boldsymbol{g}_r \tag{4.5.24}$$

4.5.6 Curl of the Laplacian of a Vector

The curl of the Laplacian of a vector is $\nabla \times \nabla^2 \boldsymbol{u}$, and it can be developed by means of the Grassmann formula

$$\nabla^2 \boldsymbol{u} = \operatorname{div} \operatorname{grad} \boldsymbol{u} = \nabla \nabla \cdot \boldsymbol{u} - \nabla \times \nabla \times \boldsymbol{u} \tag{4.5.25}$$

or

$$\nabla \times \nabla \times \boldsymbol{u} = \nabla \nabla \cdot \boldsymbol{u} - \nabla^2 \boldsymbol{u}$$

The curl of this expression is given by

$$\nabla \times \nabla \times \nabla \times \boldsymbol{u} = \nabla \times \nabla \nabla \cdot \boldsymbol{u} - \nabla \times \nabla^2 \boldsymbol{u} \tag{4.5.26}$$

or

$$\nabla \times \nabla \times \nabla \times \boldsymbol{u} = \nabla \nabla \cdot (\nabla \times \boldsymbol{u}) - \nabla \cdot \nabla (\nabla \times \boldsymbol{u}) \tag{4.5.27}$$

Expressions (4.5.4) and (4.5.6) show, respectively, that

$$\nabla \cdot (\nabla \times \boldsymbol{u}) = 0$$

$$\nabla \times \nabla \phi = 0$$

whereby the result for expression (4.5.26) is

$$\nabla \times \nabla \times \nabla \times \boldsymbol{u} = \nabla \times \nabla\nabla \cdot \boldsymbol{u} - \nabla \times \nabla^2 \boldsymbol{u} = \nabla \times \nabla\phi - \nabla \times \nabla^2 \boldsymbol{u}$$
$$= -\nabla \times \nabla^2 \boldsymbol{u}$$

and for expression (4.5.27)

$$\nabla \times \nabla \times \nabla \times \boldsymbol{u} = -\nabla \cdot \nabla(\nabla \times \boldsymbol{u})$$

The result of these two expressions is

$$\nabla \times \nabla^2 \boldsymbol{u} = \nabla^2(\nabla \times \boldsymbol{u}) \tag{4.5.28}$$

or

$$\operatorname{rot} \operatorname{lap} \boldsymbol{u} = \operatorname{lap} \operatorname{rot} \boldsymbol{u} \tag{4.5.29}$$

It is concluded that the operators ∇^2 and $\nabla\times$ are commutative when applied to vector $\boldsymbol{u}$.

4.5.7 Laplacian of a Second-Order Tensor

The gradient of a second-order tensor is given by

$$\nabla \otimes \boldsymbol{T} = \partial_m T^{ij} \boldsymbol{g}^m \otimes \boldsymbol{g}_i \otimes \boldsymbol{g}_p$$

and the divergence of the tensor defined by the previous expression stays

$$\nabla \cdot \nabla \otimes \boldsymbol{T} = \boldsymbol{g}^k \cdot \partial_k \partial_m T^{ij} \boldsymbol{g}^m \otimes \boldsymbol{g}_i \otimes \boldsymbol{g}_p = \partial_k \partial_m T^{ij} \boldsymbol{g}^m \otimes \boldsymbol{g}_i \otimes \boldsymbol{g}_p \cdot \boldsymbol{g}^k$$
$$= \partial_k \partial_m T^{ij} \boldsymbol{g}^m \otimes \boldsymbol{g}_i \delta_p^k$$

whereby

$$\nabla \cdot \nabla \otimes \boldsymbol{T} = \partial_p \partial_m T^{ij} \boldsymbol{g}^m \otimes \boldsymbol{g}_i \tag{4.5.30}$$

is a second-order tensor.

Exercise 4.17 Calculate $\nabla^2\phi$ for the scalar function $\phi(x^i)$ expressed in cylindrical coordinates.

The tensorial expression that defines the Laplacian is

$$\nabla^2\phi = \frac{1}{\sqrt{g}}\frac{\partial}{\partial x^k}\left(\sqrt{g}g^{kj}\frac{\partial\phi}{\partial x^j}\right)$$

and for the cylindrical coordinates

$$g^{11} = 1 \quad g^{22} = \frac{1}{r^2} \quad g^{33} = 1$$

$$\nabla\phi = \frac{\partial\phi}{\partial r}\boldsymbol{g}_r + \frac{1}{r}\frac{\partial\phi}{\partial\theta}\boldsymbol{g}_\theta + \frac{\partial\phi}{\partial z}\boldsymbol{g}_z$$

it follows that

$$\nabla^2\phi = \frac{1}{r}\left[\frac{\partial}{\partial r}\left(r\frac{\partial\phi}{\partial r}\right) + \frac{\partial}{\partial\theta}\left(\frac{1}{r}\frac{\partial\phi}{\partial\phi}\right) + \frac{\partial}{\partial z}\left(r\frac{\partial\phi}{\partial z}\right)\right]$$

then

$$\nabla^2\phi = \frac{1}{r}\frac{\partial\phi}{\partial r} + \frac{\partial^2\phi}{\partial r^2} + \frac{1}{r^2}\frac{\partial^2\phi}{\partial\theta^2} + \frac{\partial^2\phi}{\partial z^2}$$

Exercise 4.18 Calculate $\nabla^2\phi$ for the scalar function $\phi(x^i)$ expressed in spherical coordinates.

The tensorial expression that defines the Laplacian is

$$\nabla^2\phi = \frac{1}{\sqrt{g}}\frac{\partial}{\partial x^k}\left(\sqrt{g}g^{kj}\frac{\partial\phi}{\partial x^j}\right)$$

and for the spherical coordinates

$$g^{11} = 1 \quad g^{22} = \frac{1}{r^2} \quad g^{33} = \frac{1}{r^2\sin^2\phi}$$

$$\nabla\phi = \frac{\partial\phi}{\partial r}\boldsymbol{g}_r + \frac{1}{r}\frac{\partial\phi}{\partial\phi}\boldsymbol{g}_\phi + \frac{1}{r\sin\phi}\frac{\partial\phi}{\partial x\theta}\boldsymbol{g}_\theta$$

it follows that

$$\nabla^2\phi = \frac{1}{r^2\sin\phi}\left[\frac{\partial}{\partial r}\left(r^2\sin\phi\frac{\partial\phi}{\partial r}\right) + \frac{\partial}{\partial\phi}\left(r^2\sin\phi\frac{1}{r^2}\frac{\partial\phi}{\partial\phi}\right) + \frac{\partial}{\partial\theta}\left(r^2\sin\phi\frac{1}{r^2\sin^2\phi}\frac{\partial\phi}{\partial\theta}\right)\right]$$

then

$$\nabla^2\phi = \frac{2}{r}\frac{\partial\phi}{\partial r} + \frac{\partial^2\phi}{\partial r^2} + \frac{1}{r^2 \sin\phi}\frac{\partial}{\partial\phi}\left(\sin\phi\frac{\partial\phi}{\partial\phi}\right) + \frac{1}{r^2\sin^2\phi}\frac{\partial^2\phi}{\partial\theta^2}$$

Exercise 4.19 Let $\boldsymbol{r}$ be the position vector of the point in space E_3, show that: (a) $\nabla^2\left(\frac{x}{r^3}\right) = 0$; (b) $\nabla^2(r^n\boldsymbol{r}) = n(n+3)\,r^{n-2}\boldsymbol{r}$; (c) $\nabla^2\phi(r) = \phi''(r) + \frac{2}{r}\phi'(r)$; (d) for $\nabla^2\phi(r) = 0$ the result is $\phi(r) = \frac{m_1}{r} + m_2$, where m_1, m_2 are constant.

(a) With the definition of Laplacian

$$\nabla^2\left(\frac{x}{r^3}\right) = \left(\frac{\partial^2}{\partial x^2} + \frac{\partial^2}{\partial y^2} + \frac{\partial^2}{\partial z^2}\right)\left(\frac{x}{r^3}\right)$$

and for the derivative with respect to the variable x

$$\frac{\partial^2}{\partial x^2}\left(\frac{x}{r^3}\right) = \frac{\partial}{\partial x}\left[\frac{\partial}{\partial x}\left(\frac{x}{r^3}\right)\right] = \frac{\partial}{\partial x}\left[\frac{1}{r^3} - \frac{3x}{r^4}\frac{\partial r}{\partial x}\right]$$

but

$$2r\frac{\partial r}{\partial x} = 2x$$

then

$$\frac{\partial^2}{\partial x^2}\left(\frac{x}{r^3}\right) = \frac{\partial}{\partial x}\left[\frac{1}{r^3} - \frac{3x}{r^4}\frac{x}{r}\right] = -\frac{3}{r^4}\frac{\partial r}{\partial x} - \frac{6x}{r^5} + \frac{15x^2}{r^6}\frac{\partial r}{\partial x} = -\frac{9x}{r^5} + \frac{15x^3}{r^2}$$

In an analogous way for the other derivatives it follows that

$$\frac{\partial^2}{\partial y^2}\left(\frac{y}{r^3}\right) = -\frac{3x}{r^5} + \frac{15xy}{r^7} \quad \frac{\partial^2}{\partial z^2}\left(\frac{y}{r^3}\right) = -\frac{3x}{r^5} + \frac{15xz}{r^7}$$

Then

$$\nabla^2\left(\frac{x}{r^3}\right) = -\frac{9x}{r^5} + \frac{15x^3}{r^2} - \frac{3x}{r^5} + \frac{15xy}{r^7} - \frac{3x}{r^5} + \frac{15xz}{r^7}$$

$$\nabla^2\left(\frac{x}{r^3}\right) = 0 \qquad \text{Q.E.D.}$$

(b) Putting

$$\nabla^2(r^n\boldsymbol{r}) = \nabla[\nabla\cdot(r^n\boldsymbol{r})]$$

it follows that

$$\begin{aligned}\nabla^2(r^n\boldsymbol{r}) &= \nabla[\nabla(r^n)\cdot\boldsymbol{r} + r^n\nabla\cdot\boldsymbol{r}] = \nabla\left[\left(nr^{n-3}\boldsymbol{r}\right)\cdot\boldsymbol{r} + 3r^n\right]\\ &= \nabla\left[\left(nr^{n-3}r^2\right) + 3r^n\right] = (n+3)\nabla r^n\end{aligned}$$

then

$$\nabla^2(r^n\boldsymbol{r}) = (n+3)\,nr^{-2}\boldsymbol{r} \quad \text{Q.E.D.}$$

(c) With the definition of Laplacian

$$\nabla^2\phi(r) = \nabla\cdot[\nabla\phi(r)]$$

it follows that

$$\nabla\cdot[\nabla\phi(r)] = \nabla\cdot\left[\phi'(r)\nabla r\right] = \nabla\cdot\left[\frac{1}{r}\phi'(r)\boldsymbol{r}\right] = \frac{1}{r}\phi'(r)\nabla\cdot\boldsymbol{r} + \boldsymbol{r}\cdot\nabla\left[\frac{1}{r}\phi'(r)\right]$$

but

$$\nabla\cdot\boldsymbol{r} = 3$$

so

$$\begin{aligned}\nabla\cdot[\nabla\phi(r)] &= \frac{3}{r}\phi'(r)\nabla\cdot\boldsymbol{r} + \boldsymbol{r}\cdot\left\{\frac{d}{dr}\left[\frac{1}{r}\phi'(r)\right]\nabla r\right\}\\ &= \frac{3}{r}\phi'(r) + \boldsymbol{r}\cdot\left\{\left[-\frac{1}{r^2}\phi'(r) + \frac{1}{r}\phi''(r)\right]\frac{1}{r}\boldsymbol{r}\right\}\\ &= \frac{3}{r}\phi'(r) + \left\{\left[-\frac{1}{r^2}\phi'(r) + \frac{1}{r}\phi''(r)\right]\frac{1}{r}\right\}\boldsymbol{r}\cdot\boldsymbol{r}\\ &= \frac{3}{r}\phi'(r) + \left\{\left[-\frac{1}{r^2}\phi'(r) + \frac{1}{r}\phi''(r)\right]\frac{1}{r}\right\}r^2\end{aligned}$$

then

$$\nabla\cdot[\nabla\phi(r)] = \frac{2}{r}\phi'(r) + \phi''(r)$$

(d) Let

$$\nabla^2\phi(r) = \phi''(r) + \frac{2}{r}\phi'(r) \Rightarrow \phi''(r) + \frac{2}{r}\phi'(r) = 0 \Rightarrow \frac{\phi''(r)}{\phi'(r)} = -\frac{2}{r}$$

Putting

$$u = \phi'(r) \Rightarrow \frac{du}{dr} = \phi'(r)$$

then

$$\frac{du}{u} = -\frac{2}{r}dr$$

and integrating

$$\int \frac{du}{u} = -\int \frac{2}{r}dr$$

it follows that

$$\ell n\,(u) = -\ell n\,(r^2) + \ell n\,(m_1) = \ell n\left(\frac{m_1}{r^2}\right)$$

or

$$\ell n\,\phi'(r) = \ell n\left(\frac{m_1}{r^2}\right) \Rightarrow \phi'(r) = \frac{m_1}{r^2}$$

Integrating

$$\int d\phi(r) = \int \frac{m_1}{r^2}dr$$

then

$$\phi(r) = \frac{m_1}{r} + m_2 \quad \text{Q.E.D.}$$

4.6 Other Differential Operators

4.6.1 Hesse Operator

The operator defined on a scalar field, given by the tensorial product of two *nabla* operators applied to the scalar function that field represents

$$H(\cdots) = g^i \nabla \otimes g^j \nabla(\cdots) \tag{4.5.31}$$

In matrix form the scalar function $\phi(x^i)$ in Cartesian coordinates in the space E_3 is

$$H(\phi) = \left\{ \frac{\partial}{\partial x^1} \frac{\partial}{\partial x^2} \frac{\partial}{\partial x^3} \right\} g^i \otimes g^j \left\{ \frac{\partial}{\partial x^1} \frac{\partial}{\partial x^2} \frac{\partial}{\partial x^3} \right\} \phi$$

$$= \begin{bmatrix} \frac{\partial^2 \phi}{\partial x^1 \partial x^1} & \frac{\partial^2 \phi}{\partial x^1 \partial x^2} & \frac{\partial^2 \phi}{\partial x^1 \partial x^3} \\ \frac{\partial^2 \phi}{\partial x^2 \partial x^1} & \frac{\partial^2 \phi}{\partial x^2 \partial x^2} & \frac{\partial^2 \phi}{\partial x^2 \partial x^3} \\ \frac{\partial^2 \phi}{\partial x^3 \partial x^1} & \frac{\partial^2 \phi}{\partial x^3 \partial x^2} & \frac{\partial^2 \phi}{\partial x^3 \partial x^3} \end{bmatrix} g^i \otimes g^j \tag{4.5.32}$$

This operator is a symmetric second-order tensor, is called Hessian or Hesse operator in homage to Ludwig Otto Hesse (1881–1874).

4.6.2 D'Alembert Operator

The differential operator defined by the expression

$$\square = \nabla^2 \cdots + \frac{1}{c^2} \frac{\partial \cdots}{\partial t^2} \tag{4.5.33}$$

where c is a scalar and $\frac{\partial \cdots}{\partial t^2}$ denotes the differentiation with respect to the time t, is called D'Alembert or D'Alembertian operator in homage to Jean Le Rond d'Alembert (1717–1783).

The applying of this operator to a field represented by the scalar function that depends on the position vector and the time provides as a result the scalar function

$$\square\ \phi(x^i;\, t) = \nabla^2 \phi(x^i;\, t) + \frac{1}{c^2} \frac{\partial \phi(x^i;\, t)}{\partial t^2} \tag{4.5.34}$$

If the field is represented by a vectorial function the result is the vector

$$\square\ u(x^i;\, t) = \nabla^2 u(x^i;\, t) + \frac{1}{c^2} \frac{\partial u(x^i;\, t)}{\partial t^2} \tag{4.5.35}$$

The notation $\square \ldots$ was initially applied by Cauchy to represent the Laplacian. The D'Alembertian is the four-dimensional equivalent to the Laplacian.

Problems

4.1 Calculate the gradient of the scalar functions:
(a) $\phi(x^i) = xy + yz + xz$; (b) $\phi(x^i) = xe^{x^2+y^2}$.
Answer: (a) $(y+z)\boldsymbol{i} + (x+z)\boldsymbol{j} + (x+y)\boldsymbol{k}$; (b) $(1+2x^2)e^{x^2+y^2}\boldsymbol{i} + 2xye^{x^2+y^2}\boldsymbol{j}$

4.2 Calculate the directional derivative of the scalar function $\phi = 2(x^1)^2 + 3(x^2)^2 + (x^3)^2$ at the point (2; 1; 3) in the direction of vector $\boldsymbol{u}(1;\ 0; -2)$.
Answer: -1.789.

4.3 Calculate $\operatorname{div} u_i$ with the vector $\boldsymbol{u}$ expressed in cylindrical coordinates by its covariant components $(r;\ z\sin\theta;\ e^{\theta}\cos z)$.
Answer: $\operatorname{div} u_i = 2 + \frac{1}{(r)^2} z\cos\theta - e^{\theta}\sin z$

4.4 Show that $\nabla\cdot(\phi\psi\boldsymbol{u}) = \phi\nabla\psi\cdot\boldsymbol{u} + \psi\nabla\phi\cdot\boldsymbol{u} + \phi\psi\nabla\cdot\boldsymbol{u}$, where ϕ, ψ are scalar functions and $\boldsymbol{u}$ is a vectorial function.

4.5 Calculate the curl of the following vectorial fields:
(a) $y^2\boldsymbol{i} + z^2\boldsymbol{j} + x^2\boldsymbol{k}$; (b) $xyz(x\boldsymbol{i} + y\boldsymbol{j} + z\boldsymbol{k})$.
Answer: (a) $-2(z\boldsymbol{i} + x\boldsymbol{j} + y\boldsymbol{k})$; (b) $(xz^2 - xy^2)\boldsymbol{i} + (x^2y - yz^2)\boldsymbol{j} + (y^2z - x^2z)\boldsymbol{k}$.

4.6 Calculate the Laplacian of the function $\phi(x^i) = x^4z + 3xy^2 - zxy + 1$.
Answer: $\nabla^2\phi = 12x^2z - 6x$.

Chapter 5
Riemann Spaces

5.1 Preview

The space provided with metric is called Riemann space, for which the tensorial formalism is based on the study in its first fundamental form, being complemented by the definition of curvature and by the concept of geodesics, which allows expanding the basic conceptions of the Euclidian geometry for this type of space with N dimensions.

In the Riemann spaces the covariant derivatives of tensors are equal to the partial derivatives when the coordinates are Cartesian, but the problem arises of researching how these derivatives behave when the coordinate system is curvilinear. The analysis of this derivative leads to the definition of curvature of the space, which is the fundamental parameter for the development of a consistent study of the Riemann spaces E_N.

The concepts and expressions of Tensor Calculus are essential for the formulation of the Theory of General Relativity, and it is for this theory just as the Integral and Differential Calculus is for the Classic Mechanics.

5.2 The Curvature Tensor

The Euclidian geometry is grounded on the basic concepts of point, straight line, and plane, and in various axioms. In this geometry a curved line is defined in the Euclidian space E_2 as the one that is not a straight line, and in the Euclidian space E_3 a curved surface is defined as the one that is not a plane. The curvature is an intrinsic characteristic of the space, so it is not a property measurable by comparison between distinct spaces.

© Springer International Publishing Switzerland 2016

E. de Souza Sánchez Filho, *Tensor Calculus for Engineers and Physicists*,
DOI 10.1007/978-3-319-31520-1_5

The conception of a Riemann geometry for the space E_N is grounded on the basic concepts of the Euclidian geometry in space E_3, which generalization is carried out by means of defining the metric for the space E_N, given by

$$ds^2 = \varepsilon g_{ij} dx^i dx^j$$

where $\varepsilon = \pm 1$ is a functional indicator.

The space in which metric can be writed as an Euclidian metric, positive and definite, is called a flat space, otherwise it is called space with curvature. The concept of curvature of the space E_N was firstly conceived by Riemann as a generalization of the study of a surface's curvature developed by Gauß. Riemann presented his results in a paper in 1861, published only in 1876. Christoffel in 1869 and R. Lipschitz in four papers published in 1869, 1870 (two articles), and 1877 obtained the same results as Riemann when studying the transformation of the quadratic differential formula $g_{ij} dx^i dx^j$ to the Euclidian metric $ds^2 = \sum_i (dx^i)^2$. The curvature analysis of the Riemann space E_N was carried out by Ricci-Curbastro and Levi-Civita who deducted the expression of the curvature tensor in a very formal and concise approach, which was also obtained by Christoffel, whose deduction has an extensive algebrism. In 1917 Tulio Levi-Civita, and after Jan Arnoldus Schouten (1918) and Karl Hessenberg found independently an interpretation for the curvature tensor associating it to the concept of parallel transport of vectors.

5.2.1 Formulation

The covariant derivative of a tensor is a tensor, just as when repeating this differentiation will provide a new tensor. However, the differentiation order with respect to the variables must be considered in this analysis.

For a function $\phi(x^i)$ of class C^2 that represents a scalar field exists the derivative $\frac{\partial \phi(x^i)}{\partial x^k}$ that represents a covariant vector. Differentiating again with respect to the variable x^j results by means of the partial differentiation rule of Differential Calculus

$$\frac{\partial^2 \phi(x^i)}{\partial x^k \partial x^j} = \frac{\partial^2 \phi(x^i)}{\partial x^j \partial x^k}$$

In this case, the covariant derivative is commutative. However, for tensors which components are functions of class C^2 represented in curvilinear coordinate systems this independence of the differentiation order in general is not verified. It is concluded that only the condition of the functions being class C^2 is not enough to ensure this independence.

For the case of a covariant vector u_i the result for its covariant derivative is the tensor with variance $(0,2)$:

$$\partial_j u_i = \frac{\partial u_i}{\partial x^j} - u_\ell \Gamma^\ell_{ij} \tag{5.2.1}$$

and with

$$\partial_j u_i = T_{ij}$$

it follows that for the covariant derivative of this tensor with respect to the variable x^k

$$\partial_k T_{ij} = \frac{\partial T_{ij}}{\partial x^k} - T_{\ell j}\Gamma^\ell_{ik} - T_{i\ell}\Gamma^\ell_{jk} \tag{5.2.2}$$

The substitution of expression (5.2.1) provides

$$\begin{aligned}\partial_k T_{ij} &= \frac{\partial(\partial_j u_i)}{\partial x^k} - (\partial_j u_\ell)\Gamma^\ell_{ik} - (\partial_\ell u_i)\Gamma^\ell_{jk}\\ &= \frac{\partial}{\partial x^k}\left(\frac{\partial u_i}{\partial x^j} - u_\ell \Gamma^\ell_{ij}\right) - \left(\frac{\partial u_\ell}{\partial x^j} - u_m \Gamma^m_{\ell j}\right)\Gamma^\ell_{ik} - \left(\frac{\partial u_i}{\partial x^\ell} - u_m \Gamma^m_{i\ell}\right)\Gamma^\ell_{jk}\end{aligned}$$

it follows that

$$\begin{aligned}\partial_k T_{ij} = \partial_j \partial_k u_i &= \frac{\partial^2 u_i}{\partial x^k \partial x^j} - \frac{\partial u_\ell}{\partial x^k}\Gamma^\ell_{ij} - u_\ell \frac{\partial \Gamma^\ell_{ij}}{\partial x^j} - \frac{\partial u_\ell}{\partial x^j}\Gamma^\ell_{ik}\\ &\quad + u_m \Gamma^m_{\ell j}\Gamma^\ell_{ik} - \frac{\partial u_i}{\partial x^\ell}\Gamma^\ell_{jk} + u_m \Gamma^m_{i\ell}\Gamma^\ell_{jk}\end{aligned} \tag{5.2.3}$$

that represents a tensor with variance (0, 3).

The inversion of the differentiation order provides

$$\begin{aligned}\partial_k \partial_j u_i = \frac{\partial^2 u_i}{\partial x^j \partial x^k} - \frac{\partial u_\ell}{\partial x^j}\Gamma^\ell_{ik} - u_\ell \frac{\partial \Gamma^\ell_{ik}}{\partial x^j} - \frac{\partial u_\ell}{\partial x^k}\Gamma^\ell_{ij} + u_m \Gamma^m_{\ell k}\Gamma^\ell_{ij} - \frac{\partial u_i}{\partial x^\ell}\Gamma^\ell_{kj}\\ + u_m \Gamma^m_{i\ell}\Gamma^\ell_{kj}\end{aligned} \tag{5.2.4}$$

In Differential Calculus the differentiation order does not change the result obtained then

$$\frac{\partial^2 u_i}{\partial x^j \partial x^k} = \frac{\partial^2 u_i}{\partial x^k \partial x^j}$$

Subtracting expression (5.2.4) from expression (5.2.3) and considering the symmetry of the Christoffel symbols

$$\partial_j \partial_k u_i - \partial_k \partial_j u_i = u_\ell \left(\frac{\partial \Gamma^\ell_{ik}}{\partial x^j} - \frac{\partial \Gamma^\ell_{ij}}{\partial x^k} \right) + u_m \left(\Gamma^m_{\ell j} \Gamma^\ell_{ik} - \Gamma^m_{\ell k} \Gamma^\ell_{ij} \right)$$

and with the permutation of the dummy indexes $m \leftrightarrow \ell$ in the second term to the right

$$\partial_j \partial_k u_i - \partial_k \partial_j u_i = u_\ell \left[\left(\frac{\partial \Gamma^\ell_{ik}}{\partial x^j} - \frac{\partial \Gamma^\ell_{ij}}{\partial x^k} \right) + \left(\Gamma^\ell_{mj} \Gamma^m_{ik} - \Gamma^\ell_{mk} \Gamma^m_{ij} \right) \right]$$

Putting

$$R^\ell_{ijk} = \frac{\partial \Gamma^\ell_{ik}}{\partial x^j} - \frac{\partial \Gamma^\ell_{ij}}{\partial x^k} + \Gamma^\ell_{mj} \Gamma^m_{ik} - \Gamma^\ell_{mk} \Gamma^m_{ij} \tag{5.2.5}$$

results in

$$\partial_j \partial_k u_i - \partial_{kj} u_i = u_\ell R^\ell_{ijk}$$

The quotient law is used for verifying if the variety R^ℓ_{ijk} is a tensor, carrying out the inner product of vector u_ℓ by R^ℓ_{ijk}:

$$R^\ell_{ijk} u_\ell = R^\ell_{ijk\ell} = R_{ijk}$$

The transformation law of tensors to the variety R_{ijk} is given by

$$\overline{R}_{pqr} = \frac{\partial x^i}{\partial \overline{x}^p} \frac{\partial x^j}{\partial \overline{x}^q} \frac{\partial x^k}{\partial \overline{x}^r} R_{ijk}$$

for the vector u_ℓ the result of the transformation law is

$$u_m = \frac{\partial \overline{x}^\ell}{\partial x^m} \overline{u}_\ell$$

that substituted in previous expression provides

$$\overline{R}_{pqr} = \frac{\partial x^i}{\partial \overline{x}^p} \frac{\partial x^j}{\partial \overline{x}^q} \frac{\partial x^k}{\partial \overline{x}^r} \frac{\partial \overline{x}^\ell}{\partial x^m} R^\ell_{ijk} \overline{u}_\ell$$

In the coordinate system $\overline{X}^i$ the variety $\overline{R}_{pqr}$ is given by

$$\overline{R}_{pqr} = \overline{R}^{\ell}_{pqr}\overline{u}_{\ell}$$

whereby

$$\left(\overline{R}^{\ell}_{pqr} - \frac{\partial x^i}{\partial \overline{x}^p}\frac{\partial x^j}{\partial \overline{x}^q}\frac{\partial x^k}{\partial \overline{x}^r}\frac{\partial \overline{x}^{\ell}}{\partial x^m}R^{\ell}_{ijk}\right)\overline{u}_{\ell} = 0$$

As $\bar{u}_{\ell}$ is an arbitrary vector it results in

$$\overline{R}^{\ell}_{pqr} = \frac{\partial x^i}{\partial \overline{x}^p}\frac{\partial x^j}{\partial \overline{x}^q}\frac{\partial x^k}{\partial \overline{x}^r}\frac{\partial \overline{x}^{\ell}}{\partial x^m}R^{\ell}_{ijk}$$

that represents the transformation law of tensor with variances $(1,3)$, as R^{ℓ}_{ijk} is a tensor. The tensor defined by expression (5.2.5) is called Riemann–Christoffel curvature tensor, Riemann–Christoffel mixed tensor, or Riemann–Christoffel tensor of the second kind, or simply curvature tensor. This tensor defines a tensorial field that depends only on the metric tensor and its derivatives up to the second order, and classifies the space, for thus $R^{\ell}_{ijk} \neq 0$ the result is a space with curvature.

5.2.2 *Differentiation Commutativity*

The formulation of an analogous analysis for a contravariant vector u^i, which generates a mixed tensor with variance $(1,1)$, is carried out by calculating firstly the covariant derivative of this vector with respect to the coordinate x^j:

$$\partial_j u^i = \frac{\partial u^i}{\partial x^j} + u^{\ell}\Gamma^i_{j\ell} = T^i_j \tag{5.2.6}$$

The covariant derivative of the second order of this vector with respect to the coordinate x^k is given by

$$\partial_k(\partial_j u^i) = \partial_k T^i_j = \frac{\partial T^i_j}{\partial x^k} + T^{\ell}_j\Gamma^i_{\ell k} - T^i_{\ell}\Gamma^{\ell}_{jk}$$

Substituting expression (5.2.6) in this expression

$$\partial_k(\partial_j u^i) = \frac{\partial}{\partial x^k}\left(\frac{\partial u^i}{\partial x^j} + u^{\ell}\Gamma^i_{j\ell}\right) + \left(\frac{\partial u^{\ell}}{\partial x^j} + u^m\Gamma^{\ell}_{jm}\right)\Gamma^i_{\ell k} - \left(\frac{\partial u^i}{\partial x^{\ell}} + u^m\Gamma^i_{\ell m}\right)\Gamma^{\ell}_{jk}$$

whereby

$$\partial_k(\partial_j u^i) = \frac{\partial^2 u^i}{\partial x^k \partial x^j} + \frac{\partial u^\ell}{\partial x^k}\Gamma^i_{j\ell} + u^\ell \frac{\partial \Gamma^i_{j\ell}}{\partial x^k} + \frac{\partial u^\ell}{\partial x^j}\Gamma^i_{\ell k} + u^m \Gamma^\ell_{jm}\Gamma^i_{\ell k} - \frac{\partial u^i}{\partial x^\ell}\Gamma^\ell_{jk} - u^m \Gamma^i_{\ell m}\Gamma^\ell_{jk} \tag{5.2.7}$$

The inversion of the differentiation is obtained interchanging indexes $j \leftrightarrow k$, so

$$\partial_j(\partial_k u^i) = \frac{\partial^2 u^i}{\partial x^j \partial x^k} + \frac{\partial u^\ell}{\partial x^j}\Gamma^i_{k\ell} + u^\ell \frac{\partial \Gamma^i_{k\ell}}{\partial x^j} + \frac{\partial u^\ell}{\partial x^k}\Gamma^i_{\ell j} + u^m \Gamma^\ell_{km}\Gamma^i_{\ell j} - \frac{\partial u^i}{\partial x^\ell}\Gamma^\ell_{kj} - u^m \Gamma^i_{\ell m}\Gamma^\ell_{kj} \tag{5.2.8}$$

As in the partial derivative the order of differentiation does not change the result

$$\frac{\partial^2 u^i}{\partial x^k \partial x^j} = \frac{\partial^2 u^i}{\partial x^j \partial x^k}$$

and subtracting expression (5.2.7) from expression (5.2.8)

$$\begin{aligned}\partial_k(\partial_j u^i) - \partial_j(\partial_k u^i) = &\left(\frac{\partial^2 u^i}{\partial x^k \partial x^j} + \frac{\partial u^\ell}{\partial x^k}\Gamma^i_{j\ell} + u^\ell \frac{\partial \Gamma^i_{j\ell}}{\partial x^k} + \frac{\partial u^\ell}{\partial x^j}\Gamma^i_{\ell k}\right.\\ &\left. + u^m \Gamma^\ell_{jm}\Gamma^i_{\ell k} - \frac{\partial u^i}{\partial x^\ell}\Gamma^\ell_{jk} - u^m \Gamma^i_{\ell m}\Gamma^\ell_{jk}\right)\\ &-\left(\frac{\partial^2 u^i}{\partial x^j \partial x^k} + \frac{\partial u^\ell}{\partial x^j}\Gamma^i_{k\ell} + u^\ell \frac{\partial \Gamma^i_{k\ell}}{\partial x^j} + \frac{\partial u^\ell}{\partial x^k}\Gamma^i_{\ell j}\right.\\ &\left. + u^m \Gamma^\ell_{km}\Gamma^i_{\ell j} - \frac{\partial u^i}{\partial x^\ell}\Gamma^\ell_{kj} - u^m \Gamma^i_{\ell m}\Gamma^\ell_{kj}\right)\end{aligned}$$

and with the symmetry of the Christoffel symbols

$$\partial_k(\partial_j u^i) - \partial_j(\partial_k u^i) = u^\ell\left(\frac{\partial \Gamma^i_{j\ell}}{\partial x^k} - \frac{\partial \Gamma^i_{k\ell}}{\partial x^j}\right) + u^m\left(\Gamma^\ell_{jm}\Gamma^i_{\ell k} - \Gamma^\ell_{km}\Gamma^i_{\ell j}\right)$$

The permutation of indexes $\ell \leftrightarrow m$ in the last two terms provides

$$\partial_k(\partial_j u^i) - \partial_j(\partial_k u^i) = \left(\frac{\partial \Gamma^i_{j\ell}}{\partial x^k} - \frac{\partial \Gamma^i_{k\ell}}{\partial x^j} +^m_{j\ell} \Gamma^i_{mk} - \Gamma^m_{k\ell}\Gamma^i_{mj}\right)u^\ell$$

and putting

$$R^i_{\ell kj} = \frac{\partial \Gamma^i_{j\ell}}{\partial x^k} - \frac{\partial \Gamma^i_{k\ell}}{\partial x^j} + \Gamma^m_{j\ell}\Gamma^i_{mk} - \Gamma^m_{k\ell}\Gamma^i_{mj} \tag{5.2.9}$$

it results in

$$\partial_k\left(\partial_j u^i\right) - \partial_j\left(\partial_k u^i\right) = R^i_{\ell kj} u^\ell \tag{5.2.10}$$

The permutation of indexes $j \leftrightarrow k$ provides

$$\partial_j\left(\partial_k u^i\right) - \partial_k\left(\partial_j u^i\right) = R^i_{\ell jk} u^\ell$$

where

$$R^i_{\ell jk} = \frac{\partial \Gamma^i_{k\ell}}{\partial x^j} - \frac{\partial \Gamma^i_{j\ell}}{\partial x^k} + \Gamma^m_{k\ell}\Gamma^i_{mj} - \Gamma^m_{j\ell}\Gamma^i_{mk} \tag{5.2.11}$$

This analysis shows that $R^i_{\ell jk} = 0 \Rightarrow \partial_j \partial_k u^k = \partial_k \partial_j u^k$, i.e., the space is flat. The necessary and sufficient condition so that the differentiation commutativity be valid is that the tensor $R^i_{\ell jk}$ be null.

5.2.3 *Antisymmetry of Tensor* $\mathbf{R}^i_{\ell jk}$

The comparison of expressions (5.2.9) and (5.2.11) shows that the Riemann–Christoffel curvature tensor is antisymmetric with respect to the last two indexes

$$R^i_{\ell kj} = -R^i_{\ell jk}$$

5.2.4 *Notations for Tensor* $\mathbf{R}^i_{\ell jk}$

Putting the indexes in the sequence i, j, k, ℓ the result in tensorial notation is

$$\boldsymbol{R} = R^\ell_{ijk} \boldsymbol{g}_\ell \otimes \boldsymbol{g}^k \otimes \boldsymbol{g}^j \otimes \boldsymbol{g}^i \tag{5.2.12}$$

and rewriting the Riemann–Christoffel curvature tensor as

$$R^{\ell}_{ijk} = \frac{\partial \Gamma^{\ell}_{ik}}{\partial x^j} - \frac{\partial \Gamma^{\ell}_{ij}}{\partial x^k} + \Gamma^{m}_{ik}\Gamma^{\ell}_{mj} - \Gamma^{m}_{ij}\Gamma^{\ell}_{mk}$$

the result in symbolic form by means of determinants is

$$R^{\ell}_{ijk} = \begin{vmatrix} \dfrac{\partial}{\partial x^j} & \dfrac{\partial}{\partial x^k} \\ \Gamma^{\ell}_{ij} & \Gamma^{\ell}_{ik} \end{vmatrix} + \begin{vmatrix} \Gamma^{m}_{ik} & \Gamma^{m}_{ij} \\ \Gamma^{\ell}_{mk} & \Gamma^{\ell}_{mj} \end{vmatrix} \tag{5.2.13}$$

5.2.5 *Uniqueness of Tensor* $\mathbf{R}^{\ell}_{ijk}$

The metric tensor g_{ij} and its conjugated tensor g^{ij} are unique in a Riemann space, then their partial derivatives of the first and second order the Christoffel symbols of this space are unique at point $x^i \in E_N$. Thus it is verified that expression (5.2.11) does not ensure that tensor $R^{i}_{\ell jk}$ is the only tensor that can be expressed by the derivatives of the first and second order of the metric tensor.

However, the covariant derivatives of a contravariant vector with respect to the coordinates of a referential system are unique at point $x^i \in E_N$, and having the Riemann–Christoffel curvature tensor with variance (1, 3) obtained by means of these derivatives, it is concluded that it is unique in the point being considered. Expressions (5.2.5) and (5.2.11) obtained in distinct manners indicate this tensor's uniqueness.

For the points $x^i \in E_N$ in which the Christoffel symbols are null, it is verified that $R^{i}_{\ell jk}$ is expressed by means of a linear combination of the derivatives of the second order of the metric tensor.

5.2.6 *First Bianchi Identity*

The Riemann–Christoffel curvature tensor

$$R^{\ell}_{ijk} = \frac{\partial \Gamma^{\ell}_{ik}}{\partial x^j} - \frac{\partial \Gamma^{\ell}_{ij}}{\partial x^k} + \Gamma^{m}_{ik\Gamma}\Gamma^{\ell}_{mj} - \Gamma^{m}_{ij}\Gamma^{\ell}_{mk}$$

and the cyclic permutations of indexes i, j, k generate the expressions

$$R^{\ell}_{jki} = \frac{\partial \Gamma^{\ell}_{ji}}{\partial x^k} - \frac{\partial \Gamma^{\ell}_{jk}}{\partial x^i} + \Gamma^{m}_{ji}\Gamma^{\ell}_{mk} - \Gamma^{m}_{jk}\Gamma^{\ell}_{mi}$$

$$R^{\ell}_{kij} = \frac{\partial \Gamma^{\ell}_{kj}}{\partial x^i} - \frac{\partial \Gamma^{\ell}_{ki}}{\partial x^j} + \Gamma^{m}_{kj}\Gamma^{\ell}_{mi} - \Gamma^{m}_{ki}\Gamma^{\ell}_{mj}$$

The sum of these three expressions provides the first Bianchi identity for the Riemann–Christoffel curvature tensor

$$R^{\ell}_{ijk} + R^{\ell}_{jki} + R^{\ell}_{kij} = 0 \tag{5.2.14}$$

5.2.7 Second Bianchi Identity

The covariant derivative of a tensor with variance (1, 3) is given by

$$\partial_k T^{j}_{p\ell m} = \frac{\partial T^{j}_{p\ell m}}{\partial x^k} - T^{j}_{q\ell m}\Gamma^{q}_{pk} - T^{j}_{pqm}\Gamma^{q}_{\ell k} - T^{j}_{p\ell q}\Gamma^{q}_{mk} + T^{q}_{p\ell m}\Gamma^{j}_{kq}$$

whereby for the Riemann–Christoffel curvature tensor yields it follows that

$$R^{\ell}_{ijk} = \frac{\partial \Gamma^{\ell}_{ik}}{\partial x^j} - \frac{\partial \Gamma^{\ell}_{ij}}{\partial x^k} + \Gamma^{m}_{ik}\Gamma^{\ell}_{mj} - \Gamma^{m}_{ij}\Gamma^{\ell}_{mk}$$

The covariant derivative with respect to the coordinate x^p is given by

$$\begin{aligned}\partial_p R^{\ell}_{ijk} = {} & \frac{\partial^2 \Gamma^{\ell}_{ik}}{\partial x^p \partial x^j} - \frac{\partial^2 \Gamma^{\ell}_{ij}}{\partial x^p \partial x^k} + \frac{\partial \Gamma^{m}_{ik}}{\partial x^p}\Gamma^{\ell}_{mj} + \Gamma^{m}_{ik}\frac{\partial \Gamma^{\ell}_{mj}}{\partial x^p} - \frac{\partial \Gamma^{m}_{ij}}{\partial x^p}\Gamma^{\ell}_{mk} - \Gamma^{m}_{ij}\frac{\partial \Gamma^{\ell}_{mk}}{\partial x^p} \\ & + R^{m}_{ijk}\Gamma^{\ell}_{mp} - R^{\ell}_{mjk}\Gamma^{m}_{ip} - R^{\ell}_{imk}\Gamma^{m}_{jp} - R^{\ell}_{ijm}\Gamma^{m}_{kp}\end{aligned}$$

and with the cyclic permutation of indexes j, k, p it follows that

$$\begin{aligned}\partial_j R^{\ell}_{ikp} = {} & \frac{\partial^2 \Gamma^{\ell}_{ip}}{\partial x^j \partial x^k} - \frac{\partial^2 \Gamma^{\ell}_{ik}}{\partial x^j \partial x^p} + \frac{\partial \Gamma^{m}_{ip}}{\partial x^j}\Gamma^{\ell}_{mk} + \Gamma^{m}_{ip}\frac{\partial \Gamma^{\ell}_{mk}}{\partial x^j} - \frac{\partial \Gamma^{m}_{ik}}{\partial x^j}\Gamma^{\ell}_{mp} - \Gamma^{m}_{ik}\frac{\partial \Gamma^{\ell}_{mp}}{\partial x^j} \\ & + R^{m}_{ikp}\Gamma^{\ell}_{mj} - R^{\ell}_{mkp}\Gamma^{m}_{ij} - R^{\ell}_{imp}\Gamma^{m}_{kj} - R^{\ell}_{ikm}\Gamma^{m}_{pj} \\ \partial_k R^{\ell}_{ipj} = {} & \frac{\partial^2 \Gamma^{\ell}_{ij}}{\partial x^k \partial x^p} - \frac{\partial^2 \Gamma^{\ell}_{ip}}{\partial x^k \partial x^j} + \frac{\partial \Gamma^{m}_{ij}}{\partial x^k}\Gamma^{\ell}_{mp} + \Gamma^{m}_{ij}\frac{\partial \Gamma^{\ell}_{mp}}{\partial x^k} - \frac{\partial \Gamma^{m}_{ip}}{\partial x^k}\Gamma^{\ell}_{mj} - \Gamma^{m}_{ip}\frac{\partial \Gamma^{\ell}_{mj}}{\partial x^k} \\ & + R^{m}_{ipj}\Gamma^{\ell}_{mk} - R^{\ell}_{mpj}\Gamma^{m}_{ik} - R^{\ell}_{imj}\Gamma^{m}_{pk} - R^{\ell}_{ipm}\Gamma^{m}_{jk}\end{aligned}$$

The sum of these three expressions provides

$$\begin{aligned}
\partial_p R^{\ell}_{ikj} + \partial_j R^{\ell}_{ikp} + \partial_k R^{\ell}_{ipj} = & \frac{\partial^2 \Gamma^{\ell}_{ik}}{\partial x^p \partial x^j} - \frac{\partial^2 \Gamma^{\ell}_{ij}}{\partial x^p \partial x^k} + \frac{\partial \Gamma^m_{ik}}{\partial x^p} \Gamma^{\ell}_{mj} + \Gamma^m_{ik} \frac{\partial \Gamma^{\ell}_{mj}}{\partial x^p} - \frac{\partial \Gamma^m_{ij}}{\partial x^p} \Gamma^{\ell}_{mk} \\
& - \Gamma^m_{ij} \frac{\partial \Gamma^{\ell}_{mk}}{\partial x^p} + R^m_{ijk} \Gamma^{\ell}_{mp} - R^{\ell}_{mjk} \Gamma^m_{ip} - R^{\ell}_{imk} \Gamma^m_{jp} - R^{\ell}_{ijm} \Gamma^m_{kp} \\
& + \frac{\partial^2 \Gamma^{\ell}_{ip}}{\partial x^j \partial x^k} - \frac{\partial^2 \Gamma^{\ell}_{ik}}{\partial x^j \partial x^p} + \frac{\partial \Gamma^m_{ip}}{\partial x^j} \Gamma^{\ell}_{mk} + \Gamma^m_{ip} \frac{\partial \Gamma^{\ell}_{mk}}{\partial x^j} - \frac{\partial \Gamma^m_{ik}}{\partial x^j} \Gamma^{\ell}_{mp} \\
& - \Gamma^m_{ik} \frac{\partial \Gamma^{\ell}_{mp}}{\partial x^j} + R^m_{ikp} \Gamma^{\ell}_{mj} - R^{\ell}_{mkp} \Gamma^m_{ij} - R^{\ell}_{imp} \Gamma^m_{kj} - R^{\ell}_{ikm} \Gamma^m_{pj} \\
& + \frac{\partial^2 \Gamma^{\ell}_{ij}}{\partial x^k \partial x^p} - \frac{\partial^2 \Gamma^{\ell}_{ip}}{\partial x^k \partial x^j} + \frac{\partial \Gamma^m_{ij}}{\partial x^k} \Gamma^{\ell}_{mp} + \Gamma^m_{ij} \frac{\partial \Gamma^{\ell}_{mp}}{\partial x^k} - \frac{\partial \Gamma^m_{ip}}{\partial x^k} \Gamma^{\ell}_{mj} \\
& - \Gamma^m_{ip} \frac{\partial \Gamma^{\ell}_{mj}}{\partial x^k} + R^m_{ipj} \Gamma^{\ell}_{mk} - R^{\ell}_{mpj} \Gamma^m_{ik} - R^{\ell}_{imj} \Gamma^m_{pk} - R^{\ell}_{ipm} \Gamma^m_{jk}
\end{aligned}$$

and with the equalities

$$\frac{\partial^2 \Gamma^{\ell}_{ik}}{\partial x^p \partial x^j} = \frac{\partial^2 \Gamma^{\ell}_{ik}}{\partial x^j \partial x^p} \quad \frac{\partial^2 \Gamma^{\ell}_{ij}}{\partial x^p \partial x^k} = \frac{\partial^2 \Gamma^{\ell}_{ij}}{\partial x^k \partial x^p} \quad \frac{\partial^2 \Gamma^{\ell}_{ip}}{\partial x^j \partial x^k} = \frac{\partial^2 \Gamma^{\ell}_{ip}}{\partial x^k \partial x^j}$$

the previous expression stays

$$\begin{aligned}
\partial_p R^{\ell}_{ikj} + \partial_j R^{\ell}_{ikp} + \partial_k R^{\ell}_{ipj} = & \frac{\partial \Gamma^m_{ik}}{\partial x^p} \Gamma^{\ell}_{mj} + \Gamma^m_{ik} \frac{\partial \Gamma^{\ell}_{mj}}{\partial x^p} - \frac{\partial \Gamma^m_{ij}}{\partial x^p} \Gamma^{\ell}_{mk} - \Gamma^m_{ij} \frac{\partial \Gamma^{\ell}_{mk}}{\partial x^p} \\
& + R^m_{ijk} \Gamma^{\ell}_{mp} - R^{\ell}_{mjk} \Gamma^m_{ip} - R^{\ell}_{imk} \Gamma^m_{jp} - R^{\ell}_{ijm} \Gamma^m_{kp} \\
& + \frac{\partial \Gamma^m_{ip}}{\partial x^j} \Gamma^{\ell}_{mk} + \Gamma^m_{ip} \frac{\partial \Gamma^{\ell}_{mk}}{\partial x^j} - \frac{\partial \Gamma^m_{ik}}{\partial x^j} \Gamma^{\ell}_{mp} - \Gamma^m_{ik} \frac{\partial \Gamma^{\ell}_{mp}}{\partial x^j} \\
& + R^m_{ikp} \Gamma^{\ell}_{mj} - R^{\ell}_{mkp} \Gamma^m_{ij} - R^{\ell}_{imp} \Gamma^m_{kj} - R^{\ell}_{ikm} \Gamma^m_{pj} \\
& + \frac{\partial \Gamma^m_{ij}}{\partial x^k} \Gamma^{\ell}_{mp} + \Gamma^m_{ij} \frac{\partial \Gamma^{\ell}_{mp}}{\partial x^k} - \frac{\partial \Gamma^m_{ip}}{\partial x^k} \Gamma^{\ell}_{mj} - \Gamma^m_{ip} \frac{\partial \Gamma^{\ell}_{mj}}{\partial x^k} \\
& + R^m_{ipj} \Gamma^{\ell}_{mk} - R^{\ell}_{mpj} \Gamma^m_{ik} - R^{\ell}_{imj} \Gamma^m_{pk} - R^{\ell}_{ipm} \Gamma^m_{jk}
\end{aligned}$$

Putting the Christoffel symbols in evidence and considering the antisymmetry of the Riemann–Christoffel curvature tensor, i.e., $R^{\ell}_{imk} = -R^{\ell}_{ikm}$, $R^{\ell}_{ijm} = -R^{\ell}_{imj}$, $R^{\ell}_{imp} = -R^{\ell}_{ipm}$, and the symmetry of the Christoffel symbols, i.e., $\Gamma^m_{jp} = \Gamma^m_{pj}$, $\Gamma^m_{kp} = \Gamma^m_{pk}$, $\Gamma^m_{kj} = \Gamma^m_{jk}$ it follows that

$$
\begin{aligned}
\partial_p R^{\ell}_{ikj} + \partial_j R^{\ell}_{ikp} + \partial_k R^{\ell}_{ipj} = {} & \Gamma^{\ell}_{mp}\left(R^m_{ijk} - \frac{\partial \Gamma^m_{ik}}{\partial x^j} + \frac{\partial \Gamma^m_{ij}}{\partial x^k}\right) + \Gamma^{\ell}_{mj}\left(R^m_{ikp} + \frac{\partial \Gamma^m_{ik}}{\partial x^p} - \frac{\partial \Gamma^m_{ip}}{\partial x^k}\right) \\
& + \Gamma^{\ell}_{mk}\left(R^m_{ipj} - \frac{\partial \Gamma^m_{ij}}{\partial x^p} + \frac{\partial \Gamma^m_{ip}}{\partial x^j}\right) - \Gamma^m_{ip}\left(R^{\ell}_{mjk} - \frac{\partial \Gamma^{\ell}_{mk}}{\partial x^j} + \frac{\partial \Gamma^{\ell}_{mj}}{\partial x^k}\right) \\
& - \Gamma^m_{ij}\left(R^{\ell}_{mkp} + \frac{\partial \Gamma^{\ell}_{mk}}{\partial x^p} - \frac{\partial \Gamma^{\ell}_{mp}}{\partial x^k}\right) - \Gamma^m_{ik}\left(R^{\ell}_{mpj} - \frac{\partial \Gamma^{\ell}_{mj}}{\partial x^p} + \frac{\partial \Gamma^{\ell}_{mp}}{\partial x^j}\right)
\end{aligned}
$$

The expressions of the tensors are given by

$$
\begin{aligned}
R^m_{ijk} &= \frac{\partial \Gamma^m_{ik}}{\partial x^j} - \frac{\partial \Gamma^m_{ij}}{\partial x^k} + \Gamma^q_{ik}\Gamma^m_{qj} - \Gamma^q_{ij}\Gamma^m_{qk} \\
R^m_{ikp} &= \frac{\partial \Gamma^m_{ip}}{\partial x^k} - \frac{\partial \Gamma^m_{ik}}{\partial x^p} + \Gamma^q_{ip}\Gamma^m_{qk} - \Gamma^q_{ik}\Gamma^m_{qp} \\
R^m_{ipj} &= \frac{\partial \Gamma^m_{ij}}{\partial x^p} - \frac{\partial \Gamma^m_{ip}}{\partial x^j} + \Gamma^q_{ij}\Gamma^m_{qp} - \Gamma^q_{ip}\Gamma^m_{qj} \\
R^{\ell}_{mjk} &= \frac{\partial \Gamma^{\ell}_{mk}}{\partial x^j} - \frac{\partial \Gamma^{\ell}_{mj}}{\partial x^k} + \Gamma^q_{mk}\Gamma^{\ell}_{qj} - \Gamma^q_{mj}\Gamma^{\ell}_{qk} \\
R^{\ell}_{mkp} &= \frac{\partial \Gamma^{\ell}_{mp}}{\partial x^k} - \frac{\partial \Gamma^{\ell}_{mk}}{\partial x^p} + \Gamma^q_{mp}\Gamma^{\ell}_{qk} - \Gamma^q_{mk}\Gamma^{\ell}_{qp} \\
R^{\ell}_{mpj} &= \frac{\partial \Gamma^{\ell}_{mj}}{\partial x^p} - \frac{\partial \Gamma^{\ell}_{mp}}{\partial x^j} + \Gamma^q_{mj}\Gamma^{\ell}_{qp} - \Gamma^q_{mp}\Gamma^{\ell}_{qj}
\end{aligned}
$$

that substituted in previous expression provide

$$
\begin{aligned}
\partial_p R^{\ell}_{ikj} + \partial_j R^{\ell}_{ikp} + \partial_k R^{\ell}_{ipj} = {} & \Gamma^{\ell}_{mp}\left(\frac{\partial \Gamma^m_{ik}}{\partial x^j} - \frac{\partial \Gamma^m_{ij}}{\partial x^k} + \Gamma^q_{ik}\Gamma^m_{qj} - \Gamma^q_{ij}\Gamma^m_{qk} - \frac{\partial \Gamma^m_{ik}}{\partial x^j} + \frac{\partial \Gamma^m_{ij}}{\partial x^k}\right) \\
& + \Gamma^{\ell}_{mj}\left(\frac{\partial \Gamma^m_{ip}}{\partial x^k} - \frac{\partial \Gamma^m_{ik}}{\partial x^p} + \Gamma^q_{ip}\Gamma^m_{qk} - \Gamma^q_{ik}\Gamma^m_{qp} + \frac{\partial \Gamma^m_{ik}}{\partial x^p} - \frac{\partial \Gamma^m_{ip}}{\partial x^k}\right) \\
& + \Gamma^{\ell}_{mk}\left(\frac{\partial \Gamma^m_{ij}}{\partial x^p} - \frac{\partial \Gamma^m_{ip}}{\partial x^j} + \Gamma^q_{ij}\Gamma^m_{qp} - \Gamma^q_{ip}\Gamma^m_{qj} - \frac{\partial \Gamma^m_{ij}}{\partial x^p} + \frac{\partial \Gamma^m_{ip}}{\partial x^j}\right) \\
& - \Gamma^m_{ip}\left(\frac{\partial \Gamma^{\ell}_{mk}}{\partial x^j} - \frac{\partial \Gamma^{\ell}_{mj}}{\partial x^k} + \Gamma^q_{mk}\Gamma^{\ell}_{qj} - \Gamma^q_{mj}\Gamma^{\ell}_{qk} - \frac{\partial \Gamma^{\ell}_{mk}}{\partial x^j} + \frac{\partial \Gamma^{\ell}_{mj}}{\partial x^k}\right) \\
& - \Gamma^m_{ij}\left(\frac{\partial \Gamma^{\ell}_{mp}}{\partial x^k} - \frac{\partial \Gamma^{\ell}_{mk}}{\partial x^p} + \Gamma^q_{mp}\Gamma^{\ell}_{qk} - \Gamma^q_{mk}\Gamma^{\ell}_{qp} + \frac{\partial \Gamma^{\ell}_{mk}}{\partial x^p} - \frac{\partial \Gamma^{\ell}_{mp}}{\partial x^k}\right) \\
& - \Gamma^m_{ik}\left(\frac{\partial \Gamma^{\ell}_{mj}}{\partial x^p} - \frac{\partial \Gamma^{\ell}_{mp}}{\partial x^j} + \Gamma^q_{mj}\Gamma^{\ell}_{qp} - \Gamma^q_{mp}\Gamma^{\ell}_{qj} - \frac{\partial \Gamma^{\ell}_{mj}}{\partial x^p} + \frac{\partial \Gamma^{\ell}_{mp}}{\partial x^j}\right)
\end{aligned}
$$

Simplifying

$$\begin{aligned}\partial_p R^{\ell}_{ikj} + \partial_j R^{\ell}_{ikp} + \partial_k R^{\ell}_{ipj} = {} & \Gamma^{\ell}_{mp}\left(\Gamma^{q}_{ik}\Gamma^{m}_{qj} - \Gamma^{q}_{ij}\Gamma^{m}_{qk}\right) \\ & + \Gamma^{\ell}_{mj}\left(\Gamma^{q}_{ip}\Gamma^{m}_{qk} - \Gamma^{q}_{ik}\Gamma^{m}_{qp}\right) \\ & + \Gamma^{\ell}_{mk}\left(\Gamma^{q}_{ij}\Gamma^{m}_{qp} - \Gamma^{q}_{ip}\Gamma^{m}_{qj}\right) \\ & - \Gamma^{m}_{ip}\left(\Gamma^{q}_{mk}\Gamma^{\ell}_{qj} - \Gamma^{q}_{mj}\Gamma^{\ell}_{qk}\right) \\ & - \Gamma^{m}_{ij}\left(\Gamma^{q}_{mp}\Gamma^{\ell}_{qk} - \Gamma^{q}_{mk}\Gamma^{\ell}_{qp}\right) \\ & - \Gamma^{m}_{ik}\left(\Gamma^{q}_{mj}\Gamma^{\ell}_{qp} - \Gamma^{q}_{mp}\Gamma^{\ell}_{qj}\right)\end{aligned}$$

and with the permutation of the dummy indexes $m \leftrightarrow q$ in the first six terms it follows that

$$\begin{aligned}\partial_p R^{\ell}_{ikj} + \partial_j R^{\ell}_{ikp} + \partial_k R^{\ell}_{ipj} = {} & \Gamma^{\ell}_{qp}\Gamma^{m}_{ik}\Gamma^{q}_{mj} - \Gamma^{\ell}_{qp}\Gamma^{m}_{ij}\Gamma^{q}_{mk} \\ & + \Gamma^{\ell}_{qj}\Gamma^{m}_{ip}\Gamma^{q}_{mk} - \Gamma^{\ell}_{qj}\Gamma^{m}_{ik}\Gamma^{q}_{mp} \\ & + \Gamma^{\ell}_{qk}\Gamma^{m}_{ij}\Gamma^{q}_{mp} - \Gamma^{\ell}_{qk}\Gamma^{m}_{ip}\Gamma^{q}_{mj} \\ & - \Gamma^{m}_{ip}\Gamma^{q}_{mk}\Gamma^{\ell}_{qj} + \Gamma^{m}_{ip}\Gamma^{q}_{mj}\Gamma^{\ell}_{qk} \\ & - \Gamma^{m}_{ij}\Gamma^{q}_{mp}\Gamma^{\ell}_{qk} + \Gamma^{m}_{ij}\Gamma^{q}_{mk}\Gamma^{\ell}_{qp} \\ & - \Gamma^{m}_{ik}\Gamma^{q}_{mj}\Gamma^{\ell}_{qp} + \Gamma^{m}_{ik}\Gamma^{q}_{mp}\Gamma^{\ell}_{qj}\end{aligned}$$

whereby

$$\partial_p R^{\ell}_{ikj} + \partial_j R^{\ell}_{ikp} + \partial_k R^{\ell}_{ipj} = 0 \tag{5.2.15}$$

that is called second Bianchi identity.

5.2.8 Curvature Tensor of Variance (0, 4)

The Riemann–Christoffel curvature tensor generates a curvature tensor expressed in covariant components. With the multiplying of tensor R^{ℓ}_{ijk} by the metric tensor $g_{p\ell}$ it follows that

$$g_{p\ell}R^{\ell}_{ijk} = g_{p\ell}\left(\frac{\partial \Gamma^{\ell}_{ik}}{\partial x^j} - \frac{\partial \Gamma^{\ell}_{ij}}{\partial x^k} + \Gamma^{m}_{ik}\Gamma^{\ell}_{mj} - \Gamma^{m}_{ij}\Gamma^{\ell}_{mk}\right)$$

or

$$g_{p\ell}R^{\ell}_{ijk} = \frac{\partial\left(g_{p\ell}\Gamma^{\ell}_{ik}\right)}{\partial x^j} - \frac{\partial g_{p\ell}}{\partial x^j}\Gamma^{\ell}_{ik} - \frac{\partial\left(g_{p\ell}\Gamma^{\ell}_{ij}\right)}{\partial x^k} + \frac{\partial g_{p\ell}}{\partial x^k}\Gamma^{\ell}_{ij} + g_{p\ell}\Gamma^{m}_{ik}\Gamma\Gamma^{\ell}_{mj} - g_{p\ell}\Gamma^{m}_{ij}\Gamma^{\ell}_{mk}$$

Ricci's identity allows writing

$$\frac{\partial g_{p\ell}}{\partial x^j} = \Gamma_{pj,\ell} + \Gamma_{\ell j,p} \qquad \frac{\partial g_{p\ell}}{\partial x^k} = \Gamma_{pk,\ell} + \Gamma_{\ell k,p}$$

then

$$\begin{aligned} g_{p\ell}R^{\ell}_{ijk} =& \frac{\partial\left(g_{p\ell}\Gamma^{\ell}_{ik}\right)}{\partial x^j} - \frac{\partial\left(g_{p\ell}\Gamma^{\ell}_{ij}\right)}{\partial x^k} + \Gamma^{\ell}_{ik}\left(\Gamma_{pj,\ell} + \Gamma_{\ell j,p}\right) + \Gamma^{\ell}_{ij}\left(\Gamma_{pk,\ell} + \Gamma_{\ell k,p}\right) \\ & - \Gamma_{mk,p}\Gamma^{m}_{ij} + \Gamma_{mj,p}\Gamma^{m}_{ik} \\ =& \frac{\partial \Gamma_{ik,p}}{\partial x^j} - \frac{\partial \Gamma_{ij,p}}{\partial x^k} - \Gamma^{\ell}_{ik}\left(\Gamma_{pj,\ell} + \Gamma_{\ell j,p}\right) + \Gamma^{\ell}_{ij}\left(\Gamma_{pk,\ell} + \Gamma_{\ell k,p}\right) \\ & - \Gamma_{mk,p}\Gamma^{m}_{ij} + \Gamma_{mj,p}\Gamma^{m}_{ik} \end{aligned}$$

and replacing indexes $m \to \ell$ in the last two terms

$$\begin{aligned} g_{p\ell}R^{\ell}_{ijk} =& \frac{\partial \Gamma_{ik,p}}{\partial x^j} - \frac{\partial \Gamma_{ij,p}}{\partial x^k} - \Gamma^{\ell}_{ik}\left(\Gamma_{pj,\ell} + \Gamma_{\ell j,p}\right) + \Gamma^{\ell}_{ij}\left(\Gamma_{pk,\ell} + \Gamma_{\ell k,p}\right) - \Gamma_{\ell k,p}\Gamma^{\ell}_{ij} + \Gamma_{\ell j,p}\Gamma^{\ell}_{ik} \\ =& \frac{\partial \Gamma_{ik,p}}{\partial x^j} - \frac{\partial \Gamma_{ij,p}}{\partial x^k} + \Gamma^{\ell}_{ij}\Gamma_{pk,\ell} - \Gamma^{\ell}_{ik}\Gamma_{pj,\ell} \end{aligned}$$

whereby the result for the Riemann–Christoffel curvature tensor with variance (0, 4) or Riemann–Christoffel of first tensor type is

$$R_{pijk} = \frac{\partial \Gamma_{ik,p}}{\partial x^j} - \frac{\partial \Gamma_{ij,p}}{\partial x^k} + \Gamma^{\ell}_{ij}\Gamma_{pk,\ell} - \Gamma^{\ell}_{ik}\Gamma_{pj,\ell} \tag{5.2.16}$$

which in tensorial notation is written as

$$\boldsymbol{R} = R_{pijk}\boldsymbol{g}^p \otimes \boldsymbol{g}^i \otimes \boldsymbol{g}^j \otimes \boldsymbol{g}^k \tag{5.2.17}$$

and in symbolic form by means of determinants stays

$$R_{pijk} = \begin{vmatrix} \dfrac{\partial}{\partial x^j} & \dfrac{\partial}{\partial x^k} \\ \Gamma_{ij,p} & \Gamma_{ik,p} \end{vmatrix} + \begin{vmatrix} \Gamma^{\ell}_{ij} & \Gamma^{\ell}_{ik} \\ \Gamma_{pj,\ell} & \Gamma_{pk,\ell} \end{vmatrix} \tag{5.2.18}$$

In tensorial notation the Riemann–Christoffel tensors, mixed and covariant, are represented by $\boldsymbol{R}$.

5.2.9 *Properties of Tensor* $\mathbf{R_{pijk}}$

For the Riemann–Christoffel covariant tensor the first Bianchi identity provides

$$g_{\ell p}\left(R^{\ell}_{ikj} + R^{\ell}_{jki} + R^{\ell}_{kij}\right) = 0$$

whereby the following cyclic property results

$$R_{pikj} + R_{pjki} + R_{pkij} = 0 \tag{5.2.19}$$

Considering the antisymmetry of the Riemann–Christoffel tensor with variance (1, 3) the result is

$$g_{p\ell}R^{\ell}_{ijk} = -g_{p\ell}R^{\ell}_{ikj} \Rightarrow R_{pijk} = -R_{pikj}$$

then the Riemann–Christoffel tensor with variance (0, 4) is antisymmetric in the last two indexes.

Rewriting expression (5.2.16)

$$R_{pijk} = \frac{\partial \Gamma_{ik,p}}{\partial x^j} - \frac{\partial \Gamma_{ij,p}}{\partial x^k} + \Gamma^{\ell}_{ij}\Gamma_{pk,\ell} - \Gamma^{\ell}_{ik}\Gamma_{pj,\ell}$$

and with expressions

$$\Gamma_{ik,p} = \frac{1}{2}\left(\frac{\partial g_{pk}}{\partial x^i} + \frac{\partial g_{ip}}{\partial x^k} - \frac{\partial g_{ik}}{\partial x^p}\right) \quad \Gamma_{ij,p} = \frac{1}{2}\left(\frac{\partial g_{jp}}{\partial x^i} + \frac{\partial g_{ip}}{\partial x^j} - \frac{\partial g_{ij}}{\partial x^p}\right)$$
$$\Gamma_{pk,\ell} = g_{q\ell}\Gamma^{q}_{pk} \quad \Gamma_{pj,\ell} = g_{q\ell}\Gamma^{q}_{pj}$$

it follows that

$$R_{pijk} = \frac{\partial}{\partial x^j}\left[\frac{1}{2}\left(\frac{\partial g_{pk}}{\partial x^i} + \frac{\partial g_{ip}}{\partial x^k} - \frac{\partial g_{ik}}{\partial x^p}\right)\right] - \frac{\partial}{\partial x^k}\left[\frac{1}{2}\left(\frac{\partial g_{jp}}{\partial x^i} + \frac{\partial g_{ip}}{\partial x^j} - \frac{\partial g_{ij}}{\partial x^p}\right)\right]$$
$$+ g_{q\ell}\Gamma^{q}_{pk}\Gamma^{\ell}_{ij} - g_{q\ell}\Gamma^{q}_{pj}\Gamma^{\ell}_{ik}$$

$$R_{pijk} = \frac{1}{2}\left(\frac{\partial^2 g_{ik}}{\partial x^j \partial x^p} + \frac{\partial^2 g_{pk}}{\partial x^j \partial x^i} - \frac{\partial^2 g_{ji}}{\partial x^k \partial x^p} - \frac{\partial^2 g_{pj}}{\partial x^k \partial x^i}\right) + g_{q\ell}\left(\Gamma^q_{pk}\Gamma^\ell_{ij} - \Gamma^q_{pj}\Gamma^\ell_{ik}\right) \tag{5.2.20}$$

The expression (5.2.20) allows calculating the components of the tensor R_{pijk} directly in terms of the metric tensor. With the permutation of indexes $i \leftrightarrow p$ in expression (5.2.20)

$$R_{ipjk} = \frac{1}{2}\left(\frac{\partial^2 g_{pk}}{\partial x^j \partial x^i} + \frac{\partial^2 g_{ik}}{\partial x^j \partial x^p} - \frac{\partial^2 g_{jp}}{\partial x^k \partial x^i} - \frac{\partial^2 g_{ij}}{\partial x^k \partial x^p}\right) + g_{q\ell}\left(\Gamma^q_{pk}\Gamma^\ell_{ij} - \Gamma^q_{pj}\Gamma^\ell_{ik}\right)$$

and with the permutation of the dummy indexes $q \leftrightarrow \ell$ this expression becomes

$$R_{ipjk} = \frac{1}{2}\left(\frac{\partial^2 g_{pk}}{\partial x^j \partial x^i} + \frac{\partial^2 g_{ik}}{\partial x^j \partial x^p} - \frac{\partial^2 g_{jp}}{\partial x^k \partial x^i} - \frac{\partial^2 g_{ij}}{\partial x^k \partial x^p}\right) + g_{\ell q}\left(\Gamma^\ell_{ik}\Gamma^q_{pj} - \Gamma^\ell_{ij}\Gamma^q_{pk}\right)$$

Considering the symmetry of the metric tensor it is verified that the term to the right represents the components $-R_{pijk}$, then $R_{ipjk} = -R_{pijk}$, i.e., the tensor is antisymmetric in the first two indexes. These analyses show that the tensor R_{pijk} is antisymmetric in the first two and the last two indexes.

The permutation of indexes $p \leftrightarrow j$, $i \leftrightarrow k$ in expression (5.2.20) leads to

$$R_{pijk} = \frac{1}{2}\left(\frac{\partial^2 g_{ji}}{\partial x^p \partial x^k} - \frac{\partial^2 g_{ki}}{\partial x^p \partial x^j} - \frac{\partial^2 g_{jp}}{\partial x^i \partial x^k} + \frac{\partial^2 g_{pk}}{\partial x^i \partial x^j}\right) + g_{q\ell}\left(\Gamma^q_{ji}\Gamma^\ell_{kp} - \Gamma^q_{jp}\Gamma^\ell_{ki}\right)$$

The symmetry of the metric tensor gives $R_{pijk} = R_{jkpi}$. It is concluded that the tensor R_{pijk} is symmetric for the permutation of the pair of initial indexes for the pair of final indexes.

5.2.10 *Distinct Algebraic Components of Tensor* $\mathbf{R_{pijk}}$

The number of components of tensor R_{pijk} in the Riemann space E_N cannot be obtained counting the equations $R_{pikj} + R_{pjki} + R_{pkij} = 0$ and considering the components antisymmetric $R_{pijk} = -R_{ipjk}, R_{pijk} = -R_{pikj}$ and the symmetric components $R_{pijk} = R_{jkpi}$, because these two equations overlap. The methodology used to carry out this counting is given by means of classifying the tensor components into four groups, as a function of the number of repeated indexes:

(a) The four indexes are equal R_{iiii}
(b) The initial pair of indexes is equal to the second pair R_{ipip}

(c) One index is repeated R_{ppik}
(d) The four indexes are different R_{pijk}

Case (a) must fulfill the antisymmetry of tensor R_{pijk} that provides $R_{iiii} = -R_{iiii}$ then $R_{iiii} = 0$. The components are null when the four indexes are equal.

For case (b) only two indexes are different: R_{ipip} having that these components differ from the components R_{ippi} solely in the sign, and by the antisymmetry the result is $R_{pipi} = -R_{ippi} = -(-R_{ipip}) = R_{ipip}$. There is a number of components for R_{ipip} as many as the different pair of indexes, i.e., $i \neq p$. For index i there are N distinct combinations, and for index p there are $(N-1)$ distinct combinations, and considering the antisymmetry of the tensor for these last indexes $\frac{N}{2}(N-1)$ different combinations result. This number of combinations corresponds to the number of the $\frac{N}{2}(N-1)$ distinct combinations. There is no reduction of components due to the symmetry $R_{pijk} = R_{jkpi}$. The first Bianchi identity is satisfied, for

$$R_{pipi} + R_{ppii} + R_{piip} = R_{pipi} + 0 - R_{pipi} = 0$$

does not reduce the number of components. Therefore, in this case only $\frac{N}{2}(N-1)$ independent components are non-null.

Case (c) has components of the kind R_{ppik}. In this case there are N combinations for the index p, $(N-1)$ combinations for index i, and $(N-2)$ combinations for index k. The number of combinations for the indexes provides the number of tensor components. The antisymmetry does not reduce the number of components, for $R_{ppik} = 0$ and $R_{pipk} = 0$, and the first Bianchi identity is satisfied. Considering the symmetry $R_{pipk} = R_{pkpi}$ the number of components is reduced by half, whereby there are $\frac{N}{2}(N-1)(N-2)$ independent and non-null components. Admitting the four indexes different there are, for example, the components $R_{1234}, R_{2314}, R_{3124}$.

With methodology analogous to the previous case, it is verified that the indexes p, i, j, k can be selected in $N(N-1)(N-2)(N-3)$ modes. Considering the antisymmetries $R_{pijk} = -R_{ipjk}$ and $R_{pijk} = -R_{pikj}$, the combination of indexes is reduced to $\frac{N}{4}(N-1)(N-2)(N-3)$ modes. The symmetry $R_{pijk} = R_{jkpi}$ reduces to half these combinations, then having $\frac{N}{8}(N-1)(N-2)(N-3)$ modes. The first Bianchi identity is given by

$$R_{pikj} + R_{pjki} + R_{pkij} = 0 \Rightarrow R_{pikj} = -(R_{pjki} + R_{pkij})$$

that shows that the different combinations of the indexes are related among themselves, for a component can be expressed in terms of the other two. Therefore, the total number of combinations of indexes is reduced in $\frac{2}{3}$, and the total number of non-null independent components for this case is $\frac{2}{3}\frac{N}{8}(N-1)(N-2)(N-3)$.

The consideration of all the cases that were analyzed leads to

$$0+\frac{N}{2}(N-1)+\frac{N}{2}(N-1)(N-2)+\frac{N}{12}(N-1)(N-2)(N-3)$$

whereby there are $\frac{N^2}{12}\left(N^2-1\right)$ independent and non-null components for the tensor R_{pijk}.

The expressions that provide the Christoffel symbols for the orthogonal coordinate systems are

$$\Gamma_{ij}^{k}=0 \quad \Gamma_{ii}^{k}=-\frac{1}{2g_{kk}}\frac{\partial g_{ii}}{\partial x^k}$$
$$\Gamma_{ij}^{i}=\frac{\partial\left(\ell n\sqrt{g_{ii}}\right)}{\partial x^j} \quad \Gamma_{ii}^{i}=\frac{\partial\left(\ell n\sqrt{g_{ii}}\right)}{\partial x^i}$$

and with expression (5.2.20) that defines the Riemann–Christoffel curvature tensor with variance (0, 4) it results for the components of this tensor, where the indexes p,i,j,k indicate no summation:

- Four different indexes

$$R_{pijk}=0 \tag{5.2.21}$$

- $i=j$ and the other three indexes different

$$R_{piik}=\sqrt{g_{ii}}\left(\frac{\partial^2\sqrt{g_{ii}}}{\partial x^p\partial x^k}-\frac{\partial\sqrt{g_{ii}}}{\partial x^p}\frac{\partial\left(\ell n\sqrt{g_{pp}}\right)}{\partial x^k}-\frac{\partial\sqrt{g_{ii}}}{\partial x^k}\frac{\partial\left(\ell n\sqrt{g_{kk}}\right)}{\partial x^p}\right) \tag{5.2.22}$$

- $p=k, i=j, p\neq i$ (two different indexes)

$$R_{kiik}=\sqrt{g_{ii}}\sqrt{g_{kk}}\left[\frac{\partial}{\partial x^k}\left(\frac{1}{\sqrt{g_{kk}}}\frac{\partial\sqrt{g_{ii}}}{\partial x^k}\right)+\frac{\partial}{\partial x^i}\left(\frac{1}{\sqrt{g_{ii}}}\frac{\partial\sqrt{g_{kk}}}{\partial x^i}\right)+\frac{1}{g_{mm}}\frac{\partial\sqrt{g_{ii}}}{\partial x^m}\frac{\partial\sqrt{g_{kk}}}{\partial x^m}\right] \tag{5.2.23}$$

with $m\neq p$ and $m=i$ to fulfill the condition of having two different pairs of indexes, with the summation carried out only for the index m.

Table 5.1 shows four Riemann spaces E_N and the independent and non-null components of tensor R_{pijk}.

For the Riemann space E_1 the only component of tensor R_{pijk} is R_{1111}, which by means of its antisymmetry will always be null. Expression $\frac{N^2}{12}\left(N^2-1\right)$ proves this nullity. It is concluded that this tensor express only the internal properties of the space and not the way how this space is embedded in the Riemann spaces E_N, $N>1$, for this characteristic verifies that in E_1 a curved line has null curvature, seen that $R_{1111}=0$.

Table 5.1 Independent and non-null components of tensor R_{pijk}

Dimension of space E_N	2	3	4	5
Number of components	16	81	256	625
Independent and non-null components of R_{pijk}	1	6	20	50
Kinds of components	R_{1212}	$R_{i+1\,i+2\,j+1\,j+2}$	$R_{pipi}, R_{ppik}, R_{pijk}$	$R_{pipi}, R_{ppik}, R_{pijk}$

For the Riemann space E_2 the tensor R_{pijk} has null components when three or more indexes are equal.

Only one component cannot be null: R_{1212}. By means of the symmetry and the antisymmetry it is verified that $R_{1212} = -R_{2112} = -R_{1221} = R_{2121}$. This component is given by

$$R_{1212} = \frac{1}{2}\left(2\frac{\partial^2 g_{12}}{\partial x^1 \partial x^2} - \frac{\partial^2 g_{11}}{\partial x^2 \partial x^2} - \frac{\partial^2 g_{22}}{\partial x^1 \partial x^1}\right) + g_{q\ell}\left(\Gamma^q_{12}\Gamma^\ell_{12} - \Gamma^q_{11}\Gamma^\ell_{22}\right) \quad (5.2.24)$$

For the Riemann space E_3 the six components of tensor R_{pijk} are:

– Three components with two repeated indexes

R_{1212}	R_{1313}
R_{2323}	

– Three components with only one index repeated (three indexes are different)

R_{1213}	$R_{1223}(=R_{2123})$	$R_{1323}(=R_{3132})$

For the Riemann space E_4 there are 21 non-null components of tensor $R_{\ell ijk}$ which are:

– Six components with two repeated indexes

R_{1212}	R_{1313}	R_{1414}
R_{2323}	R_{2424}	R_{3434}

– Twelve components with only one index repeated (three indexes are different)

R_{1213}	R_{1214}	R_{1223}	R_{1224}	R_{1314}	R_{1323}	R_{1334}	R_{1424}	R_{1434}
R_{2324}	R_{2334}	R_{2434}						

– Three components with only one index repeated (three indexes are different)

R_{1234}	R_{1324}	R_{1423}

having that $R_{1234} + R_{1423} - R_{1324} = 0$, then there are 20 independent non-null components.

The non-null components of tensor $R_{\ell ijk}$ for the Riemann space E_5 are:

– Ten components with two repeated indexes

R_{1212}	R_{1313}	R_{1414}	R_{1515}
R_{2323}	R_{2424}	R_{2525}	
R_{3434}	R_{3535}		
R_{4545}			

– Thirty components with only one index repeated (three indexes are different)

R_{1213}	R_{1214}	R_{1215}	R_{1314}	R_{1315}	R_{1415}
R_{2123}	R_{2124}	R_{2125}	R_{2324}	R_{2325}	R_{2425}
R_{3132}	R_{3134}	R_{3135}	R_{3234}	R_{3235}	R_{3435}
R_{4142}	R_{4143}	R_{4145}	R_{4243}	R_{4245}	R_{4345}
R_{5152}	R_{5153}	R_{5154}	R_{5253}	R_{5254}	R_{5354}

– Ten components in which all the indexes are different

R_{1234}	R_{1235}	R_{1245}	R_{1345}	R_{2345}
R_{1324}	R_{1325}	R_{1425}	R_{1435}	R_{2435}

5.2.11 Classification of Spaces

As a function of the values assumed by the Riemann–Christoffel tensors the spaces are classified as: (a) flat: $R^{\ell}_{ijk} = R_{ijkm} = 0$; (b) curved space $R^{\ell}_{ijk} \neq 0$; $R_{ijkm} \neq 0$.

The condition $R^{i}_{\ell jk} = R_{ijkm} = 0$ indicates that the space is flat with the components of its metric tensor g_{ij} being constant. If the metric $ds^2 = g_{ij} d\overline{x}^i d\overline{x}^j$ is definite positive, i.e., $\left| g_{ij} \right| > 0$, this space is Euclidian, then it is possible to carry out a linear transformation of the coordinates $\overline{x}^i$ to the coordinates x^i for which the result is $g_{ij} = \delta^i_j$, so the metric is

$$ds^2 = \delta^i_j dx^i dx^j = dx^1 dx^1 + dx^2 dx^2 + \cdots + dx^m dx^m \qquad (5.2.25)$$

The vectors of base $\boldsymbol{e}_i$ of this new coordinate system X^i form a set of orthogonal directions, thus $\delta^i_j = \boldsymbol{e}_i \cdot \boldsymbol{e}_i$, and define an Euclidian space E_M. Consider the Riemann space E_N with the coordinates $\overline{x}^i, i = 1, 2, \dots N$, $E_N \supset E_M$, with $M > N$, which coordinates are $x^k, k = 1, 2, \dots, M$. Let the functions M be independent in terms of the coordinates x^k, so as to have the metric

$$ds^2 = g_{ij} d\overline{x}^i d\overline{x}^j = \left(dx^k\right)^2 \Rightarrow g_{ij} d\overline{x}^i d\overline{x}^j = dx^k dx^k$$

By means of the transformation law for coordinates it follows that

$$dx^k = \frac{\partial x^k}{\partial \overline{x}^i} d\overline{x}^i \quad dx^k = \frac{\partial x^k}{\partial \overline{x}^j} d\overline{x}^j$$

$$g_{ij} d\overline{x}^i d\overline{x}^j = \left(\frac{\partial x^k}{\partial \overline{x}^i} d\overline{x}^i\right)\left(\frac{\partial x^k}{\partial \overline{x}^j} d\overline{x}^j\right) \Rightarrow \left(g_{ij} - \frac{\partial x^k}{\partial \overline{x}^i}\frac{\partial x^k}{\partial \overline{x}^j}\right) d\overline{x}^i d\overline{x}^j = 0$$

As $d\overline{x}^i$ and $d\overline{x}^j$ are arbitrary, provides

$$g_{ij} = \frac{\partial x^k}{\partial \overline{x}^i}\frac{\partial x^k}{\partial \overline{x}^j}$$

that defines $\frac{N}{2}(N+1)$ independent differential equations as a function of M unknowns x^k. In this case $M < \frac{N}{2}(N+1)$ is the condition in order to have $E_N \supset E_M$. For $N = 1$ the result is $M \geq \frac{N}{2}$.

5.3 Riemann Curvature

5.3.1 Definition

The study of the Riemann space E_N is carried by means of the definition of the Riemann K curvature, which is more effective for the formulations of analyses than the Riemann–Christoffel curvature tensor R_{pijk}, for it considers the directions of the space.

For establishing a general formulation, valid for the Riemann spaces E_N with undefined metric, with the unit vectors u^i and v^i, linearly independents, defined in a point $x^i \in E_N$, and the expression

$$w^i = au^i + bv^i \tag{5.3.1}$$

that defines a coplanar vector with these two unit vectors, where a, b, are scalars that assume arbitrary values. The elementary displacements in the directions defined by the vectors w^i determine a plane π that contains the point $x^i \in E_N$.

It is admitted that u^i and v^i define coplanar vectors

$$w^i = a_1 u^i + b_1 v^i \tag{5.3.2}$$

$$r^i = a_2 u^i + b_2 v^i \tag{5.3.3}$$

where a_1, b_1, a_2, b_2 are scalars, and putting $\varepsilon(u) = \pm 1$ and $\varepsilon(v) = \pm 1$ as functional indicators of these unit vectors, and having w^i and r^i vectors mutually orthogonal it follows that

$$\varepsilon(w) = g_{k\ell} w^k w^\ell = a_1^2 u^k u^\ell + b_1^2 v^k v^\ell = a_1^2 \varepsilon(u) + b_1^2 \varepsilon(v)$$
$$\varepsilon(r) = g_{k\ell} r^k r^\ell = a_2^2 \varepsilon(u) + b_2^2 \varepsilon(v)$$

and with the condition of orthogonality

$$g_{k\ell} w^k r^\ell = \varepsilon(u) a_1 a_2 + \varepsilon(v) b_1 b_2 = 0$$

whereby

$$\begin{aligned} \varepsilon(w)\varepsilon(r) &= \left[a_1^2 \varepsilon(u) + b_1^2 \varepsilon(v)\right]\left[a_2^2 \varepsilon(u) + b_2^2 \varepsilon(v)\right] - \left[\varepsilon(u) a_1 a_2 + \varepsilon(v) b_1 b_2\right]^2 \\ &= \varepsilon(u)\varepsilon(v)(a_1 b_2 - a_2 b_1)^2 \end{aligned} \tag{5.3.4}$$

As the functional indicators assume the values ± 1:

$$a_1 b_2 - a_2 b_1 = \pm 1 \tag{5.3.5}$$

whereby

$$\varepsilon(u)\varepsilon(v) = \varepsilon(w)\varepsilon(r) \tag{5.3.6}$$

Consider two orthogonal unit vectors $\boldsymbol{u}$ and $\boldsymbol{v}$ that determine the plane π that contains the point $x^i \in E_N$, thus the Riemann curvature is defined by

$$K = \varepsilon(u)\varepsilon(v) R_{k\ell mn} u^k v^\ell u^m v^n \tag{5.3.7}$$

5.3.2 *Invariance*

For the other pair of orthogonal vectors $\boldsymbol{w}$ and $\boldsymbol{r}$ coplanar with $\boldsymbol{u}$ and $\boldsymbol{v}$, there is in an analogous way for the Riemann curvature

$$\widetilde{K} = \varepsilon(w)\varepsilon(r) R_{k\ell mn} w^k r^\ell w^m r^n$$

and with expressions (5.3.4)–(5.3.6) it follows that

$$\varepsilon(u)\varepsilon(v)(a_1 b_2 - a_2 b_1) R_{k\ell mn} w^k r^\ell w^m r^n = \varepsilon(u)\varepsilon(v) R_{k\ell mn} u^k v^\ell u^m v$$
$$\widetilde{K} = K$$

thus the Riemann curvature does not depend on the pair of unit vectors used to define it, then K is an invariant.

5.3.3 Normalized Form

The obtaining of an expression for the Riemann curvature can be carried out admitting that the Riemann space E_N is isotropic, in which the isotropic tensor is defined by

$$T_{ij\ell m} = A g_{ij} g_{\ell m} + B g_{i\ell} g_{jm} + C g_{im} g_{j\ell}$$

where A, B, C are scalars that depend on the point $x^i \in E_N$.

Assuming that tensor $T_{ij\ell m}$ is the curvature tensor $R_{ij\ell m}$ the result is

$$R_{ij\ell m} = A g_{ij} g_{\ell m} + B g_{i\ell} g_{jm} + C g_{im} g_{j\ell} \tag{5.3.8}$$

and the antisymmetry of tensor $R_{ij\ell m}$ allows writing $R_{iiii} = 0$, $R_{iijj} = 0$, $R_{ii\ell m} = R_{ij\ell\ell} = 0$, and with expression (5.3.8) it follows that

$$R_{iiii} = A g_{ii} g_{ii} + B g_{ii} g_{ii} + C g_{ii} g_{ii} = g_{ii}^2 (A + B + C) = 0$$
$$A + B + C = 0 \Rightarrow B + C = -A$$
$$R_{iijj} = A g_{ii} g_{jj} + B g_{ij} g_{ij} + C g_{ij} g_{ij} = A g_{ii} g_{jj} + (B + C) g_{ij} g_{ij}$$
$$= A\left(g_{ii} g_{jj} - g_{ij}^2\right) = 0 \tag{5.3.9}$$
$$R_{ij\ell\ell} = A g_{ij} g_{\ell\ell} + B g_{i\ell} g_{j\ell} + C g_{i\ell} g_{j\ell} = A\left(g_{ij} g_{\ell\ell} - g_{i\ell} g_{j\ell}\right) = 0 \tag{5.3.10}$$

The minors of $\det g_{i\ell}$ cannot all be simultaneously null, then in expressions (5.3.9) and (5.3.10) the result is $A = 0$ and $B = -C$, whereby

$$R_{ij\ell k} = B\left(g_{i\ell} g_{jm} - g_{im} g_{j\ell}\right) \tag{5.3.11}$$

Let

$$K = \varepsilon(u)\varepsilon(v)R_{k\ell mn}u^k v^\ell u^m v^n$$

or

$$K = R_{k\ell mn}u^k v^\ell u^m v^n \tag{5.3.12}$$

and substituting expression (5.3.11) it is concluded that $B = K$ is the Riemann curvature in $x^i \in E_N$, so

$$R_{ij\ell m} = K\left(g_{i\ell}g_{jm} - g_{im}g_{j\ell}\right) \tag{5.3.13}$$

The expression of the Riemann curvature for the isotropic space E_N, with $N > 2$, in terms of the generalized Kronecker delta and the Ricci pseudotensor

$$\varepsilon_{i_1 i_2 \ldots i_m i_{m+1} \ldots i_n}\varepsilon^{p_1 p_2 \cdots p_m p_{m+1} \cdots p_n} = (N-2)!\delta^{p_1 p_2 \cdots p_n}_{i_1 i_2 \ldots i_n}$$

takes the form

$$R_{ij\ell k} = K\frac{\varepsilon_{i_1 i_2 \ldots i_m i_{m+1} \ldots i_n}\varepsilon^{p_1 p_2 \cdots p_m p_{m+1} \cdots p_n}}{(N-2)!} = K\delta^{p_1 p_2 \cdots p_n}_{i_1 i_2 \ldots i_n} \tag{5.3.14}$$

The normalized Riemann curvature is established admitting that the vectors $\boldsymbol{u}$ and $\boldsymbol{v}$ form an angle α and define a tangent plane π in point $x^i \in E_N$. The norm of the vector perpendicular to this plane is given by

$$\|\boldsymbol{u} \times \boldsymbol{v}\|^2 = \|\boldsymbol{u}\|^2\|\boldsymbol{v}\|^2 \sin^2\alpha$$

and with the square of the dot product of these two vectors it follows that

$$(\boldsymbol{u} \cdot \boldsymbol{v})^2 = \|\boldsymbol{u}\|^2\|\boldsymbol{v}\|^2 \cos^2\alpha = \cos^2\alpha$$

$$\|\boldsymbol{u} \times \boldsymbol{v}\|^2 = \|\boldsymbol{u}\|^2\|\boldsymbol{v}\|^2(1 - \cos^2\alpha) = \|\boldsymbol{u}\|^2\|\boldsymbol{v}\|^2 - (\boldsymbol{u} \cdot \boldsymbol{v})^2$$

In terms of the components of these vectors

$$\|\boldsymbol{u}\|^2 = g_{km}u^k u^m \quad \|\boldsymbol{v}\|^2 = g_{\ell n}v^\ell v^n$$

then

$$\|\boldsymbol{u} \times \boldsymbol{v}\|^2 = g_{km}u^k u^m g_{\ell n}v^\ell v^n - g_{kn}u^k v^n g_{m\ell}u^m v^\ell = u^k v^\ell u^m v^n (g_{km}g_{\ell n} - g_{kn}g_{m\ell})$$

Expression (5.3.12) in its normalized form is

$$K\left(x^i; \boldsymbol{u}, \boldsymbol{v}\right) = \frac{R_{k\ell mn} u^k v^\ell u^m v^n}{\left(g_{km} g_{\ell n} - g_{kn} g_{m\ell}\right) u^k v^\ell u^m v^n} \tag{5.3.15}$$

or

$$K\left(x^i; \boldsymbol{u}, \boldsymbol{v}\right) = \frac{R_{k\ell mn} A^{k\ell} A^{mn}}{\left(g_{km} g_{\ell n} - g_{kn} g_{m\ell}\right) A^{k\ell} A^{mn}} \tag{5.3.16}$$

where $A^{k\ell} = u^k v^\ell, A^{mn} = u^m v^n$ represent the plane π defined by the vectors $\boldsymbol{u}, \boldsymbol{v}$.

This expression highlights that the Riemann curvature $K(x^i; \boldsymbol{u}, \boldsymbol{v})$ of the Riemann space E_N relative to the plane π defined by the vectors $\boldsymbol{u}$ and $\boldsymbol{v}$ depends on the point $x^i \in \pi \subset E_N$.

In the numerator of expression (5.3.15) the product $R_{k\ell mn} u^k v^\ell u^m v^n$ is an invariant. Putting

$$G_{k\ell mn} = g_{km} g_{\ell n} - g_{kn} g_{m\ell}$$

it is verified that $G_{k\ell mn}$ is a tensor, for it is obtained by means of algebraic operations with the metric tensor. The permutation of the tensor indexes $G_{k\ell mn}$ shows that this tensor has the same properties of symmetry and antisymmetry as tensor $R_{k\ell mn}$. For an orthogonal coordinate system exists $g_{ij} = 0$ for $i \neq j$, and the non-null components of this tensor are given by $G_{ijij} = g_{ii} g_{jj}$, where the indexes do not indicate summation.

The inner product $G_{k\ell mn} u^k v^\ell u^m v^n$ generates a scalar, then expression (5.3.15) represents an invariant, highlighting the demonstration that $\widetilde{K} = K$.

5.4 Ricci Tensor and Scalar Curvature

The Riemann–Christoffel curvature tensor R_{pijk} allows obtaining tensors of lower order by means of theirs various contractions. To obtain a non-null tensor first an index of a pair of indexes are contracted with an index of another pair of indexes, being possible the contractions: 1–3; 1–4; 2–3; 2–4. The contraction of this tensor generates the Ricci tensor, thus the multiplying of tensor R_{pijk} by g^{mp} provides

$$g^{mp} R_{pijk} = g^{mp} g_{p\ell} R^\ell_{ijk} = \delta^m_\ell R^\ell_{ijk} = R^m_{ijk}$$

and with the contraction $m = k$ the result is

$$R^k_{ijk} = R_{ij}$$

Then the Riemann–Christoffel curvature tensor with variance (1, 3) provides two Ricci tensors, one of variance (0, 2) and another of variance (1, 1). The second contraction gives a scalar with important properties, called scalar curvature. The Ricci tensor is essentially the only contraction of the Riemann–Christoffel tensor.

5.4.1 Ricci Tensor with Variance (0, 2)

The contraction of the curvature tensor

$$R^{\ell}_{ijk} = \frac{\partial \Gamma^{\ell}_{ik}}{\partial x^j} - \frac{\partial \Gamma^{\ell}_{ij}}{\partial x^k} + \Gamma^m_{ik}\Gamma^{\ell}_{mj} - \Gamma^m_{ij}\Gamma^{\ell}_{mk}$$

in indexes $\ell = k$ provides

$$R_{ij} = \frac{\partial \Gamma^{\ell}_{i\ell}}{\partial x^j} - \frac{\partial \Gamma^{\ell}_{ij}}{\partial x^{\ell}} + \Gamma^m_{i\ell}\Gamma^{\ell}_{mj} - \Gamma^m_{ij}\Gamma^{\ell}_{m\ell} \tag{5.4.1}$$

In determinants form the result is

$$R_{ij} = \begin{vmatrix} \dfrac{\partial}{\partial x^j} & \dfrac{\partial}{\partial x^{\ell}} \\ \Gamma^{\ell}_{ij} & \Gamma^{\ell}_{i\ell} \end{vmatrix} + \begin{vmatrix} \Gamma^m_{i\ell} & \Gamma^m_{ij} \\ \Gamma^{\ell}_{m\ell} & \Gamma^{\ell}_{mj} \end{vmatrix} \tag{5.4.2}$$

and with the expressions

$$\Gamma^{\ell}_{i\ell} = \frac{\partial\left(\ell n\sqrt{g}\right)}{\partial x^i} \qquad \Gamma^{\ell}_{m\ell} = \frac{\partial\left(\ell n\sqrt{g}\right)}{\partial x^m}$$

it follows that

$$R_{ij} = \frac{\partial}{\partial x^j}\left[\frac{\partial\left(\ell n\sqrt{g}\right)}{\partial x^i}\right] - \frac{\partial \Gamma^{\ell}_{ij}}{\partial x^{\ell}} + \Gamma^m_{i\ell}\Gamma^{\ell}_{mj} - \Gamma^m_{ij}\frac{\partial\left(\ell n\sqrt{g}\right)}{\partial x^m}$$

whereby for the Ricci tensor with variance (0, 2) the result is

$$R_{ij} = \frac{\partial^2\left(\ell n\sqrt{g}\right)}{\partial x^j \partial x^i} - \frac{\partial \Gamma^{\ell}_{ij}}{\partial x^{\ell}} + \Gamma^m_{i\ell}\Gamma^{\ell}_{mj} - \Gamma^m_{ij}\frac{\partial\left(\ell n\sqrt{g}\right)}{\partial x^m}$$

or

$$R_{ij} = \frac{1}{2}\frac{\partial^2(\ell n g)}{\partial x^j \partial x^i} - \frac{\partial \Gamma^{\ell}_{ij}}{\partial x^{\ell}} + \Gamma^{m}_{i\ell}\Gamma^{\ell}_{mj} - \frac{1}{2}\Gamma^{m}_{ij}\frac{\partial(\ell n g)}{\partial x^m} \tag{5.4.3}$$

If $g < 0$ it is enough to change g for $-g$ in the expression (5.4.3). The permutation of indexes $j \leftrightarrow i$ leads to

$$R_{ji} = \frac{1}{2}\frac{\partial^2(\ell n g)}{\partial x^i \partial x^j} - \frac{\partial \Gamma^{\ell}_{ji}}{\partial x^{\ell}} + \Gamma^{m}_{j\ell}\Gamma^{\ell}_{mi} - \frac{1}{2}\Gamma^{m}_{ji}\frac{\partial(\ell n g)}{\partial x^m}$$

As the Christoffel symbols are symmetric and the order of differentiation in the first term of the previous expression is independent of the sequence in which it is carried out, it is concluded that the Ricci tensor R_{ij} is symmetric, so it has $\frac{N}{2}(N+1)$ distinct components.

The contractions that can be carried out in tensor R^{ℓ}_{ijk} are: $R^{\ell}_{\ell jk}, R^{\ell}_{i\ell k}, R^{\ell}_{ij\ell}$. Considering the antisymmetry of curvature tensor $R^{\ell}_{ijk} = -R^{\ell}_{ikj}$ and with $k = \ell$ the result is $R^{\ell}_{ij\ell} = -R^{\ell}_{i\ell j}$, whereby $R_{ij} = -R^{\ell}_{i\ell j}$. The contraction $R^{\ell}_{i\ell k}$ generates the Ricci tensor R_{ij} with sign changed, then it is enough to consider only the contraction $R^{\ell}_{ij\ell}$ to obtain tensor R_{ij}, which contains components independent of R^{ℓ}_{ijk} in the more adequate form of a symmetric tensor.

The contraction of tensor R^{ℓ}_{ijk} in the indexes $i = \ell$ is given by

$$R^{\ell}_{\ell jk} = \frac{\partial \Gamma^{\ell}_{\ell k}}{\partial x^j} - \frac{\partial \Gamma^{\ell}_{\ell j}}{\partial x^k} + \Gamma^{m}_{\ell k}\Gamma^{\ell}_{mj} - \Gamma^{m}_{\ell j}\Gamma^{\ell}_{mk}$$

and with the expressions

$$\Gamma^{\ell}_{\ell k} = \frac{\partial\left(\ell n\sqrt{g}\right)}{\partial x^k} \qquad \Gamma^{\ell}_{\ell j} = \frac{\partial\left(\ell n\sqrt{g}\right)}{\partial x^j}$$

the result is

$$R^{\ell}_{\ell jk} = \frac{\partial}{\partial x^j}\left[\frac{\partial\left(\ell n\sqrt{g}\right)}{\partial x^k}\right] - \frac{\partial}{\partial x^k}\left[\frac{\partial\left(\ell n\sqrt{g}\right)}{\partial x^j}\right] + \Gamma^{m}_{\ell k}\Gamma^{\ell}_{mj} - \Gamma^{m}_{\ell j}\Gamma^{\ell}_{mk}$$

The permutation of indexes $\ell \leftrightarrow m$ in the last term and the symmetry of the Christoffel symbols allow writing

$$R^{\ell}_{\ell jk} = \frac{\partial}{\partial x^j}\left[\frac{\partial\left(\ell n\sqrt{g}\right)}{\partial x^k}\right] - \frac{\partial}{\partial x^k}\left[\frac{\partial\left(\ell n\sqrt{g}\right)}{\partial x^j}\right] + \Gamma^{m}_{\ell k}\Gamma^{\ell}_{mj} - \Gamma^{\ell}_{mj}\Gamma^{m}_{\ell k}$$

and as

$$\frac{\partial}{\partial x^j}\left[\frac{\partial\left(\ell n\sqrt{g}\right)}{\partial x^k}\right] = \frac{\partial}{\partial x^k}\left[\frac{\partial\left(\ell n\sqrt{g}\right)}{\partial x^j}\right]$$

then

$$R^{\ell}_{\ell jk} = R_{jk} = 0$$

It is concluded that the contraction of the Riemann–Christoffel curvature tensor R^{ℓ}_{ijk} in the indexes $\ell = i$ generates the null tensor.

5.4.2 *Divergence of the Ricci Tensor with Variance Ricci (0, 2)*

The calculation of the divergence of tensor R^{ℓ}_{ijk} is carried out considering the second Bianchi identity

$$\partial_{\ell}R^{\ell}_{ijk} + \partial_j R^{\ell}_{ik\ell} + \partial_k R^{\ell}_{i\ell j} = 0$$

in which the contraction of the indexes $\ell = k$ provides

$$\partial_{\ell}R^{k}_{ijk} + \partial_j R^{k}_{ik\ell} + \partial_k R^{k}_{i\ell j} = 0 \Rightarrow \partial_{\ell}R_{ij} + \partial_j R_{i\ell} + div R^{k}_{i\ell j} = 0$$

whereby

$$\operatorname{div} R^{k}_{i\ell j} = -\left(\partial_{\ell}R_{ij} + \partial_j R_{i\ell}\right)$$

and with the ordination of the indexes

$$\operatorname{div} R^{\ell}_{ijk} = -\left(\partial_j R_{ik} + \partial_k R_{ij}\right) \tag{5.4.4}$$

5.4.3 *Bianchi Identity for the Ricci Tensor with Variance (0, 2)*

An identity analogous to the second Bianchi identity can be obtained for the Ricci tensor. Rewriting expression (5.2.15)

$$\partial_p R^{\ell}_{ikj} + \partial_j R^{\ell}_{ikp} + \partial_k R^{\ell}_{ipj} = 0$$

and with the relations

$$g_i^\ell R_{kj} = R^\ell_{ikj} \quad g_i^\ell R_{kp} = R^\ell_{ikp} \quad g_i^\ell R_{pj} = R^\ell_{ipj}$$

it follows that

$$g_i^\ell \partial_p R_{kj} = \partial_p R^\ell_{ikj} \quad g_i^\ell \partial_i R_{kp} = \partial_i R^\ell_{ikp} \quad g_i^\ell \partial_k R_{pj} = \partial_k R^\ell_{ipj}$$

The sum of these three expressions provides

$$g_i^\ell \left(\partial_p R_{kj} + \partial_i R_{kp} + \partial_k R_{pj}\right) = \partial_p R^\ell_{ikj} + \partial_i R^\ell_{ikp} + \partial_k R^\ell_{ipj}$$

As the term to the right is the second Ricci identity it results in

$$\partial_p R_{kj} + \partial_i R_{kp} + \partial_k R_{pj} = 0$$

The changes of the indexes $j \to i, k \to j, p \to k$ allow the ordination of the same, then

$$\partial_k R_{ij} + \partial_i R_{jk} + \partial_j R_{ki} = 0 \tag{5.4.5}$$

that is called Bianchi identity for the Ricci tensor of covariant components.

5.4.4 Scalar Curvature

The multiplying of the Ricci tensor R_{ij} by the conjugate metric tensor g^{ij} provides

$$R = g^{ij} R_{ij} \tag{5.4.6}$$

that defines the scalar curvature, which is the trace of the Ricci tensor, also called Ricci curvature or invariant curvature of the Riemann space E_N.

5.4.5 Geometric Interpretation of the Ricci Tensor with Variance (0, 2)

Let the Riemann curvature

$$K\left(x^i; \boldsymbol{u}, \boldsymbol{v}\right) = \frac{R_{k\ell mn} u^k v^\ell u^m v^n}{\left(g_{km} g_{\ell n} - g_{kn} g_{m\ell}\right) u^k v^\ell u^m v^n}$$

where $\boldsymbol{u}, \boldsymbol{v}$ are orthogonal unit vectors, the result thereof is

$$g_{km}g_{\ell n}u^k v^\ell u^m v^n = \left(g_{km}u^k u^m\right)\left(g_{\ell n}v^\ell v^n\right)$$
$$g_{kn}g_{m\ell}u^k v^\ell u^m v^n = \left(g_{kn}u^k v^n\right)\left(g_{m\ell}u^m v^\ell\right)$$

but

$$g_{km}u^k u^m = g_{\ell n}v^\ell v^n = 1 \quad g_{kn}u^k v^n = g_{m\ell}u^m v^\ell = 0$$

then

$$K_{uv} = K\left(x^i; \boldsymbol{u}, \boldsymbol{v}\right) = \frac{R_{k\ell mn}u^k v^\ell u^m v^n}{1 \times 1 - 0} = R_{k\ell mn}u^k v^\ell u^m v^n$$

where the notation K_{uv} is adopted by convenience of graphic representation. If the unit vectors $\boldsymbol{u}, \boldsymbol{v}$ are linearly dependent, the result is $K = 0$.

The summation of all the N components of vector $\boldsymbol{u}$ is given by

$$\sum_{v^j=1}^{N} K_{uv} = \sum_{v^j=1}^{N} R_{k\ell mn}u^k v^\ell u^m v^n = u^k u^m \sum_{v^j=1}^{N} R_{k\ell mn}v^\ell v^n$$

but

$$\sum_{v^j=1}^{N} v^\ell v^n = g^{\ell n}$$

whereby the contraction $R^\ell_{i\ell j}$ generates the Ricci tensor R_{ij} with the sign changed, then

$$\sum_{v^j=1}^{N} K_{uv} = -u^k u^m g^{\ell n} R_{k\ell mn} = -u^k u^m R^{\,n}_{kmn} = -u^k u^m R_{km}$$

Putting

$$\overline{K}_u = \sum_{v^j=1}^{N} K_{uv} = -u^k u^m R_{km} \tag{5.4.7}$$

where $\overline{K}_u$ is the sum of the Riemann curvature for the space E_N determined by the components of vector $\boldsymbol{u}$ and each $(N-1)$ directions which are mutually orthogonal to them. This expression is independent of these directions and defines the mean curvature of E_N in the direction of this vector.

In expression (5.4.7) when carrying out the summation on the N directions mutually orthogonal, it follows that

$$\sum_{u^i=1}^{N} \overline{K}_u = -\sum_{u^i=1}^{N} u^k u^m R_{km}$$

$$\sum_{u^i=1}^{N} u^k u^m = g^{km}$$

$$\sum_{u^i=1}^{N} \overline{K}_u = -g^{km} R_{km} = -R \tag{5.4.8}$$

Expression (5.4.8) shows that the sum of the mean curvatures in the Riemann space E_N for mutually orthogonal directions are independent of the directions defined by the vectors $\boldsymbol{u}, \boldsymbol{v}$, being equal to the scalar curvature.

5.4.6 Eigenvectors of the Ricci Tensor with Variance (0, 2)

The Ricci tensor R_{ij} is symmetric and has in each point of the Riemann space E_N a system of linearly independent equations that define principal directions (eigenvectors).

Let the Riemann curvature

$$K\left(x^i; \boldsymbol{u}, \boldsymbol{v}\right) = \frac{R_{k\ell mn} u^k v^\ell u^m v^n}{\left(g_{km} g_{\ell n} - g_{kn} g_{m\ell}\right) u^k v^\ell u^m v^n}$$

where the vectors are orthogonal and only $\boldsymbol{v}$ is a unit vector, so

$$g_{km} g_{\ell n} u^k v^\ell u^m v^n = \left(g_{km} u^k u^m\right)\left(g_{\ell n} v^\ell v^n\right)$$

$$g_{kn} g_{m\ell} u^k v^\ell u^m v^n = \left(g_{kn} u^k v^n\right)\left(g_{m\ell} u^m v^\ell\right) = 0$$

then

$$K\left(x^i; \boldsymbol{u}, \boldsymbol{v}\right) = \frac{R_{k\ell mn} u^k v^\ell u^m v^n}{\left(g_{km} u^k u^m\right)\left(g_{\ell n} v^\ell v^n\right)}$$

but as $\boldsymbol{v}$ is a unit vector the result is

$$g_{\ell n} v^\ell v^n = 1 \Rightarrow v^\ell v^n = g^{\ell n}$$

whereby

$$K\left(x^i; \boldsymbol{u}, \boldsymbol{v}\right) = -\frac{R_{k\ell mn} g^{\ell n} u^k u^m}{g_{km} u^k u^m}$$

thereof

$$K_u = K\left(x^i; \boldsymbol{u}, \boldsymbol{v}\right) = -\frac{R_{km} u^k u^m}{g_{km} u^k u^m} \tag{5.4.9}$$

is the normalized mean curvature, where the index indicates that $\boldsymbol{u}$ is not unit vector.

The calculation of the eigenvalues is carried out by means of the equations system

$$(R_{km} + K_u g_{km}) u^k u^m = 0$$

with extreme values given by the condition

$$\frac{\partial}{\partial u^k}\left[(R_{km} + K_u g_{km}) u^k u^m\right] = 0$$

which developed stays

$$2(R_{km} + K_u g_{km}) u^m + \frac{\partial R_{km}}{\partial u^k} u^k u^m + \frac{\partial K_u}{\partial u^k} g_{km} u^k u^m = 0$$

and as the Ricci tensor R_{ij} does not depend on vector u^k the result is

$$2(R_{km} + K_u g_{km}) u^m + \frac{\partial K_u}{\partial u^k} g_{km} u^k u^m = 0$$

For the extreme values of K_u the result is $\frac{\partial K_u}{\partial x^k} = 0$, whereby the equations system

$$(R_{km} + K_u g_{km}) u^m = 0$$

allows determining the principal directions (eigenvectors) of the Ricci tensor R_{ij}.

5.4.7 Ricci Tensor with Variance (1, 1)

The Ricci tensor in terms of its mixed components is given by

$$R^i_j = g^{im} R_{mj} \tag{5.4.10}$$

An important expression that relates the Ricci tensor with variance (1, 1) with the derivative of the scalar curvature can be obtained by means of the second Bianchi identity

$$\partial_p R^{\ell}_{ijk} + \partial_j R^{\ell}_{ikp} + \partial_k R^{\ell}_{ipj} = 0$$

where with the antisymmetry $R^{\ell}_{ikp} = -R^{\ell}_{ipk}$ the result is

$$\partial_p R^{\ell}_{ijk} - \partial_j R^{\ell}_{ipk} + \partial_k R^{\ell}_{ipj} = 0$$

The contraction of these tensors in indexes $\ell = k$ provides

$$\partial_p R_{ij} - \partial_j R_{ip} + \partial_k R^{k}_{ipj} = 0$$

Multiplying by g^{ip} it follows that

$$g^{ip}\partial_p R_{ij} - g^{ip}\partial_j R_{ip} + g^{ip}\partial_k R^{k}_{ipj} = 0$$
$$\partial_p g^{ip} R_{ij} - \partial_j g^{ip} R_{ip} + \partial_k g^{ip} R^{k}_{ipj} = 0$$
$$\partial_p R^{p}_{j} - \frac{\partial R}{\partial x^j} + \partial_k R^{k}_{j} = 0 \Rightarrow \frac{\partial R}{\partial x^j} = \partial_p R^{p}_{j} + \partial_k R^{k}_{j}$$

The change of the dummy indexes $p \to k$ provides

$$\frac{\partial R}{\partial x^j} = 2\partial_k R^{k}_{j}$$

whereby

$$\partial_k R^{k}_{j} = \frac{1}{2}\frac{\partial R}{\partial x^j} \tag{5.4.11}$$

For the Riemann space E_N, with $N > 2$, multiplying expression (5.4.5) by g^{ij} the result is

$$g^{ij}\partial_k R_{ij} + g^{ij}\partial_i R_{jk} + g^{ij}\partial_j R_{ki} = 0 \Rightarrow \partial_k g^{ij} R_{ij} + \partial_i g^{ij} R_{jk} + \partial_j g^{ij} R_{ki} = 0$$

and having curvature R a scalar function at its partial derivative is equal to its covariant derivative, then

$$\frac{\partial R}{\partial x^k} + \partial_i R^{i}_{k} + \partial_j R^{j}_{k} = 0$$

and with the change of indexes $j \to i$ the result is

$$\frac{\partial R}{\partial x^k} + 2\partial_i R^i_k = 0$$

and with

$$\partial_i R^i_k = \frac{1}{2}\frac{\partial R}{\partial x^k}$$

it follows that

$$\frac{\partial R}{\partial x^k} + 2 \cdot \frac{1}{2}\frac{\partial R}{\partial x^k} = 0 \Rightarrow \frac{\partial R}{\partial x^k} = 0$$

then the scalar curvature is constant for this kind of space. The purpose of the supposition $N > 2$ will be clarified by expression (5.6.10), obtained when analyzing the scalar curvature in the Riemann space E_2.

Exercise 5.1 For the tensorial expression $T^i_j = R^i_j + \delta^i_j(\alpha R + \beta)$, where α, β are scalars, calculate the value of α so that the covariant derivative $\partial_i T^i_j$ is null.

The null covariant derivative $\partial_i T^i_j$ is given by

$$\partial_i T^i_j = \partial_i R^i_j + \partial_i \left[\delta^i_j(\alpha R + \beta)\right]$$

having $\partial_i \delta^i_j = 0$ it follows that

$$\partial_i T^i_j = \partial_i R^i_j + \alpha \partial_i R = 0$$

With the expression (5.4.11)

$$\partial_i R^i_j = \frac{1}{2}\frac{\partial R}{\partial x^j} \Rightarrow \partial_i T^i_j = \left(\frac{1}{2} + \alpha\right)\frac{\partial R}{\partial x^j} = 0$$

for $\partial_i R = \frac{\partial R}{\partial x^j}$, and as this derivative assumes any values the result is $\alpha = -\frac{1}{2}$.

5.4.8 *Notations*

In Table 5.2, in which the Tulio Levi-Civita notation was inserted, there is a compilation of the evolution of the notation for the Riemann–Christoffel curvature tensors and for the Ricci tensor. The notations that make use of (,) or (;) seek to

Table 5.2 Notations for the Riemann–Christoffel curvature tensors and Ricci tensor

Author	Riemann–Christoffel curvature tensor		Ricci tensor
	Mixed variance components (1, 3)	Covariant components (0, 4)	
Brillouin	$R^i_{j,k\ell}$	$R_{ij,k\ell}$	$R_{j\ell}=\sum_m R^m_{j,m\ell}$
Appe-Thiry	$R^i_{\bullet jk\ell}$	$R_{ijk\ell}$	$R_{jk}=\sum_m R^m_{\bullet jkm}$
Weyl	$F^i_{jk\ell}$	$F_{ijk\ell}$	$R_{j\ell}=\sum_m F^m_{jm\ell}$
Eddington-Becquerel	$B^i_{jk\ell}$	$B_{jk\ell i}$	$G_{jk}=\sum_m B^m_{jmk}$
Galbrun	$R^i_{j\ell k}$	$R_{ij\ell k}$	$R_{jk}=\sum_m R^m_{jmk}$
Juvet	$R^{\bullet i}_{j\bullet \ell k}$	$R_{ji\ell k}$	$R_{jk}=\sum_m R^{\bullet m}_{j\bullet mk}$
Cartan	$R^i_{j,\ell k}$	$R_{ji,\ell k}$	$R_{j\ell}=\sum_m R^m_{j\ell m}$
Christoffel and Bianchi	$(ji;k\ell)$	$(ji,k\ell)$	–
Levi-Civita	$\{ji,k\ell\}$	$(ji,k\ell)$	$\alpha_{j\ell}$

indicate the properties of symmetry and antisymmetry of the Riemann–Christoffel tensors. In the case of using (.) it indicates the index, or the position and the index that will be lowered or raised. The only difference between the two notations of Christoffel and Bianchi is the change of the point and comma (;) for the comma (,). Currently these two forms of spelling were abandoned. It is stressed that several authors have opted for different positioning of the indexes. The Weyl notation, with the change of the letter F for R (Riemann), was the one that became consecrated in the current literature.

Exercise 5.2 In a coordinates system let $\Gamma^i_{jk}=\delta^i_j\frac{\partial\phi}{\partial x^k}+\delta^i_k\frac{\partial\psi}{\partial x^j}$, where ϕ,ψ are functions of position. Calculate: (a) $R^i_{jk\ell}$; (b) R_{jk} for $\psi=-\ell n(a_i x^i)$.

(a) Substituting the expression

$$\Gamma^i_{jk}=\delta^i_j\frac{\partial\phi}{\partial x^k}+\delta^i_k\frac{\partial\psi}{\partial x^j}$$

in the expression of the Riemann–Christoffel curvature tensor

$$R^i_{jk\ell}=\frac{\partial\Gamma^i_{j\ell}}{\partial x^k}-\frac{\partial\Gamma^i_{jk}}{\partial x^\ell}+\Gamma^i_{rk}\Gamma^r_{j\ell}-\Gamma^i_{r\ell}\Gamma^r_{jk}$$

it follows that

$$R^i_{jk\ell} = \frac{\partial}{\partial x^k}\left(\delta^i_j \frac{\partial \phi}{\partial x^\ell} + \delta^i_\ell \frac{\partial \psi}{\partial x^j}\right) - \frac{\partial}{\partial x^\ell}\left(\delta^i_j \frac{\partial \phi}{\partial x^k} + \delta^i_k \frac{\partial \psi}{\partial x^j}\right)$$
$$+\left(\delta^i_j \frac{\partial \phi}{\partial x^k} + \delta^i_k \frac{\partial \psi}{\partial x^r}\right)\left(\delta^r_j \frac{\partial \phi}{\partial x^\ell} + \delta^r_\ell \frac{\partial \psi}{\partial x^j}\right)$$
$$-\left(\delta^i_r \frac{\partial \phi}{\partial x^\ell} + \delta^i_\ell \frac{\partial \psi}{\partial x^r}\right)\left(\delta^r_j \frac{\partial \phi}{\partial x^k} + \delta^r_k \frac{\partial \psi}{\partial x^j}\right)$$
$$= \delta^i_j \frac{\partial^2 \phi}{\partial x^k \partial x^\ell} + \delta^i_\ell \frac{\partial^2 \psi}{\partial x^k \partial x^j} - \delta^i_j \frac{\partial^2 \phi}{\partial x^\ell \partial x^k} - \delta^i_k \frac{\partial^2 \psi}{\partial x^\ell \partial x^j} + \delta^i_j \delta^r_j \frac{\partial \phi}{\partial x^k}\frac{\partial \phi}{\partial x^\ell} + \delta^i_j \delta^r_\ell \frac{\partial \phi}{\partial x^\ell}\frac{\partial \psi}{\partial x^j}$$
$$+\delta^i_k \delta^r_j \frac{\partial \psi}{\partial x^r}\frac{\partial \phi}{\partial x^\ell} + \delta^i_k \delta^r_\ell \frac{\partial \psi}{\partial x^r}\frac{\partial \psi}{\partial x^j} - \delta^i_r \delta^r_j \frac{\partial \phi}{\partial x^\ell}\frac{\partial \phi}{\partial x^k} - \delta^i_r \delta^r_k \frac{\partial \phi}{\partial x^\ell}\frac{\partial \psi}{\partial x^j} - \delta^i_\ell \delta^r_j \frac{\partial \psi}{\partial x^r}\frac{\partial \phi}{\partial x^k}$$
$$-\delta^i_\ell \delta^r_k \frac{\partial \psi}{\partial x^r}\frac{\partial \psi}{\partial x^j}$$
$$R^i_{jk\ell} = \delta^i_j \frac{\partial \phi}{\partial x^k}\frac{\partial \phi}{\partial x^\ell} + \delta^i_k \frac{\partial \psi}{\partial x^j}\frac{\partial \psi}{\partial x^\ell} + \delta^i_k \frac{\partial \phi}{\partial x^\ell}\frac{\partial \psi}{\partial x^j} + \delta^i_\ell \frac{\partial \phi}{\partial x^k}\frac{\partial \psi}{\partial x^j} - \delta^i_j \frac{\partial \phi}{\partial x^\ell}\frac{\partial \phi}{\partial x^k}$$
$$-\delta^i_\ell \frac{\partial \psi}{\partial x^j}\frac{\partial \psi}{\partial x^k} - \delta^i_\ell \frac{\partial \phi}{\partial x^k}\frac{\partial \psi}{\partial x^j} - \delta^i_k \frac{\partial \phi}{\partial x^\ell}\frac{\partial \psi}{\partial x^j} + \delta^i_j \frac{\partial^2 \phi}{\partial x^\ell \partial x^k} + \delta^i_\ell \frac{\partial^2 \psi}{\partial x^j \partial x^k}$$
$$-\delta^i_j \frac{\partial^2 \phi}{\partial x^k \partial x^\ell} - \delta^i_k \frac{\partial^2 \psi}{\partial x^j \partial x^\ell}$$
$$R^i_{jk\ell} = \delta^i_k \left(\frac{\partial \psi}{\partial x^j}\frac{\partial \psi}{\partial x^\ell} - \frac{\partial^2 \psi}{\partial x^j \partial x^\ell}\right) - \delta^i_\ell \left(\frac{\partial \psi}{\partial x^j}\frac{\partial \psi}{\partial x^k} - \frac{\partial^2 \psi}{\partial x^j \partial x^k}\right)$$

then $R^i_{jk\ell}$ only depends on the function ψ.

(b) For $\psi = -\ell n(a_i x^i)$ the partial derivatives result

$$\frac{\partial \psi}{\partial x^j} = -\frac{a_j}{a_i x^i} \Rightarrow \frac{\partial^2 \psi}{\partial x^j \partial x^\ell} = \frac{a_j a_\ell}{(a_i x^i)^2}$$
$$\frac{\partial \psi}{\partial x^\ell} = -\frac{a_\ell}{a_i x^i} \Rightarrow \frac{\partial \psi}{\partial x^j}\frac{\partial \psi}{\partial x^\ell} = \frac{a_j a_\ell}{(a_i x^i)^2}$$

and substituting this derivatives in the expression obtained in item (a) it follows that

$$R^i_{jk\ell} = \delta^i_k \left(\frac{\partial \psi}{\partial x^j} \frac{\partial \psi}{\partial x^\ell} - \frac{\partial^2 \psi}{\partial x^j \partial x^\ell} \right) - \delta^i_\ell \left(\frac{\partial \psi}{\partial x^j} \frac{\partial \psi}{\partial x^k} - \frac{\partial^2 \psi}{\partial x^j \partial x^k} \right)$$

$$R^i_{jk\ell} = \delta^i_k \left[\frac{a_j a_\ell}{(a_i x^i)^2} - \frac{a_j a_\ell}{(a_i x^i)^2} \right] - \delta^i_\ell \left[\frac{a_j a_k}{(a_i x^i)^2} - \frac{a_j a_k}{(a_i x^i)^2} \right] = 0$$

whereby

$$R^i_{jki} = R_{jk} = 0 \quad Q.E.D.$$

5.5 Einstein Tensor

The tensor $R_{ijk\ell}$, the second Bianchi identity, the Ricci tensor R_{ij} and the scalar curvature R allow obtaining a second-order tensor with peculiar characteristics. Let the second Bianchi identity

$$\partial_m R_{ijk\ell} + \partial_k R_{ij\ell m} + \partial_\ell R_{ijmk} = 0$$

and with the antisymmetry of the Riemann–Christoffel curvature tensor $R_{ijk\ell}$

$$\partial_m R_{ijk\ell} - \partial_k R_{ijm\ell} - \partial_\ell R_{jimk} = 0$$

and multiplying by $g^{i\ell}$ and g^{jk} it follows that

$$g^{i\ell} g^{jk} \partial_m R_{ijk\ell} - g^{i\ell} g^{jk} \partial_k R_{ijm\ell} - g^{i\ell} g^{jk} \partial_\ell R_{jimk} = 0$$
$$g^{jk} \partial_m R^\ell_{jk\ell} - g^{jk} \partial_k R^\ell_{jm\ell} - g^{i\ell} \partial_\ell R^k_{imk} = 0$$

whereby in terms of the Ricci tensor

$$g^{jk} \partial_m R_{jk} - g^{jk} \partial_k R_{jm} - g^{i\ell} \partial_\ell R_{im} = 0$$

The change of the dummy index $\ell \to k$ in the last term provides

$$g^{jk} \partial_m R_{jk} - g^{jk} \partial_k R_{jm} - g^{ik} \partial_k R_{im} = 0 \Rightarrow \partial_m R^{jk}_{jk} - \partial_k R^{jk}_{jm} - \partial_k R^{ik}_{im} = 0$$

The contractions of the curvature tensors provide

$$\partial_m R - \partial_k R^k_m - \partial_k R^k_m = 0 \Rightarrow \partial_m R = 2 \partial_k R^k_m$$

whereby

$$\partial_k R^k_m = \frac{1}{2}\partial_m R \tag{5.5.1}$$

is the divergence of a tensor, which can be written under the form

$$\partial_k\left(R^k_m - \frac{1}{2}\delta^k_m R\right) = 0 \tag{5.5.2}$$

where the terms in parenthesis define the Einstein tensor with variance (1, 1)

$$G^k_m = R^k_m - \frac{1}{2}\delta^k_m R \tag{5.5.3}$$

The Einstein tensor can be written as a function of its covariant components, so

$$G_{ij} = g_{ik}G^k_j = g_{ik}\left(R^k_j - \frac{1}{2}\delta^k_j R\right) \tag{5.5.4}$$

thus

$$G_{ij} = R_{ij} - \frac{1}{2}g_{ij}R \tag{5.5.5}$$

By means of this expression it is verified that the Einstein tensor is generated only by the metric tensor and the Ricci tensor. As R_{ij} and g_{ij} are two symmetric tensors then Einstein tensor is symmetric. For the contravariant components of this tensor the result is

$$G^{ij} = R^{ij} - \frac{1}{2}g^{ij}R \tag{5.5.6}$$

The divergence of the Einstein tensor is given by

$$\partial_i G^j_i = \partial_i R^i_j - \delta^j_i \frac{1}{2}\partial_i R = \partial_i R^i_j - \frac{1}{2}\partial_j R$$

but

$$\partial_i R^i_j = \frac{1}{2}\partial_j R$$

then

$$\partial_i G^i_j = 0 \tag{5.5.7}$$

Thus for any Riemann space the divergence of the Einstein tensor is null, and with the contraction of this tensor it follows that

$$G_i^i = R_i^i - \frac{1}{2}\delta_i^i R = R - \frac{1}{2}NR$$
$$G = -\frac{1}{2}(N-2)R \tag{5.5.8}$$

For the Riemann space E_2 it is verified that $G = 0$.

Exercise 5.3 Show that the tensor of the kind $T_j^i = R_j^i + \delta_j^i m$, being m a scalar function, has the characteristics of an Einstein tensor.

The divergence of this tensor given by $\partial_j T_j^i = 0$ stays

$$\partial_j T_j^i = \partial_j R_j^i + \delta_j^i \partial_j m = \partial_j \left(R_j^i + m \right) = 0$$

and with expression (5.4.11)

$$\partial_j R_j^i = \frac{1}{2}\partial_j R$$

substituted in this expression

$$\partial_j T_j^i = \partial_j \left(\frac{1}{2}R + m \right) = 0 \Rightarrow \frac{1}{2}R + m = k_1 \Rightarrow m = k_1 - \frac{1}{2}R$$

where k_1 is a constant. The substitution of this expression in the expression of tensor T_j^i provides

$$T_j^i = R_j^i - \delta_j^i \left(\frac{1}{2}R + k_2 \right)$$

where $k_2 = -k_1$.

Thus this tensor has the same characteristics of the Einstein tensor defined by expression (5.5.3).

5.6 Particular Cases of Riemann Spaces

Some kinds of Riemann spaces will be analyzed in this item with specific characteristics that make them important: the Riemann space E_2, the Riemann space with constant curvature, the Minkowski space, and the conformal space.

5.6.1 *Riemann Space* $\mathbf{E}_2$

In the Riemann space E_2 the Ricci tensor R_{ij} is defined by its components

$$R_{ij} = \begin{bmatrix} R_{11} & R_{12} \\ R_{21} & R_{22} \end{bmatrix}$$

as $R_{12} = R_{21}$ and the metric tensor in matrix form is given by

$$g_{ij} = \begin{bmatrix} g_{11} & g_{12} \\ g_{21} & g_{22} \end{bmatrix}$$

where $g_{12} = g_{21}$.

The Ricci tensor written in terms of the Riemann–Christoffel curvature tensor with variance $(0, 4)$, and considering the symmetry and the metric tensor is given by

$$R_{ij} = g^{kp} R_{pijk} = g^{pk} R_{ipkj}$$

and the development provides

$$R_{ij} = g^{11} R_{i11j} + g^{12} R_{i12j} + g^{21} R_{i21j} + g^{22} R_{i22j}$$

whereby the result for component R_{11} is

$$R_{11} = g^{11} R_{1111} + g^{12} R_{1121} + g^{21} R_{1211} + g^{22} R_{1221}$$

As the tensor R_{pijk} is antisymmetric in the first two and the last two indexes, i.e., $R_{pijk} = -R_{ipjk}$ and $R_{pijk} = -R_{pikj}$ it follows that

$$R_{11} = 0 + 0 + 0 + g^{22} R_{1221}$$

Let $g = \det g_{ij}$ and G^{22} the cofactor of g^{22}:

$$g^{22} = \frac{G^{22}}{g} = \frac{g_{11}}{g}$$

whereby

$$R_{11} = \frac{g_{11}}{g} (-R_{1212}) \Rightarrow \frac{R_{11}}{g_{11}} = -\frac{R_{1212}}{g}$$

Proceeding in an analogous way for component R_{22}:

$$R_{22} = g^{11}R_{2112} + g^{12}R_{2122} + g^{21}R_{2212} + g^{22}R_{2222}$$
$$R_{22} = g^{11}R_{2112} + 0 + 0 + 0$$
$$R_{22} = -g^{11}R_{1212}$$
$$g^{11} = \frac{G^{11}}{g} = \frac{g_{22}}{g}$$
$$R_{22} = -\frac{g_{22}}{g}R_{1212}$$

whereby

$$\frac{R_{22}}{g_{22}} = -\frac{R_{1212}}{g}$$

For component R_{12}, it follows that

$$R_{12} = g^{11}R_{1112} + g^{12}R_{1122} + g^{21}R_{1212} + g^{22}R_{1222} = 0 + 0 + g^{21}R_{1212} + 0$$
$$R_{12} = g^{21}R_{1212}$$
$$g^{21} = \frac{G^{21}}{g} = \frac{g_{12}}{g}$$
$$R_{12} = -\frac{g_{12}}{g}R_{1212}$$

thus

$$\frac{R_{12}}{g_{12}} = -\frac{R_{1212}}{g}$$

and with the symmetries $R_{ij} = R_{ji}$ and $g_{ij} = g_{ji}$ the result for component R_{21} is

$$\frac{R_{21}}{g_{21}} = -\frac{R_{1212}}{g}$$

The analysis developed shows that

$$K = \frac{R_{11}}{g_{11}} = \frac{R_{22}}{g_{22}} = \frac{R_{12}}{g_{12}} = \frac{R_{21}}{g_{21}} = -\frac{R_{1212}}{g}$$

These equalities indicate that in the Riemann space E_2 the components of the Ricci tensor R_{ij} are proportional to the components of the metric tensor g_{ij} and to its derivatives, and are independent of the directions considered. It is verified that the Riemann curvature does not vary with the orientation considered, then all the points of the space E_2 are isotropic. This, in general, is not valid for spaces with dimension $N > 2$. The scalar K in Riemann space E_2 is called Gauß curvature.

This analysis allows writing the components of the Ricci tensor as a function of the component R_{1212} and of the metric tensor, thus

$$R_{ij} = -\frac{R_{1212}}{g} g_{ij} \tag{5.6.1}$$

5.6.2 *Gauß Curvature*

Expression (5.6.1) is valid only for the Riemann space E_2. The knowledge of the properties of the surfaces in the Euclidian space E_3 is not useful for understanding the properties of the Riemann spaces E_N, with $N > 3$. For $N = 2$ several simplifications are admitted in the formulation of the expression of R_{ij}, so the conclusions obtained for the Riemann space E_2 cannot be generalized for the spaces of dimensions $N > 3$.

The scalar curvature allows expressing the Riemann–Christoffel tensor R_{pijk} as a function of the components of the metric tensor.

With the non-null components $R_{1212}, = -R_{2121}, = -R_{1221} = R_{2112}$, and the expression of the scalar curvature it follows that

$$R = g^{ij} R_{ij} = -g^{ij} g_{ij} \frac{R_{1212}}{g} = -\delta_i^i \frac{R_{1212}}{g} = -\frac{2}{g} R_{1212} \Rightarrow R_{1212} = -\frac{R}{2} g$$

and the development provides

$$R_{1212} = -\frac{R}{2} \begin{vmatrix} g_{11} & g_{12} \\ g_{21} & g_{22} \end{vmatrix} = -\frac{R}{2} (g_{11} g_{22} - g_{12} g_{21})$$

The other non-null components are obtained by means of the indexes in this expression, and considering the symmetry of tensor R_{pijk} it follows that

$$R_{2121} = -\frac{R}{2} (g_{22} g_{11} - g_{21} g_{12})$$

$$R_{1221} = -\frac{R}{2} (g_{12} g_{21} - g_{11} g_{22})$$

$$R_{2121} = -\frac{R}{2} (g_{21} g_{12} - g_{22} g_{11})$$

then

$$R_{ijk\ell} = -\frac{R}{2}\left(g_{ik}g_{j\ell} - g_{i\ell}g_{jk}\right) \tag{5.6.2}$$

or

$$R_{ijk\ell} = -K\left(g_{ik}g_{j\ell} - g_{i\ell}g_{jk}\right) \tag{5.6.3}$$

The Gauß curvature, that in general depends on the coordinates of the point considered, is determined by

$$K = \frac{1}{2}R \tag{5.6.4}$$

that can be obtained as a function of the Riemann–Christoffel curvature tensor with variance (0, 4), and with the Ricci pseudotensor for the Riemann space E_2

$$\varepsilon_{ij} = \sqrt{g}e_{ij} \quad \varepsilon^{ij} = \frac{e^{ij}}{\sqrt{g}}$$

and with the expression

$$K = \frac{R_{1212}}{g}$$

then

$$R_{ijk\ell} = K\varepsilon_{ij}\varepsilon_{k\ell} \tag{5.6.5}$$

The multiplication of both members of this expression by $\varepsilon^{ij}\varepsilon^{k\ell}$ provides

$$\varepsilon^{ij}\varepsilon^{k\ell}R_{ijk\ell} = K\varepsilon_{ij}\varepsilon_{k\ell}\varepsilon^{ij}\varepsilon^{k\ell}$$

and as

$$\varepsilon_{ij}\varepsilon^{ij} = \delta_i^i = 2$$

thus

$$K = \frac{1}{4}R_{ijk\ell}\varepsilon^{ij}\varepsilon^{k\ell} \tag{5.6.6}$$

this expression shows that the Gauß curvature is an invariant.

5.6.3 Component $\mathbf{R}_{1212}$ in Orthogonal Coordinate Systems

For the orthogonal coordinate systems in the Riemann space E_N expression (5.2.24) provides the component

$$R_{1212} = \frac{1}{2}\left(2\frac{\partial^2 g_{12}}{\partial x^1 \partial x^2} - \frac{\partial^2 g_{11}}{\partial x^2 \partial x^2} - \frac{\partial^2 g_{22}}{\partial x^1 \partial x^1}\right) + g_{q\ell}\left(\Gamma^q_{12}\Gamma^\ell_{12} - \Gamma^q_{11}\Gamma^\ell_{22}\right)$$

or more explicitly

$$R_{1212} = -\frac{1}{2}\left(\frac{\partial^2 g_{11}}{\partial x^2 \partial x^2} + \frac{\partial^2 g_{22}}{\partial x^1 \partial x^1}\right) + g_{11}\left(\Gamma^1_{12}\Gamma^1_{12} - \Gamma^1_{11}\Gamma^1_{22}\right) + g_{22}\left(\Gamma^2_{12}\Gamma^2_{12} - \Gamma^2_{11}\Gamma^2_{22}\right)$$

The Christoffel symbols for these coordinates systems are given by

- $i = j = k \Rightarrow \Gamma^k_{ij} = \Gamma^i_{ii} = \frac{1}{2g_{ii}}\frac{\partial g_{ii}}{\partial x^j} \Rightarrow \begin{cases} \Gamma^1_{11} = \dfrac{1}{2g_{11}}\dfrac{\partial g_{11}}{\partial x^1} \\ \Gamma^2_{22} = \dfrac{1}{2g_{22}}\dfrac{\partial g_{22}}{\partial x^2} \end{cases}$
- $i = j \neq k \Rightarrow \Gamma^k_{ij} = \Gamma^k_{ii} = -\frac{1}{2g_{kk}}\frac{\partial g_{ii}}{\partial x^k} \Rightarrow \begin{cases} \Gamma^2_{11} = -\dfrac{1}{2g_{22}}\dfrac{\partial g_{11}}{\partial x^2} \\ \Gamma^1_{22} = -\dfrac{1}{2g_{11}}\dfrac{\partial g_{22}}{\partial x^1} \end{cases}$
- $i = k \neq j \Rightarrow \Gamma^k_{ij} = \Gamma^i_{ij} = \frac{1}{2g_{ii}}\frac{\partial g_{ii}}{\partial x^j} \Rightarrow \begin{cases} \Gamma^1_{12} = \dfrac{1}{2g_{11}}\dfrac{\partial g_{11}}{\partial x^2} \\ \Gamma^2_{12} = \dfrac{1}{2g_{22}}\dfrac{\partial g_{22}}{\partial x^1} \end{cases}$
- For $i \neq j, j \neq k, i \neq k$ it results in $\Gamma_{ij,k} = 0$

so

$$\begin{aligned} R_{1212} = & -\frac{1}{2}\left(\frac{\partial^2 g_{11}}{\partial x^2 \partial x^2} + \frac{\partial^2 g_{22}}{\partial x^1 \partial x^1}\right) + \frac{1}{4g_{11}}\left[\left(\frac{\partial g_{11}}{\partial x^2}\right)^2 + \frac{\partial g_{11}}{\partial x^1}\frac{\partial g_{22}}{\partial x^1}\right] \\ & + \frac{1}{4g_{22}}\left[\left(\frac{\partial g_{22}}{\partial x^1}\right)^2 + \frac{\partial g_{11}}{\partial x^2}\frac{\partial g_{22}}{\partial x^2}\right] \\ = & -\frac{1}{2\sqrt{g_{11}g_{22}}}\left[\frac{\partial}{\partial x^1}\left(\frac{1}{\sqrt{g_{11}g_{22}}}\frac{\partial g_{22}}{\partial x^1}\right) + \frac{\partial}{\partial x^2}\left(\frac{1}{\sqrt{g_{11}g_{22}}}\frac{\partial g_{11}}{\partial x^2}\right)\right] \end{aligned}$$

or

$$R_{1212} = -\frac{1}{2\sqrt{g}}\left[\frac{\partial}{\partial x^1}\left(\frac{1}{\sqrt{g}}\frac{\partial g_{22}}{\partial x^1}\right) + \frac{\partial}{\partial x^2}\left(\frac{1}{\sqrt{g}}\frac{\partial g_{11}}{\partial x^2}\right)\right] \tag{5.6.7}$$

Exercise 5.4 Calculate the components of tensors $R_{ijk\ell}$, R_{ij}, and the Gauß curvature for the space E_2 defined by the fundamental form $ds^2 = c^2(dx^1)^2 - f^2(t)(dx^2)^2$ where c^2 is a constant.

The metric tensor and conjugated metric tensor are given, respectively, by

$$g_{ij} = \begin{bmatrix} c^2 & 0 \\ 0 & -f^2(t) \end{bmatrix} \quad g^{ij} = \begin{bmatrix} c^{-2} & 0 \\ 0 & -f^{-2}(t) \end{bmatrix}$$

then

$$g = c^2 f^2 i^2 \Rightarrow \sqrt{g} = cfi$$

where $i^2 = -1$ is the imaginary number and with expression (5.6.8)

$$R_{1212} = -\frac{1}{2\sqrt{g}}\left[\frac{\partial}{\partial x^1}\left(\frac{1}{\sqrt{g}}\frac{\partial g_{22}}{\partial x^1}\right) + \frac{\partial}{\partial x^2}\left(\frac{1}{\sqrt{g}}\frac{\partial g_{11}}{\partial x^2}\right)\right]$$

it follows that

$$\begin{aligned} R_{1212} &= -\frac{1}{2cfi}\frac{\partial}{\partial x^1}\left(\frac{1}{cfi}\frac{\partial g_{22}}{\partial x^1}\right) = -\frac{1}{2cfi}\frac{\partial}{\partial x^1}\left[\frac{1}{cfi}\cdot(-2f\dot{f})\right] = \frac{1}{2cfi}\frac{\partial}{\partial x^1}\left(\frac{2\dot{f}}{ci}\right) \\ &= \frac{\ddot{f}}{c^2 f i^2} = -\frac{\ddot{f}}{c^2 f} \end{aligned}$$

For the components of the Ricci tensor it follows that

$$R_{ij} = g^{pk} R_{ipkj}$$

$$R_{11} = g^{22} R_{1212} = -\frac{1}{f^2}\left(-\frac{\ddot{f}}{c^2 f}\right) = \frac{\ddot{f}}{c^2 f^3}$$

$$R_{22} = g^{11} R_{1212} = \frac{1}{c^2}\left(-\frac{\ddot{f}}{c^2 f}\right) = -\frac{\ddot{f}}{c^4 f}$$

$$R_{12} = R_{21} = g^{12} R_{1212} = 0$$

and for the Gauß curvature it results in

$$K = \frac{R_{1212}}{g} = \frac{-\frac{\ddot{f}}{c^2 f}}{c^2 f^2 i^2} = \frac{\ddot{f}}{c^4 f^3}$$

5.6.4 Einstein Tensor

For the particular case in which the metric, the metric tensor, and its conjugated tensor are given, respectively, by

$$ds^2 = h\left(x^1, x^2\right)\left(dx^1\right)^2 + h\left(x^1, x^2\right)\left(dx^2\right)^2$$

$$g_{ij} = \begin{bmatrix} h & 0 \\ 0 & h \end{bmatrix} \quad g^{ij} = \begin{bmatrix} \frac{1}{h} & 0 \\ 0 & \frac{1}{h} \end{bmatrix}$$

where $h(x^1; x^2) > 0$ is a function of the coordinates, $g = \det g_{ij} = h^2$, and the Ricci tensor is expressed by

$$R_{ij} = g^{pk} R_{ipkj} = g^{11} R_{i11j} + g^{12} R_{i12j} + g^{21} R_{i21j} + g^{22} R_{i22j}$$

then

$$R_{ij} = \frac{1}{h}\left(R_{i11j} + R_{i22j}\right)$$

Developing this expression and with the symmetry of tensor R_{ipkj} it follows that

$$R_{11} = \frac{1}{h}(R_{1111} + R_{1221}) = \frac{1}{h} R_{1221} \quad R_{22} = \frac{1}{h}(R_{2112} + R_{2222}) = \frac{1}{h} R_{2112}$$

$$R_{12} = \frac{1}{h}(R_{1112} + R_{1222}) = 0 \quad R_{21} = \frac{1}{h}(R_{2111} + R_{2221}) = 0$$

Let the scalar curvature

$$R = g^{ij} R_{ij} = g^{11} R_{11} + g^{12} R_{12} + g^{21} R_{21} + g^{22} R_{22} = g^{11} R_{11} + 0 + 0 + g^{22} R_{22}$$
$$= g^{11} R_{11} + g^{22} R_{22}$$

and with the components of the Ricci tensor as a function of the components of tensor R_{ipkj} it follows that

$$R = \frac{1}{h}\frac{1}{h} R_{1221} + \frac{1}{h}\frac{1}{h} R_{2112}$$

As $R_{ipkj} = R_{pijk}$ it results for the scalar curvature

$$R = \frac{1}{h^2}(R_{1221} + R_{1221}) = \frac{2}{h^2}R_{1221}$$

then

$$R_{1221} = \frac{h^2}{2}R$$

and with the substitution of this expression in the expressions of the components of the Ricci tensor it follows that

$$R_{11} = \frac{1}{h}\frac{h^2}{2}R = \frac{h}{2}R = \frac{R}{2}g_{11}$$
$$R_{22}\frac{1}{h}\frac{h^2}{2}R = \frac{h}{2}R = \frac{R}{2}g_{22}$$
$$R_{12} = R_{21} = 0$$

These expressions allow relating the Ricci tensor with the scalar curvature and with the metric tensor, thus

$$R_{ij} = \frac{R}{2}g_{ij} \tag{5.6.8}$$

and with the definition of the scalar curvature given by expression (5.4.6) and with the previous expression it follows

$$R = g^{ij}R_{ij} = g^{ij}g_{ij}\frac{R}{2} = \delta_i^i\frac{R}{2} = N\frac{R}{2}$$

or

$$R\left(1 - \frac{N}{2}\right) = 0 \tag{5.6.9}$$

then for the Riemann space E_2 it is verified that $R_{ij} = R = 0$.

Consider the Einstein tensor given by its covariant components

$$G_{ij} = R_{ij} - \frac{1}{2}g_{ij}R = -Kg_{ij} - \frac{1}{2}g_{ij}R = -Kg_{ij} - \frac{1}{2}g_{ij}(-2K) = 0$$

then the tensor G_{ij} is null for the Riemann space E_2.

5.6.5 Riemann Space with Constant Curvature

The Riemann curvature in point $x^i \in E_N$, in general, depends on this point in which it is defined and the vectors $\boldsymbol{u}$ and $\boldsymbol{v}$ that establish the plane π with respect to which it is calculated. It is admitted that this dependency does not exist, i.e., the space is isotropic, then the relation of the isotropy of the space with the Riemann curvature is established by the following theorem.

Schur Theorem

If all the points of a neighborhood in the Riemann space E_N, being $N > 2$, are isotropic, then the curvature K is constant in all this neighborhood.

To prove the validity of this theorem, let expression (5.3.13) be rewritten as

$$R_{ijk\ell} = G_{ijk\ell} K \tag{5.6.10}$$

with

$$G_{ijk\ell} = \left(g_{ik} g_{j\ell} - g_{i\ell} g_{jk} \right) \neq 0$$

valid in the neighborhood of point x^m of Riemann space E_N.

The covariant derivative of expression (5.6.11) with respect to variable x^m is given by

$$\partial_m R_{ijk\ell} = G_{ijk\ell} \partial_m K \tag{5.6.11}$$

with $\partial_m G_{ijk\ell} = 0$, because, in general, $\frac{\partial g_{ij}}{\partial x^m} = 0$.

With the permutation of indexes in the expression (5.6.12)

$$\partial_k R_{ij\ell m} = G_{ij\ell m} \partial_k K \tag{5.6.12}$$

$$\partial_\ell R_{ijmk} = G_{ijmk} \partial_\ell K \tag{5.6.13}$$

The sum of expressions (5.6.12)–(5.6.14) provides

$$\partial_m R_{ijk\ell} + \partial_k R_{ij\ell m} + \partial_\ell R_{ijmk} = G_{ijk\ell} \partial_m K + G_{ij\ell m} \partial_k K + G_{ijmk} \partial_\ell K$$

but the left side of expression is the second Bianchi identity thus

$$G_{ijk\ell} \partial_m K + G_{ij\ell m} \partial_k K + G_{ijmk} \partial_\ell K = 0$$

and multiplying the terms of this expression by $g^{ik} g^{j\ell}$ it follows

$$g^{ik}g^{j\ell}G_{ijk\ell}\partial_m K = g^{ik}g^{j\ell}\left(g_{ik}g_{j\ell} - g_{i\ell}g_{jk}\right) = \delta_k^k\delta_\ell^\ell - \delta_\ell^k\delta_k^\ell = N^2 - N$$

$$g^{ik}g^{j\ell}G_{ij\ell m}\partial_k K = g^{ik}g^{j\ell}\left(g_{i\ell}g_{jm} - g_{im}g_{j\ell}\right) = \delta_\ell^k\delta_m^\ell - \delta_m^k\delta_\ell^\ell = \delta_m^k - N\delta_m^k$$

$$g^{ik}g^{j\ell}G_{ijmk}\partial_\ell K = g^{ik}g^{j\ell}\left(g_{im}g_{jk} - g_{ik}g_{jm}\right) = \delta_m^k\delta_k^\ell - \delta_k^k\delta_m^\ell = \delta_m^\ell - N\delta_m^\ell$$

The sum of these three terms provides

$$\left(N^2 - N\right)\partial_m K + \left(\delta_m^k - N\delta_m^k\right)\partial_k K + \left(\delta_m^\ell - N\delta_m^\ell\right)\partial_\ell K = 0$$

it follows that

$$\left(N^2 - N\right)\partial_m K + (1 - N)\partial_m K + (1 - N)\partial_m K = 0$$

whereby

$$\left[\left(N^2 - N\right) + 2(1 - N)\right]\partial_m K = 0 \tag{5.6.14}$$

For $N > 2$ this expression is null only if $\partial_m K = 0$, and as x^m is an arbitrary coordinate it is concluded that K is constant in the neighborhood of this point in the Riemann space E_N, which proves the Schur theorem. Expression (5.3.13), where K is a constant is the necessary and sufficient condition so that the curvature of the Riemann space E_N is independent of the orientation considered.

5.6.6 *Isotropy*

Another characteristic of this type of space is related with a scalar curvature. Let expression (5.3.13) be rewritten as

$$R_{ijk\ell} = K\left(g_{ik}g_{j\ell} - g_{i\ell}g_{jk}\right)$$

and multiplied by $g^{\ell i}$

$$R_{jk} = g^{\ell i}R_{ijk\ell} = Kg^{\ell i}\left(g_{ik}g_{j\ell} - g_{i\ell}g_{jk}\right) = K\left(\delta_k^\ell g_{j\ell} - \delta_\ell^\ell g_{jk}\right) = K\left(g_{jk} - Ng_{jk}\right)$$

then

$$R_{jk} = K(1 - N)g_{jk} \tag{5.6.15}$$

For the scalar curvature it follows that

$$R_k^k = g^{kj} R_{jk} = g^{kj} K(1-N) g_{jk} = K(1-N)\delta_k^k$$

whereby

$$R = K(1-N)N \tag{5.6.16}$$

This formulation shows that in the Riemann space E_2 the tensor $R_{ijk\ell}$ leads to the Gauß curvature K, which is the reason for adopting the denomination curvature tensor by extension of this particular case for Riemann spaces of N dimensions. For the Riemann space E_N, where $N > 2$, in which the Ricci tensor results from the substitution of expression (5.6.17) in expression (5.6.16), thus

$$R_{ij} = \frac{R}{N} g_{ij} \tag{5.6.17}$$

where the ratio $\frac{R}{N}$ defines a scalar. The space in which the Ricci tensor is proportional to the metric tensor is called the Einstein space.

The scalar curvature of the Einstein space is given by

$$g^{pi} R_{ij} = \frac{K}{N} g^{pi} g_{ij}$$

following for the Ricci tensor with variance (1, 1)

$$R_j^p = \frac{K}{N} \delta_i^p$$

The covariant derivative of this expression with respect to variable x^p is given by

$$\partial_p R_j^p = \frac{K}{N} \frac{\partial \delta_j^p}{\partial x^j} = 0$$

and with expression (5.4.11)

$$\partial_p R_j^p = \frac{1}{2} \frac{\partial R}{\partial x^j} = 0$$

whereby

$$\frac{\partial R}{\partial x^j} = 0 \tag{5.6.18}$$

then the Einstein space has constant curvature, i.e., is isotropic.

The multiplying of expression (5.6.18) by vector u^j allows researching the eigenvalues of the Ricci tensor, thus

$$R_{ij}u^j = \frac{R}{N}g_{ij}u^j = \frac{R}{N}u_i \Rightarrow \left(R_{ij} - \frac{R}{N}\delta_{ij}\right)u^j = 0$$

where the scalar curvature is constant then the eigenvalues are equal to $\frac{R}{N}$. In this case the eigenvectors of tensor R_{ij} are undetermined.

Exercise 5.5 Calculate the components of the curvature tensor $R_{ijk\ell}$, of the Ricci tensor R_{ij}, the scalar curvature and the Gauß curvature K for the bidimensional spherical space which metric is given by

$$ds^2 = r^2\left(d\varphi^2 + \sin^2\varphi \, d\theta^2\right)$$

The metric tensor, the determinant g, and the conjugated tensor of g_{ij} are given, respectively, by

$$g_{ij} = \begin{bmatrix} r^2 & 0 \\ 0 & r^2\sin^2\varphi \end{bmatrix} \quad g = r^4\sin^2\varphi \quad g^{ij} = \begin{bmatrix} \frac{1}{r^2} & 0 \\ 0 & \frac{1}{r^2\sin^2\varphi} \end{bmatrix}$$

For the partial derivatives of the metric tensor the result is $g_{11,1} = g_{22,2} = 0$, following for the Christoffel symbols

$$\Gamma^1_{11} = \Gamma^2_{22} = \Gamma^2_{11} = \Gamma^2_{12} = \Gamma^2_{21} = 0$$

$$\Gamma^1_{12} = \Gamma^1_{21} = \frac{g^{11}g_{11,2}}{2} = \frac{1}{2r^2\sin^2\varphi}\frac{\partial(r^2\sin^2\varphi)}{\partial\varphi} = -\frac{\cos\varphi}{\sin\varphi}$$

$$\Gamma^1_{22} = -\frac{g^{11}g_{22,1}}{2} = \frac{1}{2r^2}\frac{\partial(r^2\sin^2\varphi)}{\partial\varphi} = -\sin\varphi\cdot\cos\varphi$$

thus

$$R^i_{jk\ell} = \frac{\partial\Gamma^i_{j\ell}}{\partial x^k} - \frac{\partial\Gamma^i_{jk}}{\partial x^\ell} + \Gamma^m_{j\ell}\Gamma^i_{mk} - \Gamma^m_{jk}\Gamma^i_{m\ell}$$

$$R_{1212} = g_{1m}R^m_{212} = g_{11}R^1_{212} = g_{11}\left(\frac{\partial\Gamma^1_{22}}{\partial x^1} - \Gamma^2_{21}\Gamma^1_{22}\right)$$

$$R_{1212} = r^2\left[\frac{\partial}{\partial\varphi}(-\sin\varphi\cdot\cos\varphi) - \left(\frac{\cos\varphi}{\sin\varphi}\right)\cdot(-\sin\varphi\cdot\cos\varphi)\right] = r^2\sin^2\varphi$$

$$K = \frac{R_{1212}}{g} = r^2\sin^2\varphi\frac{1}{r^4\sin^2\varphi} = \frac{1}{r^2}$$

For the Ricci tensor it follows that

$$g^{11} = \frac{1}{r^2} \quad g^{22} = \frac{1}{r^2 \sin^2\varphi}$$

$$R_{ij} = g^{pq} R_{ipkj}$$

$$R_{11} = g^{22} R_{1212} = \frac{1}{r^2 \sin^2\varphi} r^2 \sin^2\varphi = 1$$

$$R_{22} = g^{11} R_{1212} = \frac{1}{r^2} r^2 \sin^2\varphi = \sin^2\varphi$$

$$R_{12} = R_{21} = g^{12} R_{1212} = 0$$

As $R_{1212} = Kg$ the space is curved, and with $g = r^4 \sin^2\varphi$ results in

$$K = \frac{1}{r^2}$$

then if the radius r is large $K \to 0$, i.e., the Gauß curvature is small.

Exercise 5.6 For the tensorial equation $A^i_j = R^i_j + \delta^i_j(aR + b)$, where a and b are constants, and R is the scalar curvature, calculate the value of a for which the condition $\partial_i A^i_j = 0$ exists.

The derivative of the equation given with respect to the variable x^j stays

$$\partial_i A^i_j = \partial_i R^i_j + \left(\partial_i \delta^i_j\right)(aR + b) + \delta^i_j \left(a \frac{\partial R}{\partial x^i} + 0\right)$$

where $\partial_i \delta^i_j = 0$, then

$$\partial_i A^i_j = \partial_i R^i_j + \delta^i_j a \frac{\partial R}{\partial x^i}$$

and with the condition $\partial_i A^i_j = 0$ it follows that

$$\partial_i R^i_j + \delta^i_j a \frac{\partial R}{\partial x^i} = \partial_i R^i_j + a \frac{\partial R}{\partial x^j} = 0$$

Having

$$\partial_i R^i_j = \frac{1}{2} \frac{\partial R}{\partial x^j}$$

that substituted in the previous expression provides

$$\frac{\partial R}{\partial x^j}\left(\frac{1}{2}+a\right)=0$$

As $\frac{\partial R}{\partial x^j}\neq 0$ it results in

$$a=-\frac{1}{2}$$

Exercise 5.7 Analyze the curvature of the Riemann space E_N, $N>2$, which Riemann–Christoffel curvature tensor is given by $R^{\ell}_{ijk}=\rho\left(\delta^{\ell}_j g_{ik}-\delta^{\ell}_k g_{ij}\right)$, where ρ is a constant.

The contraction of the curvature tensor in the indexes $\ell=k$ provides

$$R^{\ell}_{ij\ell}=\rho\left(\delta^{\ell}_j g_{i\ell}-\delta^{\ell}_{\ell} g_{ij}\right)$$

it follows that

$$R^{\ell}_{ij\ell}=\rho\left(g_{ij}-Ng_{ij}\right)=\rho(1-N)g_{ij}$$

With the constant

$$\sigma=\rho(1-N)$$

it results in

$$R^{\ell}_{ij\ell}=\sigma g_{ij}$$

The multiplying of the members by g^{ij} it follows that

$$g^{ij}R_{ij}=\sigma g^{ij}g_{ij}=\sigma\delta^i_i=\rho(1-N)N$$

then for an Einstein space the scalar curvature is constant.

Exercise 5.8 Calculate the components of the Riemann–Christoffel tensor for E_2, which metric is given by

$$ds^2=dx^2+G(x,y)dy^2$$

The metric tensor and the conjugated metric tensor are given, respectively, by

$$g_{ij} = \begin{bmatrix} 1 & 0 \\ 0 & G(x,y) \end{bmatrix} \quad g^{ij} = \begin{bmatrix} 1 & 0 \\ 0 & \dfrac{1}{G(x,y)} \end{bmatrix}$$

The derivatives of the metric tensor are

$$g_{11,x} = g_{11,x} = g_{12,x} = g_{21,x} = g_{11,xx} = g_{12,xx} = g_{21,xx} = g_{11,yy} = g_{12,yy} = g_{21,yy} = 0$$

$$g_{22,x} = G(x,y)_{,x} \quad g_{22,y} = G(x,y)_{,y} \quad g_{22,xx} = G(x,y)_{,xx}$$

and the Christoffel symbols stay

$$\Gamma^1_{11} = \Gamma^1_{12} = \Gamma^1_{21} = \Gamma^2_{11} = 0$$

$$\Gamma^1_{22} = g^{1k}\Gamma_{22,k} = g^{1k}\frac{1}{2}\left(-g_{22,k}\right) = g^{11}\left[-\frac{1}{2}G(x,y)_{,x}\right] = -\frac{1}{2}G(x,y)_{,x}$$

$$\Gamma^2_{12} = \Gamma^2_{21} = g^{2k}\Gamma_{12,k} = g^{2K}\left(\frac{1}{2}g_{2k,1}\right) = g^{22}\frac{1}{2}G(x,y)_{,x} = -\frac{1}{2G(x,y)}G(x,y)_{,x}$$

$$\Gamma^2_{22} = g^{2k}\Gamma_{22,k} = g^{2K}\left(\frac{1}{2}g_{2k,2}\right) = g^{22}\frac{1}{2}G(x,y)_{,y} = -\frac{1}{2G(x,y)}G(x,y)_{,y}$$

The Riemann–Christoffel curvature tensor with variance (0, 4) is given by

$$R_{pijk} = \frac{1}{2}\left(\frac{\partial^2 g_{ik}}{\partial x^j \partial x^p} + \frac{\partial^2 g_{pk}}{\partial x^j \partial x^i} - \frac{\partial^2 g_{ji}}{\partial x^k \partial x^p} - \frac{\partial^2 g_{pj}}{\partial x^k \partial x^i}\right) + g_{q\ell}\Gamma^q_{pk}\Gamma^\ell_{ij} - g_{q\ell}\Gamma^q_{pj}\Gamma^\ell_{ik}$$

In space E_2 this tensor has a single independent non-null component, then

$$R_{1212} = \frac{1}{2}\left(\frac{\partial^2 g_{12}}{\partial x^1 \partial x^2} + \frac{\partial^2 g_{21}}{\partial x^2 \partial x^1} - \frac{\partial^2 g_{22}}{\partial x^1 \partial x^1} - \frac{\partial^2 g_{11}}{\partial x^2 \partial x^2}\right) + g_{q\ell}\Gamma^q_{21}\Gamma^\ell_{12} - g_{q\ell}\Gamma^q_{22}\Gamma^\ell_{11}$$

$$R_{1212} = \frac{1}{2}\left(-\frac{\partial^2 g_{22}}{\partial x^1 \partial x^1}\right) + g_{22}\Gamma^2_{21}\Gamma^2_{12}$$

$$R_{1212} = \frac{1}{2}\left[-G(x,y)_{,xx}\right] + G(x,y)\left\{\frac{1}{4G^2(x,y)}\left[G(x,y)_{,x}\right]^2\right\}$$

$$R_{1212} = -\frac{1}{2}G(x,y)_{,xx} + \frac{1}{4G(x,y)}\left[G(x,y)_{,x}\right]^2$$

5.6.7 Minkowski Space

The Riemann space that links three coordinates defined by lengths and a fourth coordinate related to time is called the Minkowski space. The metric of this Riemann space E_4 is defined by

$$ds^2 = dx^i dx^i = \left(dx^1\right)^2 + \left(dx^2\right)^2 + \left(dx^3\right)^2 + \left(dx^4\right)^2$$

where the fourth coordinate is $x^4 = ict$, where $i^2 = -1$ is the imaginary number, c is a constant, and t is the time variable, so for the fundamental form the result is

$$ds^2 = \left(dx^1\right)^2 + \left(dx^2\right)^2 + \left(dx^3\right)^2 - c^2(dt)^2 \tag{5.6.19}$$

and the metric tensor is given by

$$g_{ij} = \begin{bmatrix} 1 & 0 & 0 & 0 \\ 0 & 1 & 0 & 0 \\ 0 & 0 & 1 & 0 \\ 0 & 0 & 0 & -c^2 \end{bmatrix} \tag{5.6.20}$$

This tensor is not positive definite, so the Minkowski space is not Euclidian. It is verified promptly that $\frac{\partial g_{ij}}{\partial x^k} = 0, \ \forall i, j, k = 1, 2, 3, 4$, then all the Christoffel symbols are null, whereby $R^{\ell}_{ijk} = 0$, which shows that this space is flat. It is stressed that every Euclidian space is flat, but not every flat space is Euclidian, as the case of the Minkowski space.

The fundamental form and the metric tensor of the Minkowski space in spherical coordinates are given, respectively, by

$$ds^2 = dr^2 + r^2 d\varphi^2 + r^2 \sin^2\varphi \, d\theta^2 - c^2 dt^2 \tag{5.6.21}$$

$$g_{ij} = \begin{bmatrix} 1 & 0 & 0 & 0 \\ 0 & r^2 & 0 & 0 \\ 0 & 0 & r^2 \sin^2\varphi & 0 \\ 0 & 0 & 0 & -c^2 \end{bmatrix} \tag{5.6.22}$$

Exercise 5.9 Calculate the components of the Riemann–Christoffel tensor of the space defined by metric $ds^2 = \left(dx^1\right)^2 + \left(dx^2\right)^2 + \left(dx^3\right)^2 - e^{-t}(dt)^2$.

The metric tensor and its conjugated tensor are given, respectively, by

$$g_{ij} = \begin{bmatrix} 1 & 0 & 0 & 0 \\ 0 & 1 & 0 & 0 \\ 0 & 0 & 1 & 0 \\ 0 & 0 & 0 & e^{-t} \end{bmatrix} \qquad g^{ij} = \begin{bmatrix} 1 & 0 & 0 & 0 \\ 0 & 1 & 0 & 0 \\ 0 & 0 & 1 & 0 \\ 0 & 0 & 0 & e^{t} \end{bmatrix}$$

The unique non-null Christoffel symbol of second kind is

$$\Gamma_{44,4} = \frac{1}{2} g_{44,4} = -\frac{1}{2} e^{-t} \Rightarrow \Gamma^4_{44} = g^{44} \Gamma_{44,4} = -\frac{1}{2}$$

As the Riemann–Christoffel curvature tensor is defined by expression

$$R^{\ell}_{ijk} = \frac{\partial \Gamma^{\ell}_{ik}}{\partial x^j} - \frac{\partial \Gamma^{\ell}_{ij}}{\partial x^k} + \Gamma^{\ell}_{mj} \Gamma^{m}_{ik} - \Gamma^{\ell}_{mk} \Gamma^{m}_{ij}$$

and Γ^4_{44} is a constant value thus $R^{\ell}_{ijk} = 0$, then this is a flat space.

5.6.8 *Conformal Spaces*

5.6.8.1 Initial Concept

A functional relation is called conformal when the domain D of a set of complex variables in a plane generates a contradomain of values of complex variables in another plane, preserving the angle and the direction between the curves that intersect. This concept is generalized for the case of the variables in the Riemann space E_N and in the conformal space $\widetilde{E}_N$. Consider these two spaces and a coordinate system X^i, with the relation between its metric tensors $g_{ij}, \widetilde{g}_{ij}$ given by

$$\widetilde{g}_{ij} = e^{2\phi} g_{ij} \tag{5.6.23}$$

where $x^i \in D \subset E_N, \phi(x^i) > 0$ is a scalar function of class C^3.

The angles between two vectors $\boldsymbol{u}, \boldsymbol{v}$ tangent to two curves in these two Riemann spaces are given by

$$\cos\alpha = \frac{g_{ij}u^i v^j}{\|\boldsymbol{u}\|\cdot\|\boldsymbol{v}\|} = \frac{g_{ij}u^i v^j}{\sqrt{\varepsilon\cdot g_{km}u^k u^m}\sqrt{\varepsilon\cdot g_{km}v^k v^m}} = \frac{g_{ij}u^i v^j}{\varepsilon\sqrt{g_{km}u^k u^m}\sqrt{g_{km}v^k v^m}}$$

$$\cos\widetilde{\alpha} = \frac{g_{ij}u^i v^j}{\|\widetilde{\boldsymbol{u}}\|\cdot\|\widetilde{\boldsymbol{v}}\|} = \frac{\widetilde{g}_{ij}u^i v^j}{\sqrt{\varepsilon\cdot \widetilde{g}_{km}u^k u^m}\sqrt{\varepsilon\cdot \widetilde{g}_{km}v^k v^m}} = \frac{e^{2\phi}g_{ij}u^i v^j}{\sqrt{\varepsilon\cdot e^{2\phi}g_{km}u^k u^m}\sqrt{\varepsilon\cdot e^{2\phi}g_{km}v^k v^m}}$$

$$= \frac{g_{ij}u^i v^j}{\varepsilon\sqrt{g_{km}u^k u^m}\sqrt{g_{km}v^k v^m}} \tag{5.6.24}$$

where $\varepsilon = \pm 1$ is a functional operator. This expression shows that $\alpha = \widetilde{\alpha}$, then the expression (5.6.24) represents a conformal transformation.

In spaces E_N and $\widetilde{E}_N$ the conjugated metric tensors are related by

$$\widetilde{g}^{ij} = e^{-2\phi}g^{ij} \tag{5.6.25}$$

and with the following expressions being valid in these spaces

– Basis vectors

$$\widetilde{\boldsymbol{e}}^i = e^{2\phi}\boldsymbol{e}^i \tag{5.6.26}$$

$$\widetilde{\boldsymbol{e}}_i = e^{-2\phi}\boldsymbol{e}_i \tag{5.6.27}$$

– Norm of a vector

$$\|\boldsymbol{u}\| = \sqrt{\widetilde{g}_{km}u^k u^m} = \sqrt{e^{2\phi}g_{km}u^k u^m} \tag{5.6.28}$$

– Dot product of vectors

$$\boldsymbol{u}\cdot\boldsymbol{v} = \widetilde{g}_{km}u^k v^m = e^{2\phi}g_{km}u^k v^m \tag{5.6.29}$$

5.6.8.2 Christoffel Symbols

Let the Christoffel symbol of first kind be $\widetilde{\Gamma}_{jk,m}$ for the conformal space $\widetilde{E}_N$, which relates with the Riemann space E_N by means of the expressions (5.6.24) and (5.6.26), then

$$\begin{aligned}\widetilde{\Gamma}_{jk,m} &= \frac{1}{2}\left(\widetilde{g}_{jm,k} + \widetilde{g}_{km,j} - \widetilde{g}_{jk,m}\right) = \frac{1}{2}\left[\left(e^{2\phi} g_{jm}\right)_{,k} + \left(e^{2\phi} g_{km}\right)_{,j} - \left(e^{2\phi} g_{jk}\right)_{,m}\right] \\ &= \frac{1}{2}\left[\left(2e^{2\phi}\phi_{,k} g_{jm} + e^{2\phi} g_{jm,k}\right) + \left(2e^{2\phi}\phi_{,j} g_{km} + e^{2\phi} g_{km,j}\right)\right. \\ &\quad \left. - \left(2e^{2\phi}\phi_{,m} g_{jk} + e^{2\phi} g_{jk,m}\right)\right] \\ &= e^{2\phi}\left[\frac{1}{2}\left(g_{jm,k} + g_{km,j} - g_{jk,m}\right) + \left(\phi_{,k} g_{jm} + \phi_{,j} g_{km} - \phi_{,m} g_{jk}\right)\right]\end{aligned}$$

whereby the expression for this affine connection in the conformal space $\widetilde{E}_N$ stays

$$\widetilde{\Gamma}_{jk,m} = e^{2\phi}\left[\Gamma_{jk,m} + \left(\phi_{,k} g_{jm} + \phi_{,j} g_{km} - \phi_{,m} g_{jk}\right)\right] \tag{5.6.30}$$

For the Christoffel symbol of second kind the result is

$$\widetilde{\Gamma}^{i}_{jk} = \widetilde{g}^{im}\widetilde{\Gamma}_{jk,m} = e^{-2\phi} g^{im}\widetilde{\Gamma}_{jk,m}$$

it follows that

$$\begin{aligned}\widetilde{\Gamma}^{i}_{jk} &= e^{-2\phi} g^{im} e^{2\phi}\left[\Gamma_{jk,m} + \left(\phi_{,k} g_{jm} + \phi_{,j} g_{km} - \phi_{,m} g_{jk}\right)\right] \\ &= g^{im}\Gamma_{jk,m} + g^{im}\left(\phi_{,k} g_{jm} + \phi_{,j} g_{km} - \phi_{,m} g_{jk}\right)\end{aligned}$$

whereby the expression for this affine connection in the conformal space $\widetilde{E}_N$ stays

$$\widetilde{\Gamma}^{i}_{jk} = \Gamma^{i}_{jk} + \left(\phi_{,k}\delta^{i}_{j} + \phi_{,j}\delta^{i}_{k} - \phi_{,m} g^{im} g_{jk}\right) \tag{5.6.31}$$

Expressions (5.6.31) and (5.6.32) show that the Christoffel symbols are not invariant for the conformal transformation given by expression (5.6.24).

5.6.8.3 Riemann–Christoffel tensor

The definition of the Riemann–Christoffel tensor in the conformal space $\widetilde{E}_N$ is given by

$$\widetilde{R}_{ijk\ell} = \frac{1}{2}\left(\widetilde{g}_{i\ell,kj} + \widetilde{g}_{jk,\ell i} - \widetilde{g}_{j\ell,ki} - \widetilde{g}_{ik,\ell j}\right) + \widetilde{g}_{mn}\left(\widetilde{\Gamma}^{m}_{jk}\widetilde{\Gamma}^{n}_{i\ell} - \widetilde{\Gamma}^{m}_{j\ell}\widetilde{\Gamma}^{n}_{ik}\right) \tag{5.6.32}$$

For the derivatives of the metric tensor it follows that

$$\widetilde{g}_{i\ell} = e^{2\phi} g_{i\ell} \Rightarrow \widetilde{g}_{i\ell,k} = 2\phi_{,k} e^{2\phi} g_{i\ell} + e^{2\phi} g_{i\ell,k}$$
$$\widetilde{g}_{i\ell,kj} = 4\phi_{,k}\phi_{,j} e^{2\phi} g_{i\ell} + 2\phi_{,kj} e^{2\phi} g_{i\ell} + 2\phi_{,k} e^{2\phi} g_{i\ell,j} + 2\phi_{,j} e^{2\phi} g_{i\ell,k} + e^{2\phi} g_{i\ell,kj}$$

and in an analogous way

$$\widetilde{g}_{jk,\ell i} = 4\phi_{,\ell}\phi_{,i} e^{2\phi} g_{jk} + 2\phi_{,\ell i} e^{2\phi} g_{jk} + 2\phi_{,\ell} e^{2\phi} g_{jk,i} + 2\phi_{,i} e^{2\phi} g_{jk,\ell} + e^{2\phi} g_{jk,\ell i}$$
$$\widetilde{g}_{j\ell,ki} = 4\phi_{,k}\phi_{,i} e^{2\phi} g_{j\ell} + 2\phi_{,ki} e^{2\phi} g_{j\ell} + 2\phi_{,k} e^{2\phi} g_{j\ell,i} + 2\phi_{,i} e^{2\phi} g_{j\ell,k} + e^{2\phi} g_{j\ell,ki}$$
$$\widetilde{g}_{ik,\ell j} = 4\phi_{,\ell}\phi_{,j} e^{2\phi} g_{ik} + 2\phi_{,\ell j} e^{2\phi} g_{ik} + 2\phi_{,\ell} e^{2\phi} g_{ik,j} + 2\phi_{,j} e^{2\phi} g_{ik,\ell} + e^{2\phi} g_{ik,\ell j}$$

For the Christoffel symbols by means of expression (5.6.32) it follows that

$$\widetilde{\Gamma}^m_{jk} = \Gamma^m_{jk} + \left(\phi_{,k}\delta^m_j + \phi_{,j}\delta^m_k - \phi_{,i} g^{mi} g_{jk}\right)$$
$$\widetilde{\Gamma}^n_{i\ell} = \Gamma^n_{i\ell} + \left(\phi_{,\ell}\delta^n_i + \phi_{,i}\delta^n_\ell - \phi_{,m} g^{nm} g_{i\ell}\right)$$
$$\widetilde{\Gamma}^m_{j\ell} = \Gamma^m_{j\ell} + \left(\phi_{,\ell}\delta^m_j + \phi_{,j}\delta^m_\ell - \phi_{,i} g^{mi} g_{j\ell}\right)$$
$$\widetilde{\Gamma}^n_{ik} = \Gamma^n_{ik} + \left(\phi_{,k}\delta^n_i + \phi_{,i}\delta^n_k - \phi_{,n} g^{mn} g_{ik}\right)$$

The substitution of these expressions in expression (5.6.33) leads to

$$\widetilde{R}_{ijk\ell} = e^{2\phi} \left\{ \begin{array}{l} R_{ijk\ell} + \left[g_{i\ell}\left(\phi_{,jk} - \phi_{,j}\phi_{,k}\right) + g_{jk}\left(\phi_{,i\ell} - \phi_{,i}\phi_{,\ell}\right) \right. \\ \quad \left. - g_{ik}\left(\phi_{,j\ell} - \phi_{,j}\phi_{,\ell}\right) - g_{j\ell}\left(\phi_{,ik} - \phi_{,i}\phi_{,k}\right) \right] \\ \quad + \left[g_{i\ell} g_{jk}\left(g^{mn}\phi_{,m}\phi_{,n}\right) - g_{ik} g_{j\ell}\left(g^{mn}\phi_{,m}\phi_{,n}\right) \right] \end{array} \right\}$$

Putting

$$\phi_{jk} = \phi_{kj} = \phi_{,jk} - \phi_{,j}\phi_{,k}$$
$$\phi_{i\ell} = \phi_{\ell i} = \phi_{,i\ell} - \phi_{,i}\phi_{,\ell}$$
$$\phi_{j\ell} = \phi_{\ell j} = \phi_{,j\ell} - \phi_{,j}\phi_{,\ell}$$
$$\phi_{ik} = \phi_{ki} = \phi_{,ik} - \phi_{,i}\phi_{,k}$$

results in

$$\widetilde{R}_{ijk\ell} = e^{2\phi}\left[R_{ijk\ell} + g_{i\ell}\phi_{jk} + g_{jk}\phi_{i\ell} - g_{ik}\phi_{j\ell} - g_{j\ell}\phi_{ik} + g^{mn}\phi_{,m}\phi_{,n}\left(g_{i\ell}g_{jk} - g_{ik}g_{j\ell}\right)\right] \tag{5.6.33}$$

then the Riemann–Christoffel tensor is not invariant for the transformation as defined by expression (5.6.24).

5.6.8.4 Ricci Tensor

The definition of the Ricci tensor in the conformal space $\widetilde{E}_N$ is given by

$$\widetilde{R}_{jk} = \widetilde{g}^{ik}\widetilde{R}_{ijk\ell} \tag{5.6.34}$$

and with the substitution of expression (5.6.26) in expression (5.6.35) it follows that

$$\widetilde{R}_{j\ell} = e^{-2\phi}g^{ik}\left[R_{ijk\ell} + g_{i\ell}\phi_{jk} + g_{jk}\phi_{j\ell} - g_{ik}\phi_{j\ell} - g_{j\ell}\phi_{ik} + g^{mn}\phi_{,m}\phi_{,n}\left(g_{i\ell}g_{jk} - g_{ik}g_{j\ell}\right)\right]$$

$$\widetilde{R}_{j\ell} = R_{j\ell} + \delta_\ell^k\phi_{jk} + \delta_j^i\phi_{j\ell} - \delta_i^i\phi_{j\ell} - g^{ik}g_{j\ell}\phi_{ik} + g^{mn}\phi_{,m}\phi_{,n}\left(\delta_\ell^k g_{jk} - \delta_i^i g_{j\ell}\right)$$

Putting

$$\phi = g^{ik}\phi_{ik}$$

then

$$\begin{aligned} \phi = g^{mn}\phi_{mn} &= g^{mn}\left(\phi_{,mn} - \phi_{,m}\phi_{,n}\right) \\ g^{mn}\phi_{,m}\phi_{,n} &= g^{mn}\phi_{,mn} - g^{mn}\phi_{mn} \end{aligned} \tag{5.6.35}$$

thus

$$\widetilde{R}_{j\ell} = R_{j\ell} + \phi_{j\ell} + \phi_{j\ell} - N\phi_{j\ell} - \phi g_{j\ell} + g^{mn}\phi_{,m}\phi_{,n}\left(g_{j\ell} - Ng_{j\ell}\right)$$

whereby

$$\widetilde{R}_{j\ell} = R_{j\ell} - (N-2)\phi_{j\ell} - g_{j\ell}g^{mn}\phi_{,mn} - (N-2)g_{j\ell}g^{mn}\phi_{,m}\phi_{,n} \tag{5.6.36}$$

is the expression for the Ricci tensor in the conformal space $\widetilde{E}_N$, which is not invariant for the transformation as defined by expression (5.6.24).

5.6.8.5 Scalar Curvature

The definition of the scalar curvature is given by

$$\widetilde{R} = \widetilde{g}^{j\ell}\widetilde{R}_{j\ell}$$

and with the substitution of expression (5.6.26) it follows that

$$\widetilde{R} = \widetilde{g}^{j\ell}\left[R_{j\ell} - (N-2)\phi_{j\ell} - g_{j\ell}g^{mn}\phi_{,mn} - (N-2)g_{j\ell}g^{mn}\phi_{,m}\phi_{,n}\right]$$
$$= e^{-2\phi}g^{j\ell}\left[R_{j\ell} - (N-2)\phi_{j\ell} - g_{j\ell}g^{mn}\phi_{,mn} - (N-2)g_{j\ell}g^{mn}\phi_{,m}\phi_{,n}\right]$$

whereby

$$\widetilde{R} = e^{-2\phi}\{R + [-2(N-1)\phi_{,mn} - (N-1)(N-2)\phi_{,m}\phi_{,n}]g^{mn}\} \tag{5.6.37}$$

is the expression for the scalar curvature in space $\widetilde{E}_N$, which is not invariant for the conformal transformation defined by expression (5.6.24).

5.6.8.6 Weyl Tensor

Formulation

The research of a variety that remains invariant when passing from the space E_N for the conformal space $\widetilde{E}_N$ led Hermann Weyl to conceive a tensor that has the same properties of the Riemann–Christoffel tensor, and were invariant when a conformal transformation defined by expression (5.6.24) takes place.

Let the Riemann–Christoffel tensor in space $\widetilde{E}_N$ be defined by

$$\widetilde{R}_{ijk\ell} = e^{2\phi}\left[R_{ijk\ell} + \left(g_{i\ell}\phi_{jk} + g_{jk}\phi_{j\ell} - g_{ik}\phi_{j\ell} - g_{j\ell}\phi_{ik}\right) + \left(g_{i\ell}g_{jk} - g_{ik}g_{j\ell}\right)g^{mn}\phi_{,m}\phi_{,n}\right]$$

and with the expressions

$$\widetilde{g}^{ip}\widetilde{R}_{ijk\ell} = \widetilde{R}^{p}_{jk\ell} \quad \widetilde{g}^{ip} = e^{-2\phi}g^{ip}$$

the result is

$$\widetilde{R}^{p}_{jk\ell} = R^{p}_{jk\ell} + \delta^{p}_{\ell}\phi_{jk} - \delta^{p}_{k}\phi_{j\ell} + g^{ip}\left(g_{jk}\phi_{i\ell} - g_{j\ell}\phi_{ik}\right)$$
$$+ \left(\delta^{p}_{\ell}g_{jk} - \delta^{p}_{k}\phi_{j\ell}\right)g^{mn}\phi_{,m}\phi_{,n} \tag{5.6.38}$$

The term $g^{mn}\phi_{,m}\phi_{,n}$ can be eliminated, and with expression (5.6.36) it is possible to obtain the parameters $\phi_{j\ell}, \phi_{jk}, \phi_{i\ell}, \phi_{ik}$ in terms of the Ricci tensor, the scalar curvature, and the metric tensor, thus

$$\phi_{j\ell} = -\frac{1}{(N-2)}\left(\widetilde{R}_{j\ell} + R_{j\ell}\right) - \frac{1}{(N-2)}g_{j\ell}g^{mn}\phi_{,mn} - g_{j\ell}g^{mn}\phi_{,m}\phi_{,n} \tag{5.6.39}$$

whereby

$$g_{j\ell}g^{mn}\phi_{,mn} = -\frac{1}{2(N-1)}\left(\widetilde{R}e^{2\phi} + R\right) - \frac{(N-2)}{2}g_{j\ell}\phi_{,m}\phi_{,n}$$

and with

$$e^{2\phi} = \frac{\widetilde{g}_{j\ell}}{g_{j\ell}}$$

the result is

$$g_{j\ell}g^{mn}\phi_{,mn} = -\frac{1}{2(N-1)}\left(\widetilde{R}\widetilde{g}_{j\ell} + Rg_{j\ell}\right) - \frac{(N-2)}{2}g_{j\ell}\phi_{,m}\phi_{,n} \qquad (5.6.40)$$

The substitution of expression (5.6.39) in expression (5.6.38) provides

$$\phi_{j\ell} = -\frac{1}{(N-2)}\left(\widetilde{R}_{j\ell} - R_{j\ell}\right) - \frac{1}{2(N-1)(N-2)}\left(\widetilde{R}\widetilde{g}_{j\ell} - Rg_{j\ell}\right) - \frac{1}{2}g_{j\ell}g^{mn}\phi_{,m}\phi_{,n} \qquad (5.6.41)$$

The other parameters analogous to this parameter stay

$$\begin{aligned}
\phi_{jk} &= -\frac{1}{(N-2)}\left(\widetilde{R}_{jk} - R_{jk}\right) - \frac{1}{2(N-1)(N-2)}\left(\widetilde{R}\widetilde{g}_{jk} - Rg_{jk}\right) - \frac{1}{2}g_{jk}g^{mn}\phi_{,m}\phi_{,n}\\
\phi_{i\ell} &= -\frac{1}{(N-2)}\left(\widetilde{R}_{i\ell} - R_{i\ell}\right) - \frac{1}{2(N-1)(N-2)}\left(\widetilde{R}\widetilde{g}_{i\ell} - Rg_{i\ell}\right) - \frac{1}{2}g_{i\ell}g^{mn}\phi_{,m}\phi_{,n}\\
\phi_{ik} &= -\frac{1}{(N-2)}\left(\widetilde{R}_{ik} - R_{ik}\right) - \frac{1}{2(N-1)(N-2)}\left(\widetilde{R}\widetilde{g}_{ik} - Rg_{ik}\right) - \frac{1}{2}g_{ik}g^{mn}\phi_{,m}\phi_{,n}
\end{aligned} \qquad (5.6.42)$$

and with the substitution of expressions (5.6.42) and (5.6.43) in expression (5.6.39), and with expressions (5.6.24) and (5.6.26) results in

$$\begin{aligned}
&\widetilde{R}^p_{jk\ell} - \frac{1}{(N-2)}\left(\delta^p_k\widetilde{R}_{j\ell} - \delta^p_\ell\widetilde{R}_{jk} + \widetilde{g}_{j\ell}\widetilde{R}^p_k - \widetilde{g}_{jk}\widetilde{R}^p_\ell\right) + \frac{\widetilde{R}}{(N-1)(N-2)}\left(\delta^p_k\widetilde{g}_{j\ell} - \delta^p_\ell\widetilde{g}_{jk}\right)\\
&= R^p_{jk\ell} - \frac{1}{(N-2)}\left(\delta^p_k R_{j\ell} - \delta^p_\ell R_{jk} + g_{j\ell}R^p_k - g_{jk}R^p_\ell\right)\\
&\quad + \frac{R}{(N-1)(N-2)}\left(\delta^p_k g_{j\ell} - \delta^p_\ell g_{jk}\right)
\end{aligned} \qquad (5.6.43)$$

Putting

$$W^p_{jk\ell} = R^p_{jk\ell} - \frac{1}{(N-2)}\left(\delta^p_k R_{j\ell} - \delta^p_\ell R_{jk} + g_{j\ell}R^p_k - g_{jk}R^p_\ell\right) + \frac{R}{(N-1)(N-2)}\left(\delta^p_k g_{j\ell} - \delta^p_\ell g_{jk}\right) \tag{5.6.44}$$

verifies that expression (5.6.44) represents an equality between tensors

$$\widetilde{W}^p_{jk\ell} = W^p_{jk\ell}$$

and shows that tensor $W^p_{jk\ell}$ is preserved when a conformal transformation, i.e., this tensor is invariant for the space $\widetilde{E}_N$.

Lowering the index of tensor $W^p_{jk\ell}$,

$$g_{pi}W^p_{jk\ell} = W_{ijk\ell}$$

whereby

$$W_{ijk\ell} = R_{ijk\ell} - \frac{1}{(N-2)}\left(g_{ik}R_{j\ell} - g_{i\ell}R_{jk} + g_{j\ell}R^p_k - g_{jk}R^p_\ell\right) + \frac{R}{(N-1)(N-2)}\left(g_{ik}g_{j\ell} - g_{i\ell}g_{jk}\right) \tag{5.6.45}$$

defines the Weyl curvature tensor, and shows that the tensor $W_{ijk\ell}$ is obtained by means of decomposing the Riemann–Christoffel tensor $R_{ijk\ell}$ in their parts comprised by the Ricci tensor and by the scalar curvature, then the Riemann–Christoffel tensor can be decomposed into irreducible components.

Properties of the Weyl Tensor

Expression (5.6.45) indicates that the tensor $W_{ijk\ell}$ has the same number of independent components as the tensor $R_{ijk\ell}$, i.e., $\frac{1}{12}N(N+1)(N+2)(N-3)$ components. The tensorial sum given by expression (5.6.45) shows that the Weyl tensor has the same properties of symmetry and antisymmetry as tensor $R_{ijk\ell}$, then

$$W_{ijk\ell} = W_{k\ell ij}$$
$$W_{ijk\ell} = -W_{jik\ell} = -W_{ij\ell k}$$

These properties indicate that the first Bianchi identity is valid for the Weyl tensor

$$W^{\ell}_{ijk} + W^{\ell}_{jki} + W^{\ell}_{kij} = 0$$
$$W_{ijk\ell} + W_{ikj\ell} + W_{ik\ell j} = 0$$

For the Riemann space E_1 the result is $R_{ijk\ell} = 0$. For the bidimensional space E_2 there is only the component R_{1212}, and the curvature is defined by the scalar curvature. For the tridimensional space E_3 the six components of the curvature tensor are defined by the Ricci tensor, having $W_{ijk\ell} = 0$. For the space $E_N, N > 3$ the components of $R_{ijk\ell}$ are determined by the Ricci tensor and by the Weyl tensor.

Uniqueness of the Weyl tensor

Let expression (5.3.13) that determines the Riemann curvature K in point x^i of the isotropic space E_N, $N > 3$, and with expression (5.6.16) the result is

$$R_{ijk\ell} = K\left(g_{ik}g_{j\ell} - g_{i\ell}g_{jk}\right)$$
$$R_{jk} = K(1 - N)g_{jk}$$

thus

$$K = \frac{R_{jk}g^{jk}}{(1-N)}$$
$$R_{ijk\ell} = \frac{R_{jk}g^{jk}}{(1-N)}\left(g_{il}g_{j\ell} - g_{i\ell}g_{jk}\right) = \frac{1}{(1-N)}\left[\left(R_{jk}g^{jk}g_{j\ell}\right)g_{ik} - \left(R_{jk}g^{jk}g_{jk}\right)g_{i\ell}\right]$$
$$R_{jk}g^{jk}g_{j\ell} = R_{jk}\delta^k_\ell = R_{j\ell} \quad R_{jk}g^{jk}g_{jk} = R_{jk}$$
$$R_{ijk\ell} = \frac{1}{(1-N)}\left(R_{j\ell}g_{ik} - R_{jk}g_{i\ell}\right) \tag{5.6.46}$$

The Weyl tensor is defined by the expression

$$W_{ijk\ell} = R_{ijk\ell} - \frac{1}{(1-N)}\left(R_{j\ell}g_{ik} - R_{jk}g_{i\ell}\right) \tag{5.6.47}$$

If $W_{ijk\ell} = 0$ in the isotropic Riemann space E_N, where $N > 3$ the expression (5.6.48) is null, then the expression (5.6.47) is valid for this space. This is the necessary condition so that this space has constant Riemann curvature.

To demonstrate that the Riemann curvature must be constant for the condition $W_{ijk\ell} = 0$ the multiplying of expression (5.6.47) by $g^{j\ell}$ is carried out, thus

$$g^{j\ell}R_{ijk\ell} = -g^{j\ell}R_{jik\ell} = -R_{ik} = \frac{1}{(1-N)}\left(g^{j\ell}R_{j\ell}g_{ik} - g^{j\ell}R_{jk}g_{i\ell}\right) = \frac{1}{(1-N)}\left(Rg_{ik} - R_{jk}\delta_i^j\right)$$
$$= \frac{1}{(1-N)}\left(Rg_{ik} - R_{ik}\right)$$

results in

$$-R_{ik}(1-N) = Rg_{ik} - R_{ik} \Rightarrow R_{ik} = \frac{R}{N}g_{ik}$$

This last expression is identical to expression (5.6.17) that defines an Einstein space (isotropic space, whereby it has constant curvature), which proves that this condition is sufficient for the Weyl tensor to be null.

Contraction of the Weyl Tensor

The contraction of index k of the Weyl tensor $W_{ijk\ell}$ stays

$$g^{m\ell}W_{ijk\ell} = g^{m\ell}R_{ijk\ell} - \frac{2}{(N-2)}g^{m\ell}\left(R_{\ell j}g_{ik} - R_{\ell i}g_{jk}\right) + \frac{2}{(N-1)(N-2)}g^{m\ell}Rg_{ik}g_{\ell j}$$
$$W_{ijk}^{m} = R_{ijk}^{m} - \frac{2}{(N-2)}\left(g^{m\ell}R_{\ell j}g_{ik} - g^{m\ell}R_{\ell i}g_{jk}\right) + \frac{2}{(N-1)(N-2)}g^{m\ell}Rg_{ik}g_{\ell j}$$
$$W_{ijk}^{m} = R_{ijk}^{m} - \frac{2}{(N-2)}\left(R_j^m g_{ik} - R_i^k g_{jk}\right) + \frac{2}{(N-1)(N-2)}Rg_j^m g_{ik}$$

and for $m = k$

$$W_{ijk}^{k} = R_{ijk}^{k} - \frac{2}{(N-2)}\left(R_j^k g_{ik} - R_i^k g_{jk}\right) + \frac{2}{(N-1)(N-2)}Rg_j^k g_{ik}$$
$$W_{ij} = R_{ij} + \frac{2}{(N-1)(N-2)}Rg_j^k g_{ik}$$

The contraction $W_{ijk}^{j} = 0$ shows that the Weyl tensor is the portion of the Riemann–Christoffel curvature tensor for which all the contractions are null, i.e., $\mathrm{tr}\,\boldsymbol{W} = 0$.

Weyl Tensor in the Riemann Space E_4

For the Riemann space E_4 the Weyl tensor defined by expression (5.6.46) stays

$$W_{ijk\ell} = R_{ijk\ell} - \frac{1}{2}\left(R_{\ell j}g_{ik} + R_{ki}g_{j\ell} - R_{kj}g_{i\ell} - R_{\ell i}g_{jk}\right) + \frac{1}{6}R\left(g_{ik}g_{\ell j} - g_{i\ell}g_{kj}\right) \quad (5.6.48)$$

The total of components of this tensor is 256, but only 10 are algebraically independent, which are a part of the 20 components of tensor $R_{ijk\ell}$, having that the other 10 are due to tensor R_{ij}.

The curvature of the Riemann space E_4 is determined by tensor $W_{ijk\ell}$, for when $R_{ij} = 0$ the result is $R_{ijk\ell} = W_{ijk\ell}$, which indicates that if the Ricci tensor is null the space is not necessarily flat. The Weyl tensor is the tensor with null trace that comprises the Ricci tensor with an extra condition of having $R_{ij} = 0$. It is, therefore, the tensor $R_{ijk\ell}$ with all the contractions removed.

Exercise 5.10 Show that the Riemann space E_4, which Riemann–Christoffel tensor is $R_{ijk\ell} = \alpha\left(g_{ik}g_{j\ell} - g_{i\ell}g_{jk}\right)$, where α is a constant, is flat.

The Riemann–Christoffel tensor is given by

$$R_{ijk\ell} = \alpha\left(g_{ik}g_{j\ell} - g_{i\ell}g_{jk}\right)$$

and its contraction stays

$$g^{ik}R_{ijk\ell} = \alpha g^{ik}\left(g_{ik}g_{j\ell} - g_{i\ell}g_{jk}\right)$$

$$R_{j\ell} = \alpha\left(\delta_i^i g_{j\ell} - g_\ell^k g_{jk}\right) = \alpha\left(4g_{j\ell} - g_{\ell j}\right) = 3\alpha g_{j\ell}$$

For the scalar curvature the result is

$$R = g^{j\ell}R_{j\ell} = 3\alpha g^{j\ell}g_{j\ell} = 12\alpha$$

Thus, with the substitution of these values in the expression for the Weyl tensor

$$W_{ijk\ell} = R_{ijk\ell} - \frac{1}{2}\left(R_{\ell j}g_{ik} + R_{ki}g_{j\ell} - R_{kj}g_{i\ell} - R_{\ell i}g_{jk}\right) + \frac{1}{6}R\left(g_{ik}g_{\ell j} - g_{i\ell}g_{kj}\right)$$

whereby it is verified that $W_{ijk\ell} = 0$. The nullity of this Weyl tensor shows that this space is flat.

5.7 Dimensional Analysis

The dimensions of the various parameters of the Riemann space E_N are determined as a function of the formula that expresses the metric $ds^2 = \varepsilon g_{ij}dx^i dx^j$ for ds being a distance, its dimension will be a length $[L]$. With this expression the result for the

Table 5.3 Dimensions of the aim parameters of the Riemann space E_N

Parameter		Definition formula	Dimensions
Metric		$ds = \sqrt{\varepsilon g_{ij} dx^i dx^j}$	$[L]$
Metric tensor		$g_{ij} = \frac{ds^2}{dx^i dx^j}$	$[L]^2[U]^{-2}$
Conjugated metric tensor		$g^{ij} = \frac{1}{g_{ij}}$	$[L]^{-2}[U]^2$
Christoffel symbols	First kind	$\Gamma_{ij,k} = \frac{1}{2}\left(g_{ik,j} + g_{jk,i} - g_{ij,k}\right)$	$[L]^2[U]^{-1}$
	Second kind	$\Gamma_{ij}^m = g^{km}\Gamma_{ij,k}$	$[U]^{-1}$
Riemann–Christoffel curvature tensor	Variance (1, 3)	$R_{\ell kj}^i = \frac{\partial \Gamma_{j\ell}^i}{\partial x^k} - \frac{\partial \Gamma_{k\ell}^i}{\partial x^j} + \Gamma_{j\ell}^m \Gamma_{mk}^i - \Gamma_{k\ell}^m \Gamma_{mj}^i$	$[U]^{-2}$
	Variance (0, 4)	$R_{ijkm} = g_{\ell m} R_{ijk}^\ell$	$[L]^2[U]^{-4}$
Ricci tensor		$R_{ij} = g^{km} R_{ijkm}$	$[U]^{-2}$
Riemann curvature		$K = \frac{R_{k\ell mn} u^k v^\ell u^m v^n}{(g_{km} g_{\ell n} - g_{kn} g_{m\ell}) u^k v^\ell u^m v^n}$	$[L]^{-2}$
Scalar curvature		$R = 2K$	$[L]^{-2}$
Weyl tensor (isotropic space)		$W_{ijk\ell} = R_{ijk\ell} + \frac{1}{(N-1)}\left(R_{j\ell} g_{ik} - R_{jk} g_{i\ell}\right)$	$[L]^2[U]^{-4}$

metric tensor is $\left[g_{ij}\right] = [L]^2[U]^{-2}$, where $[U]$ represents a dimension for the coordinates dx^i, dx^j measured in a curvilinear or Cartesian coordinate system, whereby it can be an angle or a length. The conjugated metric tensor has dimensions $[g^{ij}] = [L]^{-2}[U]^2$. For the other parameters the dimensions are shown in Table 5.3.

Problems

5.1. Calculate the components R_{1212} of the Riemann–Christoffel tensor for the spaces defined by the metrics: (a) $ds^2 = -(x^2)^2(dx^1)^2 + (dx^2)^2$; (b) $ds^2 = dr^2 + r^2 d\theta^2 - dt^2$.
Answer: (a) $R_{1212} = 0$; (b) $R_{1212} = 0$.

5.2. Calculate the Ricci tensor and the scalar curvature of the space defined by the metric

$$ds^2 = \frac{1}{(x^2)^2}(dx^1)^2 + \frac{1}{(x^2)^2}(dx^2)^2$$

Answer: $R_{ij} = -g_{ij}$, $R = -2$.

5.3. Show that the metric

$$ds^2 = \left[(x^1)^2 + (x^2)^2\right](dx^1)^2 + \left[(x^1)^2 + (x^2)^2\right](dx^2)^2$$

defines an Euclidian space.

5.4. Calculate the Gauß curvatures of the spaces E_3 defined by the metrics:

(a) $ds^2 = \left[(x^1)^2 + (x^2)^2\right](dx^1)^2 + \left[(x^1)^2 + (x^2)^2\right](dx^2)^2 + (dx^3)^2$
(b) $ds^2 = dr^2 + a^2 \sin^2\left(\frac{r}{a}\right) \cdot \left(d\varphi^2 + \sin^2\varphi \cdot d\theta^2\right)$
(c) $ds^2 = dr^2 + a^2 \sinh^2\left(\frac{r}{a}\right) \cdot \left(d\varphi^2 + \sin^2\varphi \cdot d\theta^2\right)$

Answer: (a) $K = 0$; (b) $K = \frac{1}{a^2}$; (c) $K = -\frac{1}{a^2}$.

5.5. Calculate the Riemann–Christoffel curvature tensor of the space defined by the metric

$$ds^2 = -\left(dx^1\right)^2 - \left(dx^2\right)^2 - \left(dx^3\right)^2 + e^{-x^4}\left(dx^4\right)^2$$

Answer: $R^{\ell}_{ijk} = 0, \quad \forall i, j, k, \ell = 1, 2, 3, 4.$

5.6. Calculate the Riemann–Christoffel curvature tensor and the Gauß curvature of the spaces E_4 defined by the metric

$$ds^2 = \left(dx^1\right)^2 + 4\left(x^2\right)^2\left(dx^2\right)^2 + 4\left(x^3\right)^2\left(dx^3\right)^2 - 4\left(x^4\right)^2\left(dx^4\right)^2$$

Answer: $R_{pijk} = 0, \quad K = 0$

Chapter 6
Geodesics and Parallelism of Vectors

6.1 Introduction

The shortest distance between two points located on a surface of the Riemann space E_N is related to a curve of stationary value, which equation is obtained by means of the variational calculus. This curve is called geodesic. The checking of the existence of this type of curve is carried out from the basic concepts of the elementary geometry.

In the Euclidian space E_3, the shortest distance between two points is a straight line, and in this case the geodesic is unique. In the case of a sphere, the shortest between two points located on its surface is an arc of the circle, which radius is the radius of the sphere.

The geodesic is not necessarily unique, for instance, (a) for two points diametrically opposite in the surface of a sphere, it has several geodesics, and (b) for a circular cylinder, the geodesics depend on the positions of the points on the surface and if the points are in a generatrix, the geodesic is a straight line; otherwise, the geodesic is a spiral or an arc of circle.

6.2 Geodesics

The idea of stationary length leads to the definition of geodesic as the curve which length is minimum, keeping the initial and final point fixed. The stationary length between two points A and B is calculated by the variational condition $\delta \int_A^B ds = 0$.

In parametric form a curve in the Riemann space E_N is defined by the continuous function $x^i = x^i(t)$ of class C^2, where $t_0 \le t \le t_1$, and the distance between two points is determined by

© Springer International Publishing Switzerland 2016

E. de Souza Sánchez Filho, *Tensor Calculus for Engineers and Physicists*,
DOI 10.1007/978-3-319-31520-1_6

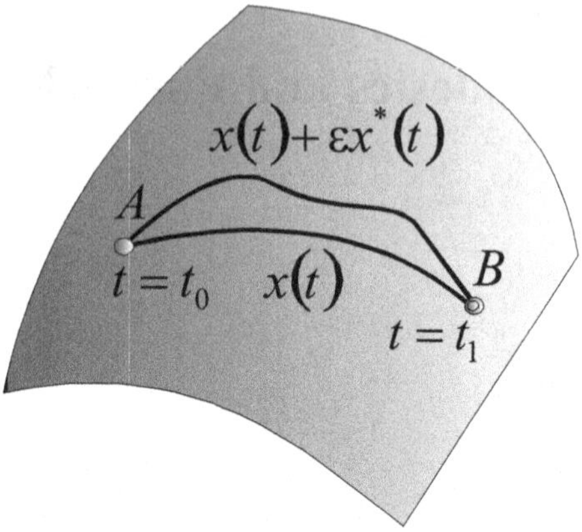

Fig. 6.1 Geodesic in the Riemann space E_N

$$s = \int_a^b \sqrt{g_{ij} \frac{dx^i}{dt} \frac{dx^j}{dt}} dt \tag{6.2.1}$$

Figure 6.1 shows a surface of the Riemann space E_N containing two curves represented by parametric equations, in which the parameter t assumes the values t_0 and t_1 in the extreme points A and B. Admitting continuous and derivable functions which cancel each other in these extreme points, neighboring curves $x(t)$ and $x(t) + \varepsilon x^*(t)$ exist, $x^*(t)$ being a continuous class C^1 parametric function that represents the change of the tracing of this curve with respect to the curve $x(t)$, and ε is a very small value. In the case of the curve with a minimum length, i.e., the geodesic, this coefficient cancels itself.

The determination of the equation for the geodesic is carried out calculating the extreme value of the functional

$$F = \sqrt{g_{ij} \frac{dx^i}{dt} \frac{dx^j}{dt}} \tag{6.2.2}$$

The variational calculus addresses this problem by means of the Euler–Lagrange formula

$$\frac{d}{dt}\left(\frac{\partial F}{\partial \dot{x}^p}\right) - \frac{\partial F}{\partial x^p} = 0 \tag{6.2.3}$$

and with expressions (6.2.1) and (6.2.2)

$$F = \frac{ds}{dt} = \sqrt{g_{ij} \frac{dx^i}{dt} \frac{dx^j}{dt}}$$

With the notation $\dot{x}^i$, $\dot{x}^j$ for the derivatives of the coordinates with respect to the parameter t, it follows that

$$\frac{\partial F}{\partial \dot{x}^p} = \frac{g_{ij}}{2F}\left(\delta^i_p \dot{x}^j + \dot{x}^i \delta^j_p\right) = \frac{1}{2F}\left(g_{ij}\delta^i_p \dot{x}^j + g_{ij}\dot{x}^i \delta^j_p\right) = \frac{1}{2F}\left(g_{pj}\dot{x}^j + g_{ip}\dot{x}^i\right)$$

and the change of the indexes $i \to j$ and the symmetry of the metric tensor $g_{pj} = g_{jp}$ provide

$$\frac{\partial F}{\partial \dot{x}^p} = \frac{1}{2F}\left(g_{pj}\dot{x}^j + g_{jp}\dot{x}^j\right) = \frac{g_{jp}\dot{x}^j}{F}$$

For the other derivatives it follows that

$$\frac{d}{dt}\left(\frac{\partial F}{\partial \dot{x}^p}\right) = \frac{d\left(\frac{1}{F}\right)}{dt} g_{jp}\dot{x}^j + \frac{1}{F}\frac{\partial g_{jp}}{\partial x^m}\dot{x}^m \dot{x}^j + \frac{1}{F} g_{jp}\ddot{x}^j$$

$$\frac{d}{dt}\left(\frac{\partial F}{\partial \dot{x}^p}\right) = \frac{1}{F}\left[-\frac{d(\ell n F)}{dt} g_{jp}\dot{x}^j + \frac{\partial g_{jp}}{\partial x^m}\dot{x}^m \dot{x}^j + g_{jp}\ddot{x}^j\right] \tag{6.2.4}$$

$$\frac{\partial F}{\partial x^p} = \frac{1}{2F}\frac{\partial g_{ij}}{\partial x^p}\dot{x}^i \dot{x}^j \tag{6.2.5}$$

Thus with the substitution of expressions (6.2.4) and (6.2.5) in expression (6.2.3)

$$\frac{1}{F}\left[-\frac{d(\ell n F)}{dt} g_{jp}\dot{x}^j + \frac{\partial g_{jp}}{\partial x^m}\dot{x}^m \dot{x}^j + g_{jp}\ddot{x}^j\right] - \frac{1}{2F}\frac{\partial g_{ij}}{\partial x^p}\dot{x}^i \dot{x}^j = 0$$

it follows that

$$\frac{1}{2F}\left[2\frac{d(\ell n F)}{dt} g_{jp}\dot{x}^j - 2\frac{\partial g_{jp}}{\partial x^m}\dot{x}^m \dot{x}^j - 2g_{jp}\ddot{x}^j + \frac{\partial g_{ij}}{\partial x^p}\dot{x}^i \dot{x}^j\right] = 0$$

and the change of the indexes $m \to i$ provides

$$2\frac{d(\ell n F)}{dt} g_{jp}\dot{x}^j - 2\frac{\partial g_{jp}}{\partial x^i}\dot{x}^i \dot{x}^j - 2g_{jp}\ddot{x}^j + \frac{\partial g_{ij}}{\partial x^p}\dot{x}^i \dot{x}^j = 0$$

whereby

$$2\frac{d(\ell n F)}{dt} g_{jp}\dot{x}^j - \left(2\frac{\partial g_{jp}}{\partial x^i} - \frac{\partial g_{ij}}{\partial x^p}\right)\dot{x}^i \dot{x}^j - 2g_{jp}\ddot{x}^j = 0 \tag{6.2.6}$$

but with expression (2.4.30)

$$\frac{\partial g_{jp}}{\partial x^i} = g_{rp}\Gamma^r_{ji} + g_{jr}\Gamma^r_{pi} \tag{6.2.7}$$

$$\frac{\partial g_{ij}}{\partial x^p} = g_{rj}\Gamma^r_{ip} + g_{ir}\Gamma^r_{jp} \tag{6.2.8}$$

The substitution of expressions (6.2.7) and (6.2.8) in expression (6.2.6) provides

$$2\frac{d(\ell nF)}{dt} g_{jp}\,\dot{x}^j - 2g_{rp}\Gamma^r_{ji}\dot{x}^i\dot{x}^j - 2g_{rj}\Gamma^r_{ip}\dot{x}^i\dot{x}^j - g_{jr}\Gamma^r_{pi}\dot{x}^i\dot{x}^j + g_{ir}\Gamma^r_{jp}\,\dot{x}^i\dot{x}^j - 2g_{jp}\,\ddot{x}^j$$
$$= 0$$

then

$$2\frac{d(\ell nF)}{dt} g_{jp}\,\dot{x}^j - \left(2g_{rp}\Gamma^r_{ji} - g_{rj}\Gamma^r_{ip} + g_{ir}\Gamma^r_{jp}\right)\dot{x}^i\dot{x}^j - 2g_{jp}\,\ddot{x}^j = 0$$

In term $g_{ir}\Gamma^r_{jp}\dot{x}^i\dot{x}^j$ the indexes i, j are dummies and then can be permutated:

$$g_{ir}\Gamma^r_{jp}\dot{x}^i\dot{x}^j = g_{jr}\Gamma^r_{ip}\dot{x}^j\dot{x}^i$$

thus

$$2\frac{d(\ell nF)}{dt} g_{jp}\dot{x}^j - 2g_{rp}\Gamma^r_{ji}\dot{x}^i\dot{x}^j - 2g_{jp}\ddot{x}^j = 0$$

and multiplying by $\frac{1}{2}g^{pm}$

$$\frac{d(\ell nF)}{dt}\delta^m_j\dot{x}^j - \delta^m_r\Gamma^r_{ji}\dot{x}^i\dot{x}^j - \delta^m_j\,\ddot{x}^j = 0$$

but with $\Gamma^m_{ji} = \Gamma^m_{ij}$, it results in

$$\frac{d^2x^m}{dt^2} + \Gamma^m_{ij}\frac{dx^i}{dt}\frac{dx^j}{dt} = \frac{d(\ell nF)}{dt}\frac{dx^m}{dt} \tag{6.2.9}$$

The N equations given by expression (6.2.9) are ordinary differential equations of the second order of the functions $x^m(s)$, and their solutions have $2N$ constants of integration. The solution $x^m(s)$ of this differential equation provides the expression of the geodesic in a surface of the Riemann space E_N, which is determined if the initial values of x^m and $\frac{dx^m}{ds}$ are known, i.e., the coordinates of the point and the direction of the tangent vector to the geodesics in this point. Then the geodesic is unique.

6.2.1 Representation by Means of Curves in the Surfaces

The calculation of the geodesics can be carried out in an alternative way in a bidimensional space if the surface is represented by coordinates ξ^1, ξ^2, then with expression (6.2.9)

$$\frac{d^2\xi^1}{dt^2} + \Gamma^1_{ij}\frac{d\xi^i}{dt}\frac{d\xi^j}{dt} = \frac{d(\ell n F)}{dt}\frac{d\xi^1}{dt} \tag{6.2.10}$$

$$\frac{d^2\xi^2}{dt^2} + \Gamma^2_{ij}\frac{d\xi^i}{dt}\frac{d\xi^j}{dt} = \frac{d(\ell n F)}{dt}\frac{d\xi^2}{dt} \tag{6.2.11}$$

Consider the functions $\xi^2 = f(\xi^1)$ or $\xi^1 = g(\xi^2)$ related with a curve in the surface, for example, with the function $\xi^2 = f(\xi^1)$, it follows that

$$\xi^1 = t \Rightarrow \frac{d\xi^1}{dt} = 1 \Rightarrow \frac{d\xi^2}{dt} = \frac{d\xi^2}{d\xi^1}\frac{d\xi^1}{dt}$$

$$\xi^2 = \xi^2(t) \Rightarrow \frac{d^2\xi^1}{dt^2} = 0 \Rightarrow \frac{d^2\xi^2}{dt^2} = \frac{d^2\xi^2}{\left(d\xi^1\right)^2}$$

The development of expressions (6.2.10) and (6.2.11) and the substitution of expressions of the respective derivatives lead to the following differential equations:

$$\Gamma^1_{11} + 2\Gamma^1_{12}\frac{d\xi^2}{d\xi^1} + \Gamma^1_{22}\left(\frac{d\xi^2}{d\xi^1}\right)^2 = \frac{d(\ell n F)}{d\xi^1} \tag{6.2.12}$$

$$\frac{d^2\xi^2}{\left(d\xi^1\right)^2} + \Gamma^1_{22} + 2\Gamma^1_{21} - \Gamma^2_{22}\frac{d\xi^1}{d\xi^2} + \left(\Gamma^1_{11} - 2\Gamma^1_{21}\right)\left(\frac{d\xi^1}{d\xi^2}\right)^2 - \Gamma^2_{11}\left(\frac{d\xi^1}{d\xi^2}\right)^3 = 0 \tag{6.2.13}$$

The solutions of these differential equations provide the expression of the curve that represents the geodesic.

6.2.2 Constant Direction

In expression (6.2.9) the parameter t is arbitrary, being plausible to admit $t = s$, then $\left\|g_{ij}\dot{x}^i\dot{x}^j\right\| = 1$, i.e., $F = 1$, thus $\frac{d(\ell n F)}{dt} = 0$, and the differential equation of the geodesics is simplified as

$$\frac{d^2x^m}{ds^2}+\Gamma^m_{ij}\frac{dx^i}{ds}\frac{dx^j}{ds}=0 \tag{6.2.14}$$

that can be expressed in another way. Let $\frac{dx^i}{ds}$ a unit tangent vector in each point of the geodesic, and then

$$g_{ij}\frac{dx^i}{ds}\frac{dx^j}{ds}=1$$

will be a solution of the differential equation (6.2.14), which multiplied by $2g_{mr}\frac{dx^r}{ds}$ and when carrying out the sum with respect to the index m takes the form

$$2g_{mr}\frac{dx^r}{ds}\frac{d^2x^m}{ds^2}+2g_{mr}\frac{dx^r}{ds}\Gamma^m_{ij}\frac{dx^i}{ds}\frac{dx^j}{ds}=0 \tag{6.2.15}$$

As

$$\frac{d}{ds}\left(g_{mr}\frac{dx^m}{ds}\frac{dx^r}{ds}\right)=\frac{dg_{mr}}{ds}\frac{dx^m}{ds}\frac{dx^r}{ds}+g_{mr}\frac{d^2x^m}{ds^2}\frac{dx^r}{ds}+g_{mr}\frac{dx^m}{ds}\frac{d^2x^r}{ds^2}$$

and with the permutation of the indexes $m \leftrightarrow r$ in the last term to the right, it follows that

$$2g_{mr}\frac{d^2x^m}{ds^2}\frac{dx^r}{ds}=\frac{d}{ds}\left(g_{mr}\frac{dx^m}{ds}\frac{dx^r}{ds}\right)-\frac{dg_{mr}}{ds}\frac{dx^m}{ds}\frac{dx^r}{ds} \tag{6.2.16}$$

Putting

$$2g_{mr}\Gamma^m_{ij}\frac{dx^r}{ds}\frac{dx^i}{ds}\frac{dx^j}{ds}=2\Gamma_{ij,r}\frac{dx^r}{ds}\frac{dx^i}{ds}\frac{dx^j}{ds}$$

and with the cyclic permutation of the indexes of the Christoffel symbol of first kind $\Gamma_{ij,r}=\Gamma_{ri,j}$, thus

$$2g_{mr}\Gamma^m_{ij}\frac{dx^r}{ds}\frac{dx^i}{ds}\frac{dx^j}{ds}=2\Gamma_{ij,r}\frac{dx^r}{ds}\frac{dx^i}{ds}\frac{dx^j}{ds}$$

$$\frac{\partial g_{ir}}{\partial x^j}=\Gamma_{ij,r}+\Gamma_{ri,j}$$

$$2g_{mr}\Gamma^m_{ij}\frac{dx^r}{ds}\frac{dx^i}{ds}\frac{dx^j}{ds}=2\left(\frac{\partial g_{ir}}{\partial x^j}\frac{dx^r}{ds}\right)\frac{dx^i}{ds}\frac{dx^j}{ds}$$

whereby

$$2g_{mr}\Gamma^{m}_{ij}\frac{dx^r}{ds}\frac{dx^i}{ds}\frac{dx^j}{ds}=\frac{dg_{ir}}{ds}\frac{dx^i}{ds}\frac{dx^j}{ds} \tag{6.2.17}$$

The substitution of expressions (6.2.16) and (6.2.17) in expression (6.2.15) provides

$$\frac{d}{ds}\left(g_{mr}\frac{dx^m}{ds}\frac{dx^r}{ds}\right)-\frac{dg_{mr}}{ds}\frac{dx^m}{ds}\frac{dx^r}{ds}+\frac{dg_{ir}}{ds}\frac{dx^i}{ds}\frac{dx^j}{ds}=0$$

and with the change of the indexes $i \longrightarrow m$ in the last term

$$\frac{d}{ds}\left(g_{mr}\frac{dx^m}{ds}\frac{dx^r}{ds}\right)=0 \tag{6.2.18}$$

that is another way of writing the differential equation of the geodesics, then its first integral is given by

$$g_{mr}\frac{dx^m}{ds}\frac{dx^r}{ds}=\text{constant} \tag{6.2.19}$$

with constant $=1$. The terms $\frac{dx^m}{ds}$ and $\frac{dx^r}{ds}$ represent unit tangent vectors to the geodesic whereby this curve always maintains its direction. This is the necessary and sufficient condition so that this curve is a geodesic.

6.2.3 *Representation by Means of the Unit Tangent Vector*

The previous ascertaining allows writing the geodesic's equation in another manner. Let the unit tangent vector to the geodesics $\xi^i=\frac{dx^i}{ds}$, and with expression (6.2.14), it follows that

$$\frac{d^2x^k}{ds^2}+\Gamma^{k}_{\ell m}\frac{dx^\ell}{ds}\frac{dx^m}{ds}=0\Rightarrow\frac{d}{ds}\left(\frac{dx^k}{ds}\right)+\Gamma^{k}_{\ell m}\frac{dx^\ell}{ds}\frac{dx^m}{ds}=0\Rightarrow\frac{d\xi^k}{ds}+\Gamma^{k}_{\ell m}\xi^\ell\xi^m=0$$

and with

$$\frac{d\xi^k}{ds}=\frac{\partial\xi^k}{\partial x^\ell}\frac{dx^\ell}{ds}=\frac{\partial\xi^k}{\partial x^\ell}\xi^\ell$$

thus

$$\left(\frac{\partial\xi^k}{\partial x^\ell}+\Gamma^{k}_{\ell m}\xi^m\right)\xi^\ell=0$$

and the geodesic's equation can be written as a function of the unit tangent vector and its covariant derivative as follows:

$$\left(\partial_\ell \xi^k\right) \xi^\ell = 0 \tag{6.2.20}$$

6.2.4 Representation by Means of an Arbitrary Parameter

The calculation of the geodesics can be carried out considering a parameter $\zeta(s)$, and then

$$\frac{dx^m}{ds} = \frac{dx^m}{d\zeta}\frac{d\zeta}{ds} \Rightarrow \frac{d^2x^m}{ds^2} = \frac{d^2x^m}{d\zeta^2}\left(\frac{d\zeta}{ds}\right)^2 + \frac{dx^m}{d\zeta}\frac{d^2\zeta}{ds^2}$$

and the substitution of these derivatives in expression (6.2.14) provides

$$\frac{d^2x^m}{d\zeta^2} + \Gamma^m_{ij}\frac{dx^i}{d\zeta}\frac{dx^j}{d\zeta} = -\frac{\frac{d^2\zeta}{ds^2}}{\left(\frac{d\zeta}{ds}\right)^2}\frac{dx^m}{d\zeta} \tag{6.2.21}$$

which is valid for any parameter $\zeta(s)$.

If $\zeta(s)$ is a linear function, it results in $\frac{d\zeta}{ds} = 1$ and $\frac{d^2\zeta}{ds^2} = 0$; then the term to the right of expression (6.2.21) is null, and the result is expression (6.2.14).

Exercise 6.1 Determine the geodesic in the Riemann space E_N, with metric $ds = \sqrt{\left[(dx^1)^2 + (dx^2)^2 \cdots (dx^N)^2\right]}$ given by the Cartesian coordinates.

The differential equation of the geodesic is given by

$$\frac{d^2x^m}{ds^2} + \Gamma^m_{ij}\frac{dx^i}{ds}\frac{dx^j}{ds} = 0$$

and the metric tensor of the Riemann space E_N *is* $g = \delta_{ij}$, but $\frac{\partial g_{ij}}{\partial x^p} = 0$, thus Christoffel symbols are null $\Gamma^m_{ij} = 0$, so

$$\frac{d^2x^m}{ds^2} = 0$$

The solution of this equation is the straight line

$$x^m = a^m s + b^m$$

where a^m and b^m are constants.

Exercise 6.2 Determine the differential equations of the geodesic in the Riemann space defined by the metric $ds^2 = e^{-2kt}(dx^2 + dy^2 + dz^2 - dt^2)$.

The coordinates are $x^1 = x, x^2 = y, x^3 = z$, and $x^4 = t$; thus, the metric tensor and the metric conjugated tensor are given, respectively, by

$$g_{ij} = \begin{bmatrix} e^{-2kt} & 0 & 0 & 0 \\ 0 & e^{-2kt} & 0 & 0 \\ 0 & 0 & e^{-2kt} & 0 \\ 0 & 0 & 0 & 1 \end{bmatrix} \quad g^{ij} = \begin{bmatrix} e^{2kt} & 0 & 0 & 0 \\ 0 & e^{2kt} & 0 & 0 \\ 0 & 0 & e^{2kt} & 0 \\ 0 & 0 & 0 & 1 \end{bmatrix}$$

and the non-null Christoffel symbols are

$$\Gamma^1_{14} = \Gamma^1_{41} = \frac{1}{2}e^{2kt}g_{11,4} = -k \quad \Gamma^2_{24} = \Gamma^2_{42} = \frac{1}{2}e^{2kt}g_{22,4} = -k$$
$$\Gamma^3_{34} = \Gamma^3_{43} = \frac{1}{2}e^{2kt}g_{33,4} = -k \quad \Gamma^4_{11} = g^{44}\left[\frac{1}{2}(g_{11,4})\right] = \Gamma^4_{22} = \Gamma^4_{33} = ke^{2kt}$$

The first differential equation of the geodesics

$$\frac{d^2x}{ds^2} + \Gamma^1_{14}\frac{dt}{ds}\frac{dx}{ds} + \Gamma^1_{41}\frac{dx}{ds}\frac{dt}{ds} = 0$$

stays

$$\frac{d^2x}{ds^2} - 2k\frac{dx}{ds}\frac{dt}{ds} = 0$$

In an analogous way, for the other variables, it follows that

$$\frac{d^2y}{ds^2} - 2k\frac{dy}{ds}\frac{dt}{ds} = 0 \quad \frac{d^2z}{ds^2} - 2k\frac{dz}{ds}\frac{dt}{ds} = 0$$

and

$$\frac{d^2t}{ds^2} + \Gamma^4_{11}\frac{dx}{ds}\frac{dx}{ds} + \Gamma^4_{22}\frac{dy}{ds}\frac{dy}{ds} + \Gamma^4_{33}\frac{dz}{ds}\frac{dz}{ds} = 0 \Rightarrow$$

$$\frac{d^2t}{ds^2} + ke^{-2kt}\left[\left(\frac{dx}{ds}\right)^2 + \left(\frac{dy}{ds}\right)^2 + \left(\frac{dz}{ds}\right)^2\right] = 0$$

Exercise 6.3 Determine the geodesics on the circular cylinder of radius r, represented by the parametric equations $x^1 = r\cos\xi^1$, $x^2 = r\sin\xi^1$, and $x^3 = \xi^2$.

The bidimensional space defined by the surface of the cylinder which components of the metric tensor are

$$g_{\ell m} = \frac{\partial x^i}{\partial \xi^\ell} \frac{\partial x^j}{\partial \xi^m}$$

$$g_{11} = \left(\frac{\partial x^1}{\partial \xi^1}\right)^2 = r^2 \sin^2 \xi^1 + r^2 \cos^2 \xi^1 = r^2 \quad g_{12} = g_{21} = 0$$

$$g_{22} = \left(\frac{\partial x^2}{\partial \xi^2}\right)^2 = 1$$

and determine the metric

$$ds^2 = r^2 \left(d\xi^1\right)^2 + \left(d\xi^2\right)^2$$

The Christoffel symbols are all null, whereby expressions (6.2.12) and (6.2.13) stay

$$\frac{d^2 \xi^2}{\left(d\xi^1\right)^2} = 0 \quad \frac{d^2 \xi^1}{\left(d\xi^2\right)^2} = 0$$

The solutions of these differential equations are

$$\xi^2 = k_1 \xi^1 + k_2 \quad \xi^1 = k_3 \xi^2 + k_4$$

where k_1, k_2, k_3, k_4 are constants and represent a circular helix.

For $k_1 = 0$, $k_3 \neq 0$ there is $\xi^2 = k_2$, and $\xi^1 = k_3 \xi^2 + k_4$; thus,

$$x^1 = r \cos\left(k_3 \xi^2 + k_4\right) \quad x^2 = r \sin\left(k_3 \xi^2 + k_4\right) \quad x^3 = k_2$$

and then this curve is a circle.

For $k_1 \neq 0$, $k_3 \neq 0$ there is $\xi^1 = k_3 \xi^1 + k_4$, and $\xi^2 = k_1 \xi^1 + k_2$; thus,

$$x^1 = r \cos\left(k_3 \xi^2 + k_4\right) \quad x^2 = r \sin\left(k_3 \xi^2 + k_4\right) \quad x^3 = k_1 \xi^1 + k_2$$

and then this curve is the generatrix of the cylinder.

For $k_1 \neq 0$, $k_3 = 0$ there is $\xi^1 = k_4$, and $\xi^2 = k_1 \xi^2 + k_2$; thus,

$$x^1 = r \cos k_4 \quad x^2 = r \sin k_4 \quad x^3 = k_1 \xi^2 + k_2$$

and then the geodesics is a straight line.

Exercise 6.4 Determine the geodesics on the sphere of radius r, which metric is given by $ds^2 = r^2\, d\varphi^2 + r^2 \sin^2 \varphi\, d\theta^2$, with $x^1 = \varphi$ and $x^2 = \theta$.

The metric tensor and the conjugated metric tensor are given, respectively, by

$$g_{ij} = \begin{bmatrix} r^2 & 0 \\ 0 & r^2 \sin^2\varphi \end{bmatrix} \quad g^{ij} = \begin{bmatrix} \dfrac{1}{r^2} & 0 \\ 0 & \dfrac{1}{r^2 \sin^2\varphi} \end{bmatrix}$$

Expression (6.2.9)

$$\frac{d^2x^m}{dt^2} + \Gamma^m_{ij}\frac{dx^i}{dt}\frac{dx^j}{dt} = \frac{d(\ell n F)}{dt}\frac{dx^m}{dt}$$

with

$$F = \left(g_{ij}\frac{dx^i}{dt}\frac{dx^j}{dt}\right)^{\frac{1}{2}}$$

$$\frac{d(\ell n F)}{dt} = \frac{d}{dt}\left[\ell n\left(g_{ij}\frac{dx^i}{dt}\frac{dx^j}{dt}\right)^{\frac{1}{2}}\right] = \frac{1}{2}\frac{d}{dt}\left[\ell n\left(g_{ij}\frac{dx^i}{dt}\frac{dx^j}{dt}\right)\right]$$
$$= \frac{1}{2}\frac{d}{dt}\left(\ell n F^2\right) = \frac{1}{2F^2}\dot{F}^2$$

stays

$$\ddot{x}^m + \Gamma^m_{ij}\dot{x}^i\dot{x}^j = \frac{1}{2F^2}\dot{F}^2$$

and with the non-null Christoffel symbols

$$\Gamma^1_{22} = -\sin\varphi\cos\varphi \quad \Gamma^2_{12} = \Gamma^2_{21} = \cot\varphi$$

it follows that

$$F^2 = g_{11}\left(\dot{x}^1\right)^2 + g_{22}\left(\dot{x}^2\right)^2 = r^2\left(\dot{\varphi}^2 + \sin^2\varphi\cdot\dot{\theta}^2\right)$$

$$\dot{F}^2 = 2r^2\left(\dot{\varphi}\,\ddot{\varphi} + \sin\varphi\cos\varphi\cdot\dot{\varphi}\,\dot{\theta}^2 + \sin^2\varphi\cdot\dot{\theta}\,\ddot{\theta}\right)$$

$$2F^2\left(\ddot{x}^m + \Gamma^m_{ij}\dot{x}^i\dot{x}^j\right) = F^2\dot{x}^m$$

then

$$\left(\dot{\varphi}^2 + \sin^2\varphi\cdot\dot{\theta}^2\right)\left(\ddot{x}^m + \Gamma^m_{ij}\dot{x}^i\dot{x}^j\right) - \left(\dot{\varphi}\,\ddot{\varphi} + \sin\varphi\cos\varphi\cdot\dot{\varphi}\,\dot{\theta}^2 + \sin^2\varphi\cdot\dot{\theta}\,\ddot{\theta}\right)\dot{\varphi} = 0$$

– $m = 1$

$$2F^2\left(\ddot{x}^1 + \Gamma^1_{22}\dot{x}^2\dot{x}^2\right) = F^2\dot{x}^1$$

$$\left(\dot{\varphi}^2 + \sin^2\varphi\cdot\dot{\theta}^2\right)\left(\ddot{\varphi} - \sin\varphi\cos\varphi\cdot\dot{\theta}^2\right) - \left(\dot{\varphi}\ddot{\varphi} + \sin\varphi\cos\varphi\cdot\dot{\varphi}\dot{\theta}^2 + \sin^2\varphi\cdot\dot{\theta}\ddot{\theta}\right)\dot{\varphi} = 0$$

$$\ddot{\varphi} - 2\frac{\cos\varphi}{\sin\varphi}\dot{\varphi}^2 - \sin\varphi\cos\varphi\cdot\dot{\theta}^2 - \frac{\dot{\varphi}}{\dot{\theta}}\ddot{\theta} = 0$$

– $m = 2$

$$2F^2\left(\ddot{x}^2 + \Gamma^2_{21}\dot{x}^1\dot{x}^2 + \Gamma^2_{12}\dot{x}^1\dot{x}^2\right) = F^2\dot{x}^2$$

$$\left(\dot{\varphi}^2 + \sin^2\varphi\cdot\dot{\theta}^2\right)\left(\ddot{\theta} + 2\cot\varphi\cdot\dot{\varphi}\dot{\theta}\right) - \left(\dot{\varphi}\ddot{\varphi} + \sin\varphi\cos\varphi\cdot\dot{\varphi}\dot{\theta}^2 + \sin^2\varphi\cdot\dot{\theta}\ddot{\theta}\right)\dot{\theta} = 0$$

Let $\varphi = \varphi(\theta)$ and $\theta \equiv t$, thus $\dot{\theta} = 1,\ \ddot{\theta} = 0$, whereby a differential equation for $m = 2$ stays

$$\left(\dot{\varphi}^2 + \sin^2\varphi\right)2\cot\varphi\cdot\dot{\varphi} - \left(\dot{\varphi}\ddot{\varphi} + \sin\varphi\cos\varphi\cdot\dot{\varphi}\right) = 0$$

follows

$$2\dot{\varphi}^3\cdot\cot\varphi + 2\dot{\varphi}\cdot\sin^2\varphi\cot\varphi - \dot{\varphi}\ddot{\varphi} - \sin\varphi\cos\varphi\cdot\dot{\varphi} = 0$$

$$\ddot{\varphi} - 2\frac{\cos\varphi}{\sin\varphi}\dot{\varphi}^2 - \sin\varphi\cos\varphi = 0$$

Putting

$$F(\theta) = \cot\varphi$$

it follows that

$$\frac{dF}{d\theta} = -\frac{1}{\sin^2\varphi}\frac{d\varphi}{d\theta}$$

$$\frac{d\varphi}{d\theta} = -\sin^2\varphi\frac{dF}{d\theta}$$

$$\frac{d^2\varphi}{d\theta^2} = -2\sin\varphi\cos\varphi\frac{d\varphi}{d\theta}\frac{dF}{d\theta}\sin^2\varphi\frac{d^2F}{d\theta^2} = 2\sin^3\varphi\cos\varphi\left(\frac{dF}{d\theta}\right)^2 - \sin^2\varphi\frac{d^2F}{d\theta^2}$$

then

$$2\sin^3\varphi\cos\varphi\left(\frac{dF}{d\theta}\right)^2 - \sin^2\varphi\frac{d^2F}{d\theta^2} - 2\frac{\cos\varphi}{\sin\varphi}\sin^4\varphi\left(\frac{dF}{d\theta}\right)^2 - \sin\varphi\cos\varphi = 0$$

The division by $\sin^2\varphi$ provides

$$\frac{d^2F}{d\theta^2} - \frac{\cos\varphi}{\sin\varphi} = 0 \Rightarrow \frac{d^2F}{d\theta^2} + F(\theta) = 0$$

which solution is

$$F = k_1 \cos\theta + k_2 \sin\theta$$

where k_1, k_2 are constants, then

$$\cot\varphi = k_1 \cos\theta + k_2 \sin\theta$$

represents the geodesics on the surfaces of the sphere, which can be rewritten in an implicit form as

$$k_1 r \sin\varphi \cos\theta + k_2 r \sin\varphi \sin\theta - r \sin\varphi \frac{\cos\varphi}{\sin\varphi} = 0$$

The relations between the spherical and Cartesian coordinates are given by

$$x = r \sin\varphi \cos\theta \quad y = r \sin\varphi \sin\theta \quad z = r \cos\varphi$$

whereby

$$k_1 x + k_2 y - z = 0$$

represents a plane that passes through the center of the sphere.

The geodesics are intersections of the sphere with the diametral planes, which normal vectors have the components $(k_1; k_2; -1)$, i.e., they are the maximum circles of the sphere defined by the expressions

$$\begin{aligned} x^2 + y^2 + z^2 &= r^2 \\ k_1 x + k_2 y - z &= 0 \end{aligned}$$

6.3 Geodesics with Null Length

The length of the geodesic between two points can be null, i.e., the fundamental form is undefined, which makes applying expression (6.2.9) invalid for calculating this curve, for its parametric representation $x^i = x^i(t)$ is not appropriate, and the variational equation $\delta \int_A^B ds = 0$ has no meaning, and the tangent vector to the curve is undefined.

With a new parametric representation $x^i = x^i(\lambda)$, which equations are continuous and class C^2, and $\lambda_0 \leq \lambda \leq \lambda_1$, where λ is an invariant, so that the tangent vector $\frac{dx^i}{d\lambda}$ exists for each point of the curve and has null modulus, thus yielding the condition

$$ds^2 = g_{ij} \frac{dx^i}{d\lambda} \frac{dx^j}{d\lambda} = 0 \tag{6.3.1}$$

and the contravariant vector $\frac{dx^i}{d\lambda}$ has the direction of the displacement along the curve. Thus, the geodesics with null length has tangent vectors in all their points, so the displacement of this vector from one point to another neighboring point keeps it parallel to the null vector, then these vectors must be equipollent.

The condition of parallelism between vectors for the case of the geodesic with null length is given by

$$\frac{d^2x^m}{d\lambda^2} + \Gamma^m_{ij} \frac{dx^i}{d\lambda} \frac{dx^j}{d\lambda} = 0 \tag{6.3.2}$$

A geodesic is null if one of its sub-arcs is null or if it has null length. In the set of values of λ for which there are null unit tangent vectors, the geodesic generates the undefined fundamental form $ds^2 = 0$. By this condition it is verified that the geodesics can be null without having null length, but if it has null length, it is necessarily null in all its points. Expressions (6.3.1) and (6.3.2) provide $(N+1)$ ordinary differential equations in which the unknown values are functions $x^i = x^i(\lambda)$ that determine the condition of geodesics with null length.

Exercise 6.5 Let the metric tensor g_{ij} with constant components. Determine the equation of geodesic with null length.

The metric tensor has its components constant, so the Christoffel symbols are all null and the equation of the geodesic

$$\frac{d^2x^m}{ds^2} + \Gamma^m_{ij} \frac{dx^i}{ds} \frac{dx^j}{ds} = 0 \Rightarrow \frac{d^2x^m}{ds^2} = 0$$

with general solution

$$x^m = a^m s + b^m$$

where a^m, b^m are constants, it follows that

$$\frac{dx^m}{ds} = a^m \Rightarrow a^m = \frac{x^m - b^m}{s}$$

The geodesic with null length is determined by the equation

$$g_{ij} \frac{dx^i}{ds} \frac{dx^j}{ds} = 0 \Rightarrow g_{ij} a^i a^j = 0$$

The substitution of expression a^m provides the differential equation of the geodesic with null length when $g_{ij} = constant$:

$$g_{ij}\left(x^i - b^i\right)\left(x^j - b^j\right) = 0$$

Exercise 6.6 Calculate the geodesics with null length for the space E_4 in which the metric is $ds^2 = \left(dx^1\right)^2 + \left(dx^2\right)^2 + \left(dx^3\right)^2 - \left(dx^4\right)^2$.

The components of tensor g_{ij} are constants, then the equation of the geodesics with null length, in accordance with Exercise 6.5, is given by

$$g_{ij}\left(x^i - x_0^i\right)\left(x^j - x_0^j\right) = 0$$

which when developed provides

$$\left(x^1 - x_0^1\right)^2 + \left(x^2 - x_0^2\right)^2 + \left(x^3 - x_0^3\right)^2 - \left(x^4 - x_0^4\right)^2 = 0$$

6.4 Coordinate Systems

In tensor calculus, the choice of the coordinate system is a function of the type of problem to be solved. For the Euclidian space E_3, the simplest coordinate system is the Cartesian system. However, when analyzing a few problems of the use of curvilinear coordinates, it is more convenient. For the Riemann space E_N, the Cartesian coordinate sometimes is not adequate; then it becomes necessary to search for a few coordinate systems that have special characteristics which make the solving of specific cases easier.

6.4.1 Geodesic Coordinates

For the Cartesian coordinate system, and only for this type of coordinates, the Christoffel symbols are all null in any point of the Riemann space E_N, because the coefficients in a fundamental form $ds^2 = g_{ij}dx^i dx^j$ are constants. However, it is possible to determine a coordinate system with respect to which these symbols cancel each other in a certain point $P_0(x^i) \in E_N$, called pole. The symmetry of the Christoffel symbols $\Gamma_{ij}^k = \Gamma_{ji}^k$ allows determining a coordinate system for which all these symbols are null.

Consider point $P_0(x^i)$ be chosen as the origin of the coordinate system X^i having its coordinates $x^i = 0$, and the linear transformation

$$x^i = \bar{x}^i + \frac{1}{2} C^i_{jk} \bar{x}^j \bar{x}^k \tag{6.4.1}$$

where the constant coefficients C^i_{jk} are chosen so as to be symmetric in the indexes j and k. Differencing this expression it follows that

$$\frac{\partial x^i}{\partial \bar{x}^j} = \frac{\partial \bar{x}^i}{\partial \bar{x}^j} + \frac{1}{2} C^i_{jk} \frac{\partial \bar{x}^j}{\partial \bar{x}^j} \bar{x}^k + \frac{1}{2} C^i_{jk} \bar{x}^j \frac{\partial \bar{x}^k}{\partial \bar{x}^j} = \delta^j_i + \frac{1}{2} C^i_{jk} \delta^j_j \bar{x}^k + \frac{1}{2} C^i_{jk} \bar{x}^j \delta^k_j$$
$$= \delta^j_i + \frac{1}{2} C^i_{jk} \bar{x}^k + \frac{1}{2} C^i_{jk} \bar{x}^k = \delta^j_i + C^i_{jk} \bar{x}^k$$

$$\frac{\partial^2 x^i}{\partial \bar{x}^k \partial \bar{x}^j} = 0 + C^i_{jk} \frac{\partial \bar{x}^k}{\partial \bar{x}^k} = C^i_{jk}$$

and with the expressions

$$\frac{\partial x^i}{\partial \bar{x}^j} \frac{\partial \bar{x}^j}{\partial \bar{x}^k} = \delta^i_k \quad \frac{\partial \bar{x}^i}{\partial x^k} = \delta^i_k$$

and the transformation law of the Christoffel symbols

$$\bar{\Gamma}^i_{jk} = \frac{\partial \bar{x}^i}{\partial x^p} \frac{\partial^2 x^p}{\partial \bar{x}^j \partial \bar{x}^k} + \frac{\partial \bar{x}^i}{\partial x^p} \frac{\partial x^q}{\partial \bar{x}^j} \frac{\partial x^r}{\partial \bar{x}^k} \Gamma^p_{qr}$$

for the point P_0

$$\frac{\partial \bar{g}_{ik}}{\partial x^j} = \bar{\Gamma}_{ij,k} + \bar{\Gamma}_{kj,i} \Rightarrow \bar{\Gamma}^i_{jk} = C^i_{jk} + \Gamma^i_{jk}$$

Consider $\bar{\Gamma}^i_{jk} = 0$ it results in $C^i_{jk} = -\Gamma^i_{jk}$ at the point P_0, i.e., the Christoffel symbols are null at the pole. This condition leads to the definition of a coordinate system called geodesic or normal coordinate system, which condition of existence is grounded in the symmetry of the Christoffel symbols.

Ricci's lemma

$$\partial_k g_{ij} = \frac{\partial g_{ij}}{\partial x^k} - g_{pj} \Gamma^p_{ik} - g_{ip} \Gamma^p_{kj} = 0 \tag{6.4.2}$$

is valid when the Christoffel symbols are null in pole, whereby the metric is constant, and the geodesic coordinates correspond to a local Euclidian coordinate system. It is highlighted that the derivatives of the Christoffel symbols do not necessarily cancel each other in this point.

Pole P_0 has an important relation with a derivative covariant. Let the covariant vector u_i expressed in geodesic coordinates, and with its derivative covariant

$$\partial_i u_i = \frac{\partial u_i}{\partial x^j} - u_p \Gamma^p_{ij}$$

in geodesic coordinates with respect to which $\Gamma^p_{ij} = 0$, thus

$$\partial_i u_i = \frac{\partial u_i}{\partial x^j}$$

It is concluded that in geodesic coordinates, the covariant derivative of vector u_i is equal to its partial derivative in the pole P_0. The generalization of this property for the higher-order tensors is immediate. The demonstrations of the tensor relations using this type of coordinate system are simpler, and if they are valid for this special case, they will be valid for the other coordinate systems. This can be proven comparing the algebraic development carried out in item 5.2, when demonstrating the second Bianchi identity and the solution of Exercise 6.7.

Exercise 6.7 Using the geodesic coordinates, show that the curvature tensor R^p_{ijk} satisfies the second Bianchi identity $\partial_\ell R^p_{ijk} + \partial_j R^p_{ik\ell} + \partial_k R^p_{i\ell j} = 0$.

The geodesic coordinates correspond to a local Euclidian coordinate system, and then the covariant derivatives of the curvature tensor are equal to their partial derivatives; thus,

$$\partial_\ell R^p_{ijk} = \frac{\partial R^p_{ijk}}{\partial x^\ell} = \frac{\partial}{\partial x^\ell}\left(\frac{\partial \Gamma^p_{ik}}{\partial x^j} - \frac{\partial \Gamma^p_{ij}}{\partial x^k} + \Gamma^p_{ik}\Gamma^p_{qj} - \Gamma^q_{ij}\Gamma^p_{qk}\right)$$

As in the pole, the Christoffel symbols cancel each other, but their derivatives do not necessarily cancel each other in this point

$$\frac{\partial R^p_{ijk}}{\partial x^\ell} = \frac{\partial^2 \Gamma^p_{ik}}{\partial x^\ell \partial x^j} - \frac{\partial^2 \Gamma^p_{ij}}{\partial x^\ell \partial x^k}$$

and the permutations of the indexes allow writing

$$\frac{\partial R^p_{ik\ell}}{\partial x^j} = \frac{\partial^2 \Gamma^p_{i\ell}}{\partial x^j \partial x^k} - \frac{\partial^2 \Gamma^p_{ik}}{\partial x^j \partial x^\ell} \qquad \frac{\partial R^p_{i\ell j}}{\partial x^k} = \frac{\partial^2 \Gamma^p_{ij}}{\partial x^k \partial x^\ell} - \frac{\partial^2 \Gamma^p_{i\ell}}{\partial x^k \partial x^j}$$

The sum of these three expressions provides

$$\partial_\ell R^p_{ijk} + \partial_j R^p_{ik\ell} + \partial_k R^p_{i\ell j} = 0 \qquad Q.E.D.$$

6.4.2 Riemann Coordinates

The coordinate systems in which the partial derivatives of the metric tensor g_{ij} cancel each are called Riemann coordinate systems.

Consider a geodesic that passes by the point $P_0(x^i) \in E_N$ and the notation $\xi^i = \left(\frac{dx^i}{ds}\right)_0$ for the unit tangent vector to this curve in P_0, in which s is the arc measured from this point. Parameter ξ^i represents only one geodesic that contains P_0; thus, with the coordinates

$$y^i = \xi^i s \tag{6.4.3}$$

there is a set of values of ξ^i that generates the equations which define the geodesics in new coordinates, called Riemann coordinates. The geodesics that contain point P_0 are analogous to the straight lines that pass by the origin of a coordinate system in Euclidian geometry.

The quadratic form of the curve in this new coordinate system is given by

$$ds^2 = \overline{g}_{ij} dy^i dy^j \tag{6.4.4}$$

and with the Christoffel symbols $\overline{\Gamma}^i_{jk}$, $\overline{\Gamma}_{ij,k}$, the geodesics are determined by

$$\frac{d^2 y^i}{ds^2} + \overline{\Gamma}^i_{jk} \frac{dy^i}{ds} \frac{dy^k}{ds} = 0 \tag{6.4.5}$$

Expression (6.4.3) must satisfy expression (6.4.5); then

$$\overline{\Gamma}^i_{jk} \xi^j \xi^k = 0 \tag{6.4.6}$$

or

$$\overline{\Gamma}^i_{jk} y^j y^k = 0 \tag{6.4.7}$$

This ascertaining translates the necessary and sufficient condition so that the coordinates are valid in the Riemann space E_N. It is stressed that $\overline{\Gamma}^i_{jk} = 0$ in point P_0, and with $\overline{g}_{\ell p} \overline{\Gamma}^\ell_{jk} = \overline{\Gamma}_{jk,p}$, it is concluded by means of the expression

$$\frac{\partial \overline{g}_{ik}}{\partial x^j} = \overline{\Gamma}_{ij,k} + \overline{\Gamma}_{kj,i}$$

which in Riemann coordinates the partial derivatives of the metric tensor g_{ij} are null. The Riemann coordinate system is a geodesic coordinate system.

6.5 Geodesic Deviation

The deviation between two geodesics in the Riemann space E_N is a generalization of the behavior of two straight lines $R1$ and $R2$ in the Euclidian space E_2. Let $R1$ and $R2$ two parallel straight lines (Fig. 6.2a) on which the points A, A', B, and B' are located. The distance ξ between these two geodesics remains unchanged, i.e., $\eta = \eta'$.

For the case in which $R1$ and $R2$ intersect, there are small values of the angle α which deviation between the geodesics is given by

$$\eta \cong \alpha \cdot s \tag{6.5.1}$$

where s is the distance of the point being considered to the point of interception of $R1$ with $R2$ (Fig. 6.2b).

Expression (6.5.1) shows that in this case the separation between the straight lines varies linearly with the distance from their points until the origin O; then

$$\frac{d^2\eta}{ds^2} = 0 \tag{6.5.2}$$

This behavior is not valid for geodesics in curved spaces.

Consider a sphere of unit radius in the surface of which two segments of the geodesic are considered $OA = u$ and $OB = u$, distant from each other $AB = \eta$ (arc of latitude), where point O is the origin of the distances u, measured on these curves (Fig. 6.3). Admitting a small value for the angle α, it follows that

$$\eta \cong \alpha . \sin u \tag{6.5.3}$$

$$\frac{d^2\eta}{ds^2} \neq 0 \tag{6.5.4}$$

Expression (6.5.3) shows that the deviation η between the two geodesics varies with the parameter u, measured along the same. From the origin O up to the midpoints of the geodesics, there is an increase of η; from there, this distance decreases until it cancels itself at point O' (diametrically opposite to the point O). Expression (6.5.4) shows that the variation of η is not linear and highlights the difference between the behavior of the geodesics of a flat space and a curved space.

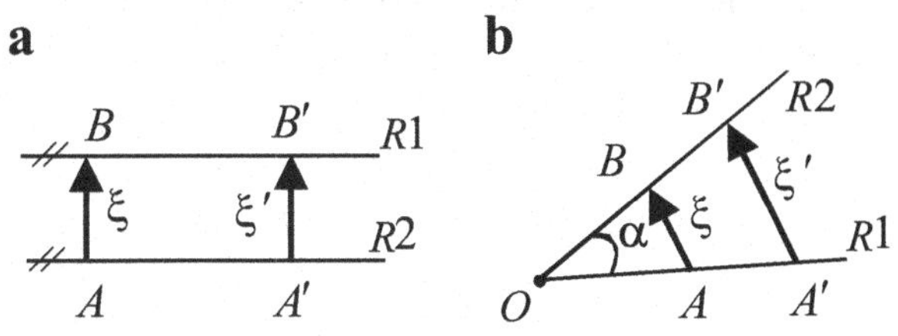

Fig. 6.2 Geodesics in the Riemann space E_2: (**a**) parallel straight lines and (**b**) converging straight lines

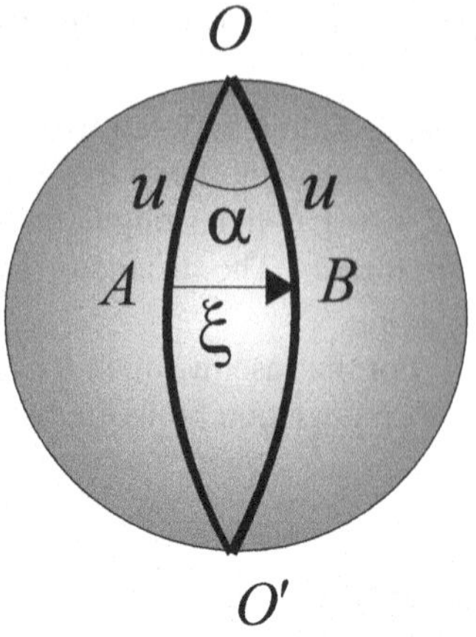

Fig. 6.3 Geodesics in spherical space E_2

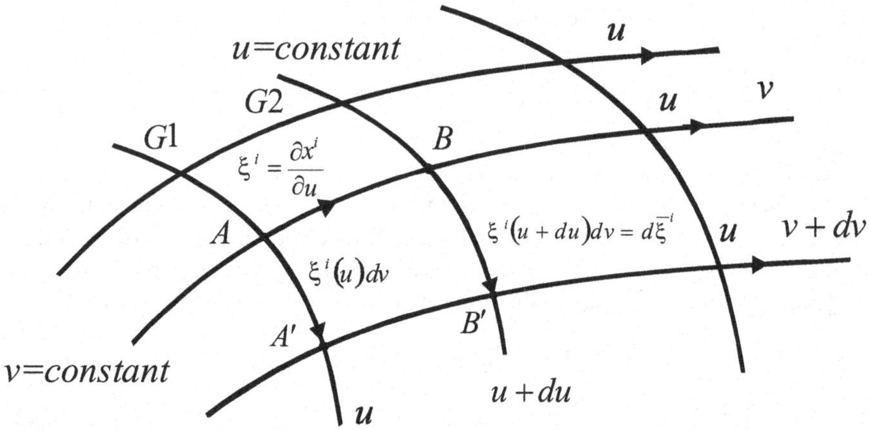

Fig. 6.4 Geodesics in the Riemann space E_N

The generalization of this ascertaining for the Riemann space E_N is carried out when admitting a family of geodesics defined by the functions $x^i = x^i(u, v)$ of class C^2, in which the parameter u (length of arc) varies along each curve fixing the points on them, and parameter v is constant along its length but varies when passing from one geodesics to another, i.e., it distinguishes the curves.

Consider the geodesics $G1$ and $G2$ which contain the points A, A', B, and B', defining the distances $\eta^i(u)$, $\overline{\eta}^i(u)$ measured orthogonally to these curves. Figure 6.4 shows these parameters with the geodesics $G1, G2 \in E_N$.

The partial derivatives

$$\xi^i = \frac{\partial x^i}{\partial u} \tag{6.5.5}$$

$$\eta^i = \frac{\partial x^i}{\partial v} \tag{6.5.6}$$

determine the tangent vector to the geodesics defined by the parameter v and the distance $\eta^i(u)$ (displacement or deviation) between two nearby geodesics, the length being measured in each curve from points A and A'.

For intrinsic derivative of the tangent vector, it follows that

$$\frac{\delta\xi^i}{\delta v} = \frac{\partial\xi^k}{\partial v} + \left(\frac{\partial\xi^i}{\partial x^j} + \Gamma^i_{kj}\xi^k\right)\frac{\partial x^j}{\partial v}$$

$$\frac{\partial\xi^k}{\partial v} = 0$$

$$\frac{\delta\xi^i}{\delta v} = \left(\frac{\partial\xi^i}{\partial x^j} + \Gamma^i_{kj}\xi^k\right)\frac{\partial x^j}{\partial v} = \partial_j\xi^i\frac{\partial x^j}{\partial v}$$

Thus,

$$\frac{\delta\xi^i}{\delta v} = \left(\frac{\partial\xi^i}{\partial x^j} + \Gamma^i_{kj}\xi^k\right)\eta^j = \frac{\partial\xi^i}{\partial x^j}\frac{\partial x^j}{\partial v} + \Gamma^i_{kj}\xi^k\eta^j = \frac{\partial\xi^i}{\partial v} + \Gamma^i_{kj}\xi^k\eta^j$$
$$= \frac{\partial}{\partial v}\left(\frac{\partial x^i}{\partial u}\right) + \Gamma^i_{kj}\xi^k\eta^j$$

Then

$$\frac{\delta\xi^i}{\delta v} = \frac{\partial^2 x^i}{\partial v\partial u} + \Gamma^i_{kj}\xi^k\eta^j$$

In an analogous way,

$$\frac{\delta\eta^i}{\delta u} = \partial_j\eta^i\frac{\partial x^j}{\partial u}$$

Thus,

$$\frac{\delta\eta^i}{\delta u} = \frac{\partial^2 x^i}{\partial u\partial v} + \Gamma^i_{kj}\eta^j\xi^k$$

These two expressions are equal, i.e.:

$$\frac{\delta\xi^i}{\delta v} = \frac{\delta\eta^i}{\delta u} = \frac{\partial^2 x^i}{\partial v\partial u} + \Gamma^i_{kj}\xi^k\eta^j \tag{6.5.7}$$

The second-order derivative of vector η^i with respect to the parameter u is given by

$$\frac{\delta^2\eta^i}{\delta u^2} = \frac{\delta}{\delta u}\left(\frac{\delta\eta^i}{\delta u}\right) = \frac{\delta}{\delta u}\left(\frac{\delta\xi^i}{\delta v}\right)$$

and follows

$$\begin{aligned}\frac{\delta^2\eta^i}{\delta u^2} &= \frac{\delta}{\delta u}\left(\partial_k\xi^i\frac{\partial x^k}{\partial v}\right) = \frac{\delta}{\delta u}\left[\left(\frac{\partial\xi^i}{\partial x^k}+\Gamma^i_{jk}\xi^j\right)\frac{\partial x^k}{\partial v}\right] = \frac{\delta}{\delta u}\left(\frac{\partial\xi^i}{\partial v}+\Gamma^i_{jk}\xi^j\frac{\partial x^k}{\partial v}\right)\\ &= \frac{\delta}{\delta u}\left(\frac{\partial\xi^i}{\partial v}+\Gamma^i_{jk}\xi^j\eta^k\right)\end{aligned}$$

that can be written as

$$\frac{\delta^2\eta^i}{\delta u^2} = \partial_\ell\left(\frac{\partial\xi^i}{\partial v}+\Gamma^i_{jk}\xi^j\eta^k\right)\frac{\partial x^\ell}{\partial u}$$

which development

$$\begin{aligned}\frac{\delta^2\eta^i}{\delta u^2} &= \frac{\partial}{\partial x^\ell}\left(\frac{\partial\xi^i}{\partial v}\right)\frac{\partial x^\ell}{\partial u}+\frac{\partial\Gamma^i_{jk}}{\partial x^\ell}\xi^j\eta^k\frac{\partial x^\ell}{\partial u}+\Gamma^i_{jk}(\partial_\ell\xi^j)\eta^k\frac{\partial x^\ell}{\partial u}+\Gamma^i_{jk}\xi^j(\partial_\ell\eta^k)\frac{\partial x^\ell}{\partial u}\\ &= \frac{\partial^2\xi^i}{\partial u\partial v}+\frac{\partial\Gamma^i_{jk}}{\partial x^\ell}\xi^j\eta^k\frac{\partial x^\ell}{\partial u}+\Gamma^i_{jk}\left(\frac{\partial\xi^j}{\partial x^\ell}+\Gamma^j_{\ell m}\xi^m\right)\eta^k\frac{\partial x^\ell}{\partial u}+\Gamma^i_{jk}\xi^j\left(\frac{\partial\eta^k}{\partial x^\ell}+\Gamma^k_{\ell n}\eta^n\right)\frac{\partial x^\ell}{\partial u}\end{aligned}$$

with

$$\frac{\partial\xi^j}{\partial x^\ell}\frac{\partial x^\ell}{\partial u}=\frac{\partial\xi^j}{\partial u}\qquad\frac{\partial\eta^k}{\partial x^\ell}\frac{\partial x^\ell}{\partial u}=\frac{\partial\eta^k}{\partial u}$$

provides

$$\begin{aligned}\frac{\delta^2\eta^i}{\delta u^2} &= \frac{\partial^2\xi^i}{\partial u\partial v}+\frac{\partial\Gamma^i_{jk}}{\partial x^\ell}\xi^j\xi^\ell\eta^k+\Gamma^i_{jk}\left(\frac{\partial\xi^j}{\partial u}\eta^k+\xi^j\frac{\partial\eta^k}{\partial u}\right)+\Gamma^i_{jk}\Gamma^j_{\ell m}\xi^m\xi^\ell\eta^k\\ &\quad+\Gamma^i_{jk}\Gamma^k_{\ell n}\xi^j\xi^\ell\eta^n\end{aligned}\tag{6.5.8}$$

The variation rate of the tangent vector along the geodesics is null, and with expression (6.2.20) that defines the geodesic, this rate can be written as

$$\left(\partial_\ell\xi^k\right)\xi^\ell=0$$

or

$$\begin{aligned}\frac{\delta}{\delta v}\left(\frac{\delta\xi^i}{\delta u}\right) &= 0 = \frac{\delta}{\delta v}\left[\left(\partial_k\xi^i+\Gamma^i_{jk}\xi^j\right)\frac{\partial x^k}{\partial u}\right] = \frac{\delta}{\delta v}\left(\partial_k\xi^i\frac{\partial x^k}{\partial u}+\Gamma^i_{jk}\xi^j\frac{\partial x^k}{\partial u}\right)\\ &= \frac{\delta}{\delta v}\left(\frac{\partial\xi^i}{\partial u}+\Gamma^i_{jk}\xi^j\xi^k\right)\end{aligned}$$

thus

$$\frac{\delta}{\delta v}\left(\frac{\partial \xi^i}{\partial u} + \Gamma^i_{jk}\xi^j\xi^k\right) = 0$$

and with expression (6.2.20), it follows that

$$\partial_\ell\left(\frac{\partial \xi^i}{\partial u} + \Gamma^i_{jk}\xi^j\xi^k\right)\frac{\partial x^\ell}{\partial v} = \partial_\ell\left(\frac{\partial \xi^i}{\partial u} + \Gamma^i_{jk}\xi^j\xi^k\right)\eta^\ell = 0$$

and

$$\frac{\partial}{\partial x^\ell}\left(\frac{\partial \xi^i}{\partial u}\right)\frac{\partial x^\ell}{\partial v} + \frac{\partial \Gamma^i_{jk}}{\partial x^\ell}\xi^j\xi^k\eta^\ell + \Gamma^i_{jk}\left(\partial_\ell \xi^j\right)\xi^k\eta^\ell + \Gamma^i_{jk}\xi^j\left(\partial_\ell \xi^k\right)\eta^\ell = 0$$

$$\frac{\partial^2 \xi^i}{\partial u \partial v} + \frac{\partial \Gamma^i_{jk}}{\partial x^\ell}\xi^j\xi^k\eta^\ell + \Gamma^i_{jk}\left(\frac{\partial \xi^j}{\partial x^\ell} + \Gamma^j_{\ell m}\xi^m\right)\xi^k\eta^\ell + \Gamma^i_{jk}\xi^j\left(\frac{\partial \xi^k}{\partial x^\ell} + \Gamma^k_{\ell n}\xi^n\right)\eta^\ell = 0$$

$$\frac{\partial^2 \xi^i}{\partial u \partial v} + \frac{\partial \Gamma^i_{jk}}{\partial x^\ell}\xi^j\xi^k\eta^\ell + \Gamma^i_{jk}\left(\frac{\partial \xi^j}{\partial v}\xi^k + \xi^j\frac{\partial \xi^k}{\partial v}\right) + \Gamma^i_{jk}\Gamma^j_{\ell m}\xi^m\xi^k\eta^\ell + \Gamma^i_{jk}\Gamma^k_{\ell n}\xi^j\xi^n\eta^\ell = 0$$

allows writing

$$\frac{\partial^2 \xi^i}{\partial u \partial v} = -\left[\frac{\partial \Gamma^i_{jk}}{\partial x^\ell}\xi^j\xi^k\eta^\ell + \Gamma^i_{jk}\left(\frac{\partial \xi^j}{\partial v}\xi^k + \xi^j\frac{\partial \xi^k}{\partial v}\right) + \Gamma^i_{jk}\Gamma^j_{\ell m}\xi^m\xi^k\eta^\ell + \Gamma^i_{jk}\Gamma^k_{\ell n}\xi^j\xi^n\eta^\ell\right]$$

and with the substitution in expression (6.5.8)

$$\frac{\delta^2 \eta^i}{\delta u^2} = -\left[\frac{\partial \Gamma^i_{jk}}{\partial x^\ell}\xi^j\xi^k\eta^\ell + \Gamma^i_{jk}\left(\frac{\partial \xi^j}{\partial v}\xi^k + \xi^j\frac{\partial \xi^k}{\partial v}\right) + \Gamma^i_{jk}\Gamma^k_{\ell m}\xi^m\xi^k\eta^\ell + \Gamma^i_{jk}\Gamma^k_{\ell n}\xi^j\xi^n\eta^\ell\right]$$
$$+ \frac{\partial \Gamma^i_{jk}}{\partial x^\ell}\xi^j\eta^k\xi^\ell + \Gamma^i_{jk}\left(\frac{\partial \xi^j}{\partial u}\eta^k + \xi^j\frac{\partial \eta^k}{\partial u}\right) + \Gamma^i_{jk}\Gamma^j_{\ell m}\xi^m\xi^\ell\eta^k + \Gamma^i_{jk}\Gamma^k_{\ell n}\xi^j\xi^\ell\eta^n$$

This expression will be analyzed in parts:

(a) Terms that cancel each other

$$A = -\Gamma^i_{jk}\left(\frac{\partial \xi^j}{\partial v}\xi^k + \xi^j\frac{\partial \xi^k}{\partial v}\right) + \Gamma^i_{jk}\left(\frac{\partial \xi^j}{\partial u}\eta^k + \xi^j\frac{\partial \eta^k}{\partial u}\right) = \Gamma^i_{jk}\left[\frac{\partial\left(\xi^j\eta^k\right)}{\partial u} - \frac{\partial\left(\xi^j\xi^k\right)}{\partial v}\right]$$

and with expressions (6.5.5) and (6.5.6)

$$\frac{\partial u}{\partial v} = \frac{\eta^i}{\xi^i} \Rightarrow \eta^i = \frac{\partial u}{\partial v}\xi^i$$

thus

$$A = \Gamma^i_{jk} \left[\frac{\partial}{\partial u} \left(\xi^j \frac{\partial u}{\partial v} \xi^k \right) - \frac{\partial (\xi^j \xi^k)}{\partial v} \right] = \Gamma^i_{jk} \left[\frac{\partial (\xi^j \xi^k)}{\partial v} - \frac{\partial (\xi^j \xi^k)}{\partial v} \right] = 0$$

(b) Terms with the derivatives of the Christoffel symbols

$$B = -\frac{\partial \Gamma^i_{jk}}{\partial x^\ell} \xi^j \xi^k \eta^\ell + \frac{\partial \Gamma^i_{jk}}{\partial x^\ell} \xi^j \eta^k \xi^\ell$$

and interchanging the indexes $\ell \leftrightarrow k$

$$B = \left(-\frac{\partial \Gamma^i_{j\ell}}{\partial x^k} + \frac{\partial \Gamma^i_{jk}}{\partial x^\ell} \right) \xi^j \xi^\ell \eta^k$$

(c) Other terms

$$C = -\Gamma^i_{jk} \Gamma^k_{\ell m} \xi^m \xi^k \eta^\ell - \Gamma^i_{jk} \Gamma^k_{\ell n} \xi^j \xi^n \eta^\ell + \Gamma^i_{jk} \Gamma^j_{\ell m} \xi^m \xi^\ell \eta^k + \Gamma^i_{jk} \Gamma^k_{\ell n} \xi^j \xi^\ell \eta^n$$

With the change of the indexes $n \longrightarrow m$, it follows that

$$C = -\Gamma^i_{jk} \Gamma^k_{\ell m} \xi^m \xi^k \eta^\ell - \Gamma^i_{jk} \Gamma^k_{\ell m} \xi^j \xi^m \eta^\ell + \Gamma^i_{jk} \Gamma^j_{\ell m} \xi^m \xi^\ell \eta^k + \Gamma^i_{jk} \Gamma^k_{\ell m} \xi^j \xi^\ell \eta^m$$

and with the permutation of the indexes $m \leftrightarrow \ell$ in the second term, the expression is reduced to

$$\begin{aligned} C &= -\Gamma^i_{jk} \Gamma^k_{\ell m} \xi^m \xi^k \eta^\ell - \Gamma^i_{jk} \Gamma^k_{m\ell} \xi^j \xi^\ell \eta^m + \Gamma^i_{jk} \Gamma^j_{\ell m} \xi^m \xi^\ell \eta^k + \Gamma^i_{jk} \Gamma^k_{\ell m} \xi^j \xi^\ell \eta^m \\ &= -\Gamma^i_{jk} \Gamma^k_{\ell m} \xi^m \xi^k \eta^\ell + \Gamma^i_{jk} \Gamma^j_{\ell m} \xi^m \xi^\ell \eta^k \end{aligned}$$

The permutation of the indexes $j \leftrightarrow m$ provides

$$C = -\Gamma^i_{mk} \Gamma^k_{\ell j} \xi^j \xi^k \eta^\ell + \Gamma^i_{mk} \Gamma^m_{\ell j} \xi^j \xi^\ell \eta^k$$

and with the permutation of the indexes $\ell \leftrightarrow k$ in the first term

$$C = -\Gamma^i_{mk} \Gamma^k_{\ell j} \xi^j \xi^\ell \eta^k + \Gamma^i_{m\ell} \Gamma^m_{kj} \xi^j \xi^\ell \eta^k$$

Joining this parcel

$$\frac{\delta^2 \eta^i}{\delta u^2} = \left(-\frac{\partial \Gamma^i_{j\ell}}{\partial x^k} + \frac{\partial \Gamma^i_{jk}}{\partial x^\ell} - \Gamma^i_{mk} \Gamma^k_{\ell j} + \Gamma^i_{m\ell} \Gamma^m_{kj} \xi^j \xi^\ell \eta^k \right) \xi^j \xi^\ell \eta^k$$

and with expression (5.2.11), the result is

$$\frac{\partial^2 \eta^i}{\partial u^2} + R^i_{jk\ell}\xi^j\xi^\ell\eta^k = 0 \tag{6.5.9}$$

This expression allows establishing N second-order ordinary differential equation for the vectors η^i that represent the deviations (distances) between the geodesics, the unit vectors ξ^i being tangents to these curves. The distances η^i are determined if the initial values of η^i and $\frac{\partial \eta^i}{\partial u}$ (or $\frac{d\eta^i}{du}$) are known.

Exercise 6.8 Show that in a flat space the deviation of the family of geodesics defined by the function $x^i(u,v) = uF^i(v) + G^i(v)$ is null.

The family of geodesics is defined by

$$x^i(u,v) = uF^i(v) + G^i(v)$$

thus

$$\frac{\partial x^i}{\partial u} = \xi^i = F^i \quad \frac{\partial x^i}{\partial v} = \eta^i = u\frac{\partial F^i}{\partial v} + \frac{\partial G^i}{\partial v}$$
$$\frac{\partial \eta^i}{\partial u} = \frac{\partial F^i}{\partial v} \quad \frac{\partial^2 \eta^i}{\partial u^2} = 0$$

and expression (6.5.9)

$$\frac{\partial^2 \eta^i}{\partial u^2} + R^i_{jk\ell}\xi^j\xi^\ell\eta^k = 0$$

stays

$$0 + R^i_{jk\ell}\xi^j\xi^\ell\eta^k = 0$$

and as the space is flat $R^i_{jk\ell} = 0$, which verifies the previous equation and shows that the deviation of this family of geodesics is null.

6.6 Parallelism of Vectors

6.6.1 *Initial Notes*

In the Euclidian space, two coplanar vectors that move with the origin over a straight line AB located in the plane of these vectors are parallel and have the same norm $\|\boldsymbol{u}\| = \|\boldsymbol{v}\|$ and maintain the same direction defined by angle α, and there is no geometric difference between these varieties, i.e., the vectors are equipollent. In Fig. 6.5 the vector $\boldsymbol{u}$ in point $P(x^i) \in E_3$ and the vector $\boldsymbol{v}$ with origin in point

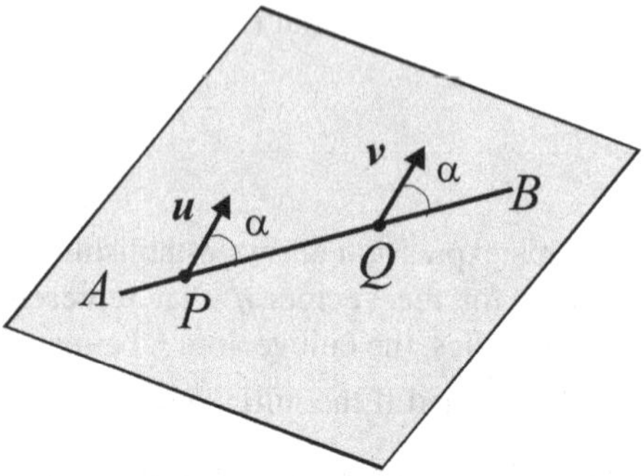

Fig. 6.5 Parallelism of vectors in the Euclidian space E_2

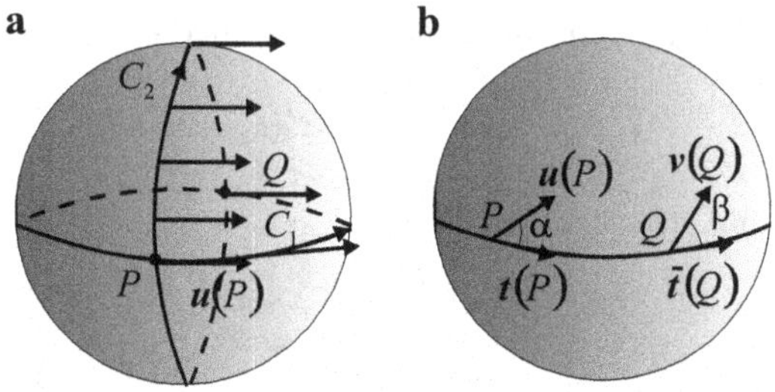

Fig. 6.6 Parallelism of vectors in the bidimensional spherical space: (a) parallel transport of vector $\boldsymbol{u}$ along different paths, (b) condition of parallelism $\alpha = \beta$

$Q(x^i + dx^i)$ neighbor to point $P(x^i)$ are equipollent. This equipollence between the two vectors indicates that the same are parallels, i.e., they shift from one point to another point of the plane.

The concept of displacement of a Cartesian vector can be generalized for the space E_N with the definition of parallel shift of the vector along a curve (Fig. 6.5). However, in general, the space is not Euclidian, which requires a few considerations more specific for study of the parallel transport of the vector.

The unit tangent vectors to the geodesic $u^i = \frac{dx^i}{ds}$ represent velocities (variation rates), and the velocity variation in the measurement unit of the independent variable s is called acceleration (variation of the variation rate); thus,

$$\partial_s u^i = \frac{\partial u^i}{\partial s} + u^j \Gamma^i_{jk} = \frac{d^2 x^i}{ds^2} + \frac{du^j}{ds} \Gamma^i_{jk}$$

and if u^i is constant $\partial_s u^i = 0$, then the displacement of the unit tangent vector is a translation, i.e., it is a displacement which trajectory keeps its direction constant.

The comparison between the varieties defined in distinct points of the Riemann space E_N is carried out from the concept of parallel transport of vector along a curve of this space.

To interpret the curvature tensor geometrically, admit the parallel transport of vector $\boldsymbol{u}$ from point $P(x^i)$ to point $Q(x^i + dx^i)$, along a curve on the spherical surface shown in Fig. 6.6a, running two different paths C_1 and C_2, having path C_1 along the equator and path C_2 along the meridian. It is observed that vector $\boldsymbol{u}$ is not kept constant, for in point Q will depend on the path being run, concluding that the parallel transport of $\boldsymbol{u}$ depends on the curvature of the space.

Consider, for instance, in point $P(x^i)$ located on the circle of maximum diameter of the sphere shown in Fig. 6.6b the vector $\boldsymbol{u}(P)$ and the tangent vector $\boldsymbol{t}(P)$ that form an angle α and are located in the plane tangent to the sphere in this point, and the point $Q(x^i + dx^i)$ neighbor to point $P(x^i)$ and also located on the circle of maximum diameter, which tangent plane contains the vector $\boldsymbol{v}(Q)$ and the tangent vector $\bar{\boldsymbol{t}}(Q)$ that form an angle β. Let $||\boldsymbol{u}(P)|| = ||\boldsymbol{v}(Q)||$. The vectors $\boldsymbol{u}(P)$ and $\boldsymbol{v}(Q)$ will be parallel if $\alpha = \beta$. The concept of parallelism between vectors is generalized for the Riemann space E_N.

6.6.2 Parallel Transport of Vectors

The parallel transport of a vector is its displacement from one point to another of the space, during which the vector is kept constant. In the curved Riemann space E_N, the result of the parallel transport of the vector depends on the path run between the two points.

The study of the parallelism of vectors in the Riemann space E_N is made according to the Levi-Civita approach by means of the elementary curved parallelogram $PQRS$ (Fig. 6.7). The vertexes of this parallelogram have coordinates $P(x^i)$, $Q(x^i + \varepsilon^i)$, $S(x^i + \delta^i)$, and $R(x^i + \varepsilon^i + \delta^i)$, ε^i and δ^i being elementary quantities.

Consider the vector $\boldsymbol{u}(P)$ embedded in the tangent space to the Riemann space E_N and shifted parallel along path $C_1 = PQR$ from P, with vector originating in point Q:

$$u^m(Q) = \left(u^m - u^p \Gamma^m_{np} \varepsilon^n\right)_{(P)}$$

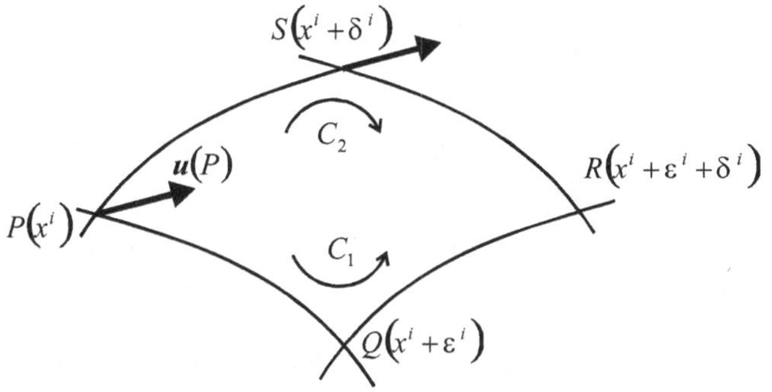

Fig. 6.7 Levi-Civita parallelism

where the letters in parenthesis indicate the point where it refers to the variety being analyzed. The transport of the vector from point $Q(x^i + \varepsilon^i)$ to point $R(x^i + \varepsilon^i + \delta^i)$ is given by

$$u^m(R) = \left(u^m - u^p\Gamma^m_{np}\varepsilon^n\right)_{(Q)}$$

$$= \left(u^m - u^p\Gamma^m_{np}\varepsilon^n\right)_{(P)} - \left(u^p - u^r\Gamma^p_{sr}\varepsilon^s\right)_{(P)}\left(\Gamma^m_{np} + \frac{\partial \Gamma^m_{np}}{\partial x^\ell}\varepsilon^\ell\right)_{(P)}\delta^n$$

where vector $u^m(R)$ appears in terms of the parameters defined in point P; then without the lower index that indicates this point, it follows that

$$u^m(R) = u^m - u^p\Gamma^m_{np}\varepsilon^n - u^p\Gamma^m_{np}\delta^n - u^r\Gamma^p_{sr}\Gamma^m_{np}\varepsilon^s\delta^n + u^p\frac{\partial \Gamma^m_{np}}{\partial x^\ell}\varepsilon^\ell\delta^n - u^r\Gamma^p_{sr} \times \frac{\partial \Gamma^m_{np}}{\partial x^\ell}\varepsilon^s\varepsilon^\ell\delta$$

and the last term can be disregarded on account of being of a higher order; thus,

$$u^m(R) \cong u^m - u^p\Gamma^m_{np}\varepsilon^n - u^p\Gamma^m_{np}\delta^n - u^r\Gamma^p_{sr}\Gamma^m_{np}\varepsilon^s\delta^n + u^p\frac{\partial \Gamma^m_{np}}{\partial x^\ell}\varepsilon^\ell\delta^n$$

The permutation of the indexes $r \leftrightarrow p$ and the change of the indexes $s \leftrightarrow \ell$ in the fourth term to the right allow writing

$$u^m(R) \cong u^m - u^p\Gamma^m_{np}\varepsilon^n - u^p\Gamma^m_{np}\delta^n + u^p\left(\frac{\partial \Gamma^m_{np}}{\partial x^\ell} - \Gamma^r_{\ell p}\Gamma^m_{nr}\right)\varepsilon^\ell\delta^n$$

and with an analogous formulation for the path $C_2 = PSR$, the result is

$$\tilde{u}^m(R) \cong u^m - u^p\Gamma^m_{np}\delta^n - u^p\Gamma^m_{np}\varepsilon^n - u^p\left(\frac{\partial \Gamma^m_{\ell p}}{\partial x^n} - \Gamma^r_{np}\Gamma^m_{\ell r}\right)\varepsilon^\ell\delta^n$$

with $u^m(R) \neq \tilde{u}^m(R)$; thus,

$$\tilde{u}^m(R) - u^m(R) \cong u^p\left(\frac{\partial \Gamma^m_{np}}{\partial x^\ell} - \frac{\partial \Gamma^m_{\ell p}}{\partial x^n} - \Gamma^r_{\ell p}\Gamma^m_{nr} + \Gamma^r_{np}\Gamma^m_{\ell r}\right)\varepsilon^\ell\delta^n = u^p R^m_{p\ell n}\varepsilon^\ell\delta^n$$

where the term to the left is a vector, and in the term to the right, there is the inner product of vector u^p by the variety $R^m_{p\ell n}$, this variety being the Riemann–Christoffel curvature tensor.

The deduction of the Riemann–Christoffel tensor by means of the concept of parallelism of vectors in space E_N is due to Levi-Civita. The approach adopted as

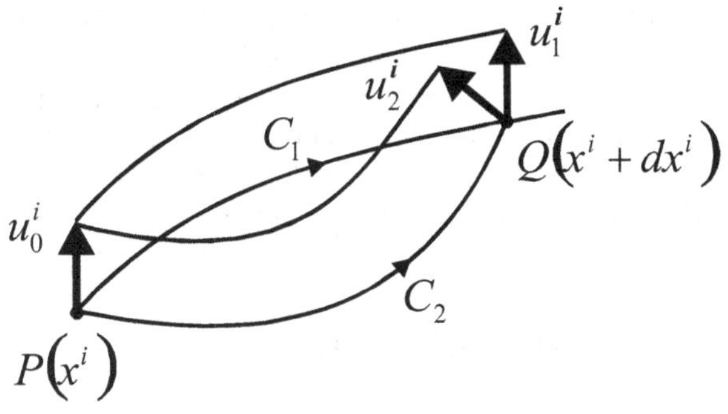

Fig. 6.8 Parallel transport to vector u^i along the paths C_1 and C_2

definition of this tensor, such as developed in item 5.2, where $R^m_{p\ell n}$ was obtained by simple algebraic formalism when calculating the covariant derivatives of the second-order tensor of variance $(0, 2)$, is due to Erwin Christoffel. The geometric interpretation of tensor $R^m_{p\ell n}$ is that the change of relative orientation between the vectors shift in parallel along different paths is measured by this tensor.

Complementing the analysis of parallelism of vectors let, in Fig. 6.8, the paths of vector u^i from point $P(x^i)$ where u_0^i exists, from point $Q(x^i + dx^i)$ along two different paths C_1 and C_2, which results, in general, in the condition $u_1^i \neq u_2^i$.

The condition for the vectors u^i or u_i being displaced in parallel to itself along the parameterized curve $x^k = x^k(t)$ is that their absolute derivatives are null:

$$\frac{\delta u^i}{dt} = \frac{\delta u_i}{dt} = 0$$

The differential equation that represents the parallel transport of the vector u^i along the curve defined by the parametric equations $x^k = x^k(t) \in E_N$ is given by

$$du^i + \Gamma^i_{jk} u^j dx^k = 0 \tag{6.6.1}$$

In an analogous way for the covariant components of the vector, in differential form is given by

$$du_i - \Gamma^j_{ik} u_j dx^k = 0 \tag{6.6.2}$$

and the simple analogy with the parallelism of vectors defined in Cartesian coordinate systems would lead to the condition $du^i = du_i = 0$, which would be the condition for the vector to be displaced in parallel to itself keeping its components constants. This analogy is incorrect, because these increments do not represent the components of the vector, and when the coordinate system is changed, these differentials du^i, du_i are not necessarily null and imply that the condition of parallelism would depend on the coordinate system. These differentials are vectors, for expressions (6.6.1) and (6.6.2) provide, respectively, $du^i = -\Gamma^i_{jk} u^j dx^k$ and $du_i = \Gamma^j_{ik} u_j dx^k$, whereby if they are null, a coordinate system will cancel in all the others.

The substitution of the parametric equations of the curve $x^k = x^k(t)$, on which will be given the path of the vector defined by expression (6.6.1), results in a system of ordinary differential equations which unknown values are the functions $u^i(t)$. The values of these functions in the final point of the path will depend on the value of parameter t in this point. The result of solving these ordinary differential equations, in general, depends on the path.

6.6.2.1 Independence of Path

The independence of path is linked to the condition of the derivatives $\frac{du^i}{dx^k}$ of expression (6.6.1) representing exact differential

$$\frac{du^i}{dx^k} = -\Gamma^i_{jk} u^j$$

and the equality

$$\frac{\partial\left(\Gamma^i_{jk} u^j\right)}{\partial x^\ell} = \frac{\partial\left(\Gamma^i_{j\ell} u^j\right)}{\partial x^k}$$

allows the analytic development of this condition, whereby

$$\frac{\partial \Gamma^i_{jk}}{\partial x^\ell} u^j + \Gamma^i_{jk} \frac{du^j}{dx^\ell} = \frac{\partial \Gamma^i_{j\ell}}{\partial x^k} u^j + \Gamma^i_{j\ell} \frac{du^j}{dx^k} \tag{6.6.3}$$

The change of the indexes $i \to j,\ k \to \ell,\ j \to p$ in expression (6.6.1) rewritten as

$$\frac{du^i}{dx^k} = -\Gamma^i_{jk} u^j$$

provides

$$\frac{du^j}{dx^\ell} = -\Gamma^j_{p\ell} u^p$$

and with the change of the indexes $i \to j, j \to p$

$$\frac{du^j}{dx^k} = -\Gamma^j_{pk} u^p$$

The substitution of these two expressions in expression (6.6.3) allows writing

$$\frac{\partial \Gamma^i_{jk}}{\partial x^\ell} u^j - \Gamma^i_{jk} \Gamma^j_{p\ell} u^p = \frac{\partial \Gamma^i_{j\ell}}{\partial x^k} u^j - \Gamma^i_{j\ell} \Gamma^j_{pk} u^p$$

whereby

$$\left(\frac{\partial \Gamma^i_{jk}}{\partial x^\ell} - \frac{\partial \Gamma^i_{j\ell}}{\partial x^k}\right) u^j + \left(\Gamma^j_{j\ell}\Gamma^i_{pk} - \Gamma^i_{jk}\Gamma^j_{p\ell}\right) u^p = 0$$

and with the change of the indexes $j \to p$ in terms between the first parenthesis

$$\left(\frac{\partial \Gamma^i_{pk}}{\partial x^\ell} - \frac{\partial \Gamma^i_{p\ell}}{\partial x^k} + \Gamma^j_{j\ell}\Gamma^i_{pk} - \Gamma^i_{jk}\Gamma^j_{p\ell}\right) u^p = 0$$

and as

$$R^i_{p\ell k} = \frac{\partial \Gamma^i_{pk}}{\partial x^\ell} - \frac{\partial \Gamma^i_{p\ell}}{\partial x^k} + \Gamma^j_{j\ell}\Gamma^i_{pk} - \Gamma^i_{jk}\Gamma^j_{p\ell}$$

results in

$$R^i_{p\ell k} u^p = 0 \tag{6.6.4}$$

Therefore, $R^i_{p\ell k} u^p = 0 \, \forall i, p, \ell, k$ expresses the conditions that must be fulfilled so that the parallel transport of the vector u^i is independent of the path. The necessary and sufficient condition so that the parallel transport is independent of the path for any vector is $R^i_{p\ell k} = 0$.

6.6.2.2 Invariance of the Modulus and the Angle Between Vectors

With expression (6.6.1) under the form

$$\frac{du^i}{dt} + \Gamma^i_{jk} u^j \frac{dx^k}{dt} = 0 \tag{6.6.5}$$

and having the vectors $u^i(t)$ and $v^i(t)$, two solutions of differential equation, the dot product between these two vectors

$$\boldsymbol{u} \cdot \boldsymbol{v} = \|\boldsymbol{u}\| \|\boldsymbol{v}\| \cos \alpha = g_{ij} u^i v^j$$

is invariant; thus,

$$\frac{d}{dt}\left(g_{ij} u^i v^j\right) = 0$$

As $g_{ij}u^i v^j$ is an invariant and by Ricci's lemma tensor g_{ij} behaves as a constant in the covariant derivation

$$\frac{d}{dt}\left(g_{ij}u^i v^j\right) = g_{ij}\frac{du^i}{dt}v^j + g_{ij}u^i\frac{dv^j}{dt}$$

By hypothesis, the vectors are solutions of the differential equation given by expression (6.6.5); then

$$\frac{du^i}{dt} = 0 \quad \frac{dv^j}{dt} = 0$$

It is concluded that $g_{ij}u^i v^j$ is constant along the curve represented by the parametric equation $x^k = x^k(t)$. If the vectors are equal $u^i = v^j$, then $g_{ij}u^i v^j = g_{ij}u^2$, which is a constant; therefore, the angle α between the vectors is constant. With this analysis it is verified that the modulus of these vectors are invariant when they move along the parameterized curve, thence the angle α between u^i and v^j also remains unchanged when varying parameter t. The modulus of the vectors $\boldsymbol{u}$ and $\boldsymbol{v}$ is maintained unchanged, just as the angle between them; thus, the dot product $\boldsymbol{u} \cdot \boldsymbol{v}$ also remains unchanged.

The straight line in the Euclidian space is the only curve for which the parallel transport of a vector is the own tangent vector to this curve.

6.6.2.3 Space with Affine Connections

The parallel transport of a vector is independent of the metric tensor, because Christoffel symbol of second kind can be determined by expression (2.3.9). The spaces with a parallel transport are called space with affine connections. For the Riemann space E_N, the affine connections are the Christoffel symbols.

6.6.2.4 Integrability

The Euclidian characteristics of the Riemann space E_N depend only of its metric. As the Christoffel symbols are linked to the metric tensor, the use of the contravariant and covariant coordinates of the vector is indifferent for determining the parallel displacement of a vector along a curve represented by parametric equations, which is governed by the differential equations given by expressions (6.6.1) and (6.6.2).

Consider the parallel displacement that takes place along a closed curve such as Fig. 6.9a shows. In this case the vectors remain unchanged along the path, and the affine connections of this space are integratable, and then the vector in a point generates the field of parallel vectors in space E_N. Figure 6.9b shows the displacement of a vector along a closed curve for non-integratable affine connections, where

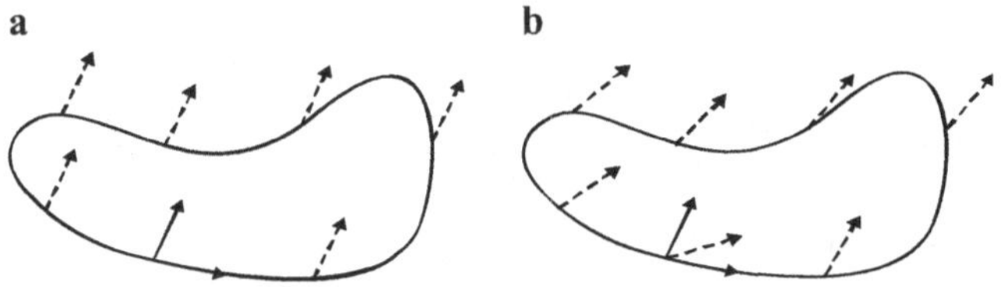

Fig. 6.9 Integrability of the affine connections: (**a**) integratable affine connections, (**b**) non-integratable affine connections

the change of the vector's direction along the path is verified. If the affine connections of the space are linked to the metric tensor, the condition of the space being Euclidian is directly related with their integrability.

For the Euclidian space the affine connections are the Christoffel symbols; then the differentials du^i and du_i given by expressions (6.6.1) and (6.6.2), respectively, are null, and the parallel vectors have the same components in all the points of the space, whereby these affine connections are integratable. The affine connections of the Euclidian spaces are always integratable, and then it is always possible to determine a Cartesian coordinate system when the affine connections are integratable.

Integrability is an invariant property of the affine connections, so it is independent of the coordinate system.

Exercise 6.9 For the Riemann space in which all the pairs of points $(x^1; x^2) \in R$, $x^2 > 0$, which metric tensor is given by

$$g_{ij} = \begin{bmatrix} \frac{1}{x^2} & 0 \\ 0 & \frac{1}{x^2} \end{bmatrix}$$

calculate the parallel transport of a vector v^j along the curve of parametric coordinates $u^i(t) = \left\{ \begin{matrix} x_0^1 \\ x_0^2 + t \end{matrix} \right\}$, having $(x_0^1; x_0^2)$ initial values of the coordinates and $x^1 = \text{constant}$.

The derivatives of the components of the metric tensor are

$$g_{11,2} = g_{22,1} = -\frac{2}{(x^2)^3}$$

and the non-null Christoffel symbols are given by

$$\Gamma_{11,2} = -\frac{1}{2} g_{11,2} = \frac{1}{(x^2)^3} \quad \Gamma_{12,1} = \frac{1}{2} g_{11,2} = -\frac{1}{(x^2)^3}$$

$$\Gamma_{21,1} = \frac{1}{2} g_{11,2} = -\frac{1}{(x^2)^3} \quad \Gamma_{22,2} = \frac{1}{2} g_{22,2} = -\frac{1}{(x^2)^3}$$

thus

$$\Gamma^2_{11} = g^{22}\Gamma_{11,2} = \frac{1}{x^2} \quad \Gamma^1_{12} = \Gamma^1_{21} = \Gamma^2_{22} = -\frac{1}{x^2}$$

The parallel transport of the vector v^j is given by

$$\dot{v}^j = -\Gamma^j_{ik}\dot{u}^i v^k$$

and having

$$\dot{u}^i(t) = \begin{Bmatrix} 0 \\ 1 \end{Bmatrix}$$

it follows that

$$\dot{v}^1 = -\Gamma^1_{21}v^1 = \frac{v^1}{x^2} \quad \dot{v}^2 = -\Gamma^2_{21}v^1 = \frac{v^1}{x^2}$$

These two differential equations have as solutions

$$v^i(t) = v^i_0 e^{\frac{t}{x^2}}$$

where v^i_0 is a constant in $t = 0$.

Exercise 6.10 Calculate the parallel displacement of vector u^i along the curve defined by the parametric equations $\xi^1 = c =$ constant and $\xi^2 = t$, located on a cone of parametric equation $\left(\xi^3\right)^2 = \left(\xi^1 \cos \xi^2\right)^2 + \left(\xi^1 \sin \xi^2\right)^2$, where the relations between the parametric coordinates (ξ^1, ξ^2) and the Cartesian coordinates are $x^1 = \xi^1 \cos \xi^2$, $x^2 = \xi^1 \sin \xi^2$, $x^3 = \xi^1$.

The fundamental form is given by

$$ds^2 = g_{\ell m} d\xi^\ell d\xi^m$$

where

$$g_{\ell m} = \frac{\partial x^i}{\partial \xi^\ell}\frac{\partial x^i}{\partial \xi^m}$$

which components are

$$g_{11} = \left(\frac{\partial x^i}{\partial \xi^1}\right)^2 = \left(\frac{\partial x^1}{\partial \xi^1}\right)^2 + \left(\frac{\partial x^2}{\partial \xi^1}\right)^2 + \left(\frac{\partial x^3}{\partial \xi^1}\right)^2 = \left(\cos \xi^2\right)^2 + \left(\sin \xi^2\right)^2 + 1$$
$$= 2$$

$$g_{12} = \frac{\partial x^i}{\partial \xi^1}\frac{\partial x^i}{\partial \xi^2} = \frac{\partial x^1}{\partial \xi^1}\frac{\partial x^1}{\partial \xi^2} + \frac{\partial x^2}{\partial \xi^1}\frac{\partial x^2}{\partial \xi^2} + \frac{\partial x^3}{\partial \xi^1}\frac{\partial x^3}{\partial \xi^2}$$
$$= \cos \xi^2 (-\xi^1 \sin \xi^2) + \sin \xi^2 (\xi^1 \cos \xi^2) + 0 = 0$$
$$g_{22} = \left(\frac{\partial x^i}{\partial \xi^2}\right)^2 = \left(\frac{\partial x^1}{\partial \xi^2}\right)^2 + \left(\frac{\partial x^2}{\partial \xi^2}\right)^2 + \left(\frac{\partial x^3}{\partial \xi^2}\right)^2$$
$$= (-\xi^1 \sin \xi^2)^2 + (\xi^1 \cos \xi^2)^2 + 0 = (\xi^1)^2$$

then

$$g_{ij} = \begin{bmatrix} 2 & 0 \\ 0 & (\xi^1)^2 \end{bmatrix} \quad g^{ij} = \begin{bmatrix} \frac{1}{2} & 0 \\ 0 & \frac{1}{(\xi^1)^2} \end{bmatrix}$$

For the non-null Christoffel symbols, it follows that

$$\Gamma_{22,1} = -\xi^1 \quad \Gamma_{12,2} = \Gamma_{21,2} = \xi^1$$
$$g^{11}\Gamma_{22,1} = -\frac{\xi^1}{2} \quad g^{22}\Gamma_{12,2} = \Gamma^2_{12} = \Gamma^2_{21} = \frac{1}{\xi^1}$$

The parametric equations of the curve are given by

$$\frac{d\xi^1}{dt} = 0 \quad \frac{d\xi^2}{dt} = 1$$

and with expression (6.6.2) written under the form

$$\frac{du^\ell}{dt} + \Gamma^\ell_{mn} u^m \frac{d\xi^m}{dt} = 0$$

it follows that

$$\frac{du^1}{dt} + \Gamma^1_{22} u^2 \frac{d\xi^2}{dt} = 0 \quad \frac{du^2}{dt} + \Gamma^2_{12} u^1 \frac{d\xi^2}{dt} = 0$$

whereby

$$\frac{du^1}{dt} - \frac{\xi^1}{2} u^2 = 0 \quad \frac{du^2}{dt} + \frac{1}{\xi^1} u^1 = 0$$

Differentiating the first differential equation and by substitution

$$\frac{d^2u^1}{dt^2} + \frac{u^1}{2} = 0$$

which solution is

$$u^1 = k_3 \cos \frac{\sqrt{2}}{2} t + k_4 \sin \frac{\sqrt{2}}{2} t$$

The derivative of this solution substituted in the differential equation

$$\frac{du^1}{dt} - \frac{\xi^1}{2} u^2 = 0$$

provides

$$u^2 = \frac{\sqrt{2}}{c} \left(-k_3 \sin \frac{\sqrt{2}}{2} t + k_4 \cos \frac{\sqrt{2}}{2} t \right)$$

For $t = 0$ the point of coordinates $(c; 0)$ exists; in this point writing the initial values of the coordinates as u_0^m, it follows that

$$u_0^1 = k_3$$

$$u_0^2 = \frac{\sqrt{2}}{c} k_4 \; \therefore \; k_4 = \frac{c}{\sqrt{2}} u_0^2$$

whereby

$$u^1 = u_0^1 \cos \frac{\sqrt{2}}{2} t + \frac{c\sqrt{2}}{2} u_0^2 \sin \frac{\sqrt{2}}{2} t \quad u^2 = -\frac{\sqrt{2}}{c} u_0^1 \sin \frac{\sqrt{2}}{2} t + u_0^2 \cos \frac{\sqrt{2}}{2} t$$

For $t = 2\pi$:

$$u^1 = u_0^1 \cos \sqrt{2}\pi + \frac{c\sqrt{2}}{2} u_0^2 \sin \sqrt{2}\pi \quad u^2 = -\frac{\sqrt{2}}{c} u_0^1 \sin \sqrt{2}\pi + u_0^2 \cos \sqrt{2}\pi$$

These expressions show that $u^1 \neq u^2$ in the interval $[0; 2\pi]$, so the direction of this vector varies along the curve.

Exercise 6.11 For the parallel displacement of a vector along a defined path in the space with metric $ds^2 = \left(d\xi^1\right)^2 + g_{22}\left(d\xi^2\right)^2$, show that upon reaching the final point of the path this vector forms with its initial position an angle equal to $\iint_S K\, dS$, where $dS = d\xi^1 d\xi^2$.

The metric tensor linked to the fundamental form is

$$g_{ij} = \begin{bmatrix} 1 & 0 \\ 0 & g_{22} \end{bmatrix}$$

and the unit vector that moves along the path has components

$$\xi^1 = \cos\alpha \quad \xi^2 = \frac{\sin\alpha}{\sqrt{g_{22}}}$$

As this unit vector moves in parallel to itself

$$\frac{\delta\xi^1}{\delta s} = 0$$

follows

$$\frac{\delta\xi^1}{\delta s} = \frac{d}{ds}(\cos\alpha) + \Gamma^1_{\ell m}\xi^\ell \frac{d\xi^2}{ds} = -\sin\alpha\frac{d\alpha}{ds} + \Gamma^1_{22}\xi^2\frac{d\xi^2}{ds}$$

$$\Gamma^1_{22} = g^{11}\Gamma_{22,1} = -\frac{1}{2}\frac{\partial g_{22}}{\partial\xi^1}$$

$$\frac{\delta\xi^1}{\delta s} = -\sin\alpha\frac{d\alpha}{ds} - \frac{1}{2}\frac{\partial g_{22}}{\partial\xi^1}\frac{\sin\alpha}{\sqrt{g_{22}}}\frac{d\xi^2}{ds} = 0$$

then

$$d\alpha = -\frac{\partial\sqrt{g_{22}}}{\partial\xi^1}\frac{d\xi^2}{ds}ds$$

The integration provides

$$\alpha = \int\left(\frac{\partial\sqrt{g_{22}}}{\partial\xi^1}\frac{d\xi^2}{ds}\right)ds$$

but

$$\alpha = \int\left(\frac{\partial\sqrt{g_{22}}}{\partial\xi^1}\frac{d\xi^2}{ds}\right)ds = \int\frac{\partial\sqrt{g_{22}}}{\partial\xi^1}d\xi^2 = \iint_S\left(\frac{\partial^2\sqrt{g_{22}}}{\partial\xi^1\partial\xi^1}d\xi^1\right)d\xi^2$$

The non-null component of the Riemann–Christoffel tensor of the first kind is

$$R_{1212} = -\frac{1}{2}\frac{\partial^2 g_{22}}{\partial\xi^1\partial\xi^1}$$

and as $\det g = g_{22}$

$$K = -\frac{1}{2g_{22}}\frac{\partial^2 g_{22}}{\partial\xi^1\partial\xi^1} = -\frac{\partial^2\sqrt{g_{22}}}{\partial\xi^1\partial\xi^1}$$

It is verified that

$$\alpha = \iint_S K\,dS \qquad Q.E.D.$$

Exercise 6.12 Given vector u^i of constant modulus that moves in parallel along a curve $x^i = x^i(s)$ in the Riemann space E_N, and the vector $v^i = \alpha u^i$ parallel to vector u^i, where α is a scalar, show that the condition that vector v^i must obey is $\frac{\delta v^i}{\delta s} = \frac{d(\ell n\,\alpha)}{ds} v^i$.

The condition of parallel displacement is given by

$$\frac{\delta u^i}{\delta s} = 0$$

and with the expression of vector $v^i = \alpha u^i$, it follows that

$$\frac{\delta v^i}{\delta s} = \alpha\frac{\delta u^i}{\delta s} + u^i\frac{d\alpha}{ds} = u^i\frac{d\alpha}{ds} = \frac{v^i}{\alpha}\frac{d\alpha}{ds}$$

whereby

$$\frac{\delta v^i}{\delta s} = \frac{d(\ell n\,\alpha)}{ds} v^i \qquad Q.E.D.$$

6.6.3 Torsion

The parallel transport of a vector along different ways can lead to two coincident points. Let point $P(x^i) \in E_N$ and the points Q and S be located in the neighborhood of this point, and determined by means of the translations of vectors $\boldsymbol{du} = d\varepsilon^i \boldsymbol{e}_i$ and $\boldsymbol{dv} = d\lambda^i \boldsymbol{e}_i$, respectively (Fig. 6.10).

Admitting the parallel transport along PS vector $\boldsymbol{SR}_1$ is obtained with components $\left(d\varepsilon^i - \Gamma^m_{n\ell} d\varepsilon^\ell d\lambda^n\right)$; then $\boldsymbol{PR}_1 = \boldsymbol{PS} + \boldsymbol{SR}_1$, which components are $\left(d\delta^i + d\varepsilon^i - \Gamma^m_{n\ell} d\varepsilon^\ell d\lambda^n\right)$. In a similar way, the parallel transport of λ^i along the segment PQ provides the vector $\boldsymbol{PR}_2 = \boldsymbol{PQ} + \boldsymbol{QR}_2$, with components $\left(d\varepsilon^i + d\lambda^i - \Gamma^m_{\ell n} d\varepsilon^\ell d\lambda^n\right)$.

Fig. 6.10 Curved space E_N with torsion

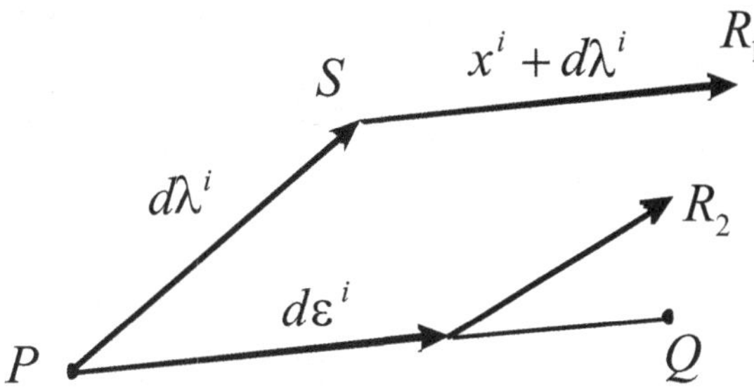

For the space E_N with affine connections not necessarily symmetric in the indexes n and ℓ, vector $\boldsymbol{R}_2\boldsymbol{R}_1 = \boldsymbol{PR}_2 - \boldsymbol{PR}_1$ is given by

$$\boldsymbol{R}_2\boldsymbol{R}_1 = \left(\Gamma^m_{n\ell} - \Gamma^m_{\ell n}\right) d\varepsilon\, d\lambda\, \boldsymbol{e_m} \neq 0 \tag{6.6.6}$$

which allows defining the tensor

$$T^m_{n\ell} = \Gamma^m_{n\ell} - \Gamma^m_{\ell n} \tag{6.6.7}$$

Expression (6.6.7) defines the torsion tensor of space E_N. This tensor measures the difference of closing the "elementary parallelogram" formed by the vectors and their transport parallels. If the connections are the Christoffel symbols, then $T^m_{n\ell} = 0$ due to the symmetry. In this space for the vectors $\boldsymbol{e_m}$ results the rotation when having the parallel transport between nearby points.

The Christoffel symbols are symmetric connections of the Riemann space E_N and form the Levi-Civita connections of this space, i.e., the torsion tensor of a Levi-Civita connection is null.

If $\Gamma^m_{n\ell} \neq \Gamma^m_{\ell n}$, the vectors $\boldsymbol{e_m}$ are submitting a torsion; thus,

$$\begin{aligned}\Omega^m &= -T^m_{n\ell} ds^{n\ell}\\ \Omega^m &= -\left(\Gamma^m_{n\ell} - \Gamma^m_{\ell n}\right) ds^{n\ell}\end{aligned} \tag{6.6.8}$$

The antisymmetric pseudotensor of the second order is $ds^{n\ell}$ and is obtained by means of the dyadic product of the vectors $d\boldsymbol{u} = d\varepsilon^i \boldsymbol{e}_i$ and $d\boldsymbol{v} = d\lambda^i \boldsymbol{e}_i$, whereby

$$ds^{n\ell} = \frac{1}{2}\left(d\varepsilon^n d\lambda^\ell - d\varepsilon^\ell d\lambda^n\right) = \frac{1}{2}\begin{vmatrix} d\varepsilon^n & d\varepsilon^\ell \\ d\lambda^n & d\lambda^\ell \end{vmatrix} \tag{6.6.9}$$

has $\frac{1}{2}N(N-1)$ independent components.

Putting

$$\Omega^m = -\frac{1}{2}\left(\Gamma^m_{n\ell} - \Gamma^m_{\ell n}\right) d\varepsilon^n d\lambda^\ell + \frac{1}{2}\left(\Gamma^m_{n\ell} - \Gamma^m_{\ell n}\right) d\varepsilon^\ell d\lambda^n$$

and with the permutation of the dummy indexes $n \leftrightarrow \ell$ in the first term to the right

$$\Omega^m = -\frac{1}{2}\left(\Gamma^m_{\ell n} - \Gamma^m_{n\ell}\right) d\varepsilon^n d\lambda^\ell + \frac{1}{2}\left(\Gamma^m_{n\ell} - \Gamma^m_{\ell n}\right) d\varepsilon^\ell d\lambda^n$$

then

$$\Omega^m = \left(\Gamma^m_{n\ell} - \Gamma^m_{\ell n}\right) d\varepsilon^\ell d\lambda^n$$

It is concluded that

$$\boldsymbol{R}_2\boldsymbol{R}_1 = \Omega^m \boldsymbol{e}_m \tag{6.6.10}$$

where Ω^m are the N components of the vector that evaluates the torsion of the space E_N. If $\Omega^m = 0$, then E_N is a Riemann space.

If there is no symmetry of the affine connections, i.e., $\Gamma^m_{n\ell} \neq \Gamma^m_{\ell n}$, then $\partial_k g_{ij} \neq 0$.

Exercise 6.13 Show that $\partial_k g_{ij} = 0$ requires that the torsion tensor is null.

The nullity of the covariant derivative of the metric tensor allows writing

$$\partial_k g_{ij} = g_{ij,k} - g_{i\ell}\Gamma^\ell_{jk} - g_{j\ell}\Gamma^\ell_{ki} = 0$$
$$g_{i\ell}\Gamma^\ell_{jk} = g_{ij,k} - g_{j\ell}\Gamma^\ell_{ki}$$

and with the cyclic permutation of the indexes

$$g_{j\ell}\Gamma^\ell_{ki} = g_{jk,i} - g_{k\ell}\Gamma^\ell_{ij}$$

thus

$$g_{i\ell}\Gamma^\ell_{jk} = g_{ij,k} - g_{jk,i} + g_{k\ell}\Gamma^\ell_{ij}$$

and in an analogous way

$$g_{k\ell}\Gamma^\ell_{ij} = g_{ki,j} - g_{i\ell}\Gamma^\ell_{jk}$$

then

$$g_{i\ell}\Gamma^\ell_{jk} = g_{ij,k} - g_{jk,i} + g_{ki,j} - g_{i\ell}\Gamma^\ell_{jk} \Rightarrow \Gamma^\ell_{jk} = \frac{1}{2}g^{ij}\left(g_{ij,k} - g_{jk,i} + g_{ki,j}\right)$$

that defines the Christoffel symbol of the first kind that is symmetrical, whereby by the definition of the torsion tensor

$$T^\ell_{jk} = \Gamma^\ell_{jk} - \Gamma^\ell_{kj} = 0 \qquad Q.E.D.$$

Problems

6.1. Show that in the space with metric $ds^2 = dx^2 + dy^2 + dz^2 - cdt^2$, the curve with parametric representation $x = c\int r \cos\varphi \cdot r \cos\theta\, ds, y = c\int r \cos\varphi \cdot r \sin\theta\, ds,\ z = c\int r \sin\varphi\, ds, t = c\int r\, ds$ has null length.

6.2. Deduce the Euler–Lagrange equation $\frac{d}{dt}\left(\frac{\partial F}{\partial \dot{x}^p}\right) - \frac{\partial F}{\partial x^p} = 0$.

6.3. Using the Euler–Lagrange equation, determine the geodesics on the sphere of radius r.

6.4. Show that the distance L between two points $P(x^i)$ and $Q(\overline{x}^i)$ in the Riemann space E_N is given by $L = \sqrt{\sum_{i=1}^{N} \left(\overline{x}^i - x^i\right)^2}$.

6.5. Demonstrate that Pythagoras theorem is valid for the Riemann space E_N.

Bibliography

Ahsan, Zafar. 2008. *Tensor Analysis with Applications*. New Delhi: Anamaya Publishers.

Akivis, Maks A., and Vladislav V. Golderg. 1977. *An Introduction to Linear Algebra & Tensors*. New York: Dover.

Akivis, Maks A., and Vladislav V. Golderg. 2003. *Tensor Calculus with Applications*. Singapore: World Scientific.

Aracil, Carlos Mataix. 1951. *Cálculo Vectorial Intrínseco*, 3rd ed. Madrid: Editorial Dossat.

Aris, Rutherford. 1989. *Vectors, Tensors and the Basic Equations of Fluid Mechanics*. New York: Dover.

Barbotte, Jean. 1948. *Le Calcul Tensoriel*. Paris: Bordas.

Bergamann, Peter Gabriel. 1976. *Introduction to the Theory of Relativity*. New York: Dover.

Betten, J. 1987. *Tensorrechnung für Ingenieure*. Stuttgart: B.G. Teubner.

Bishop, Richard L., and Samuel I. Goldemberg. 1980. *Tensor Analysis on Manifolds*. New York: Dover.

Blaschke, Wilhelm, and Hans Reichardt. 1960. *Einführung in die Differentialgeometrie*. Berlin: Springer.

Block, H.D. 1962. *Introduction to Tensor Analysis*. Columbus: Charles E. Merrill Books.

Bonola, R. 1955. *Non-Euclidian Geometry*. New York: Dover.

Borishenko, A.I., and I.E. Tarapov. 1979. *Vector and Tensor Analysis with Applications*. New York: Dover.

Bourne, D.E., and P.C. Kendall. 1996. *Vector Analysis and Cartesians Tensor*, 3rd ed. London: Chapman & Hall.

Bowen, Ray M., and C.C. Wang. 2008. *Vectors & Tensors*. New York: Dover.

Brand, Louis. 1947. *Vector and Tensor Analysis*. New York: Wiley.

Brillouin, L. 1946. *Les Tenseurs en Mécanique et en Élasticié*. New York: Dover.

Charon, Jean E. 1963. *Relativité Générale*. Paris: Éditions René Kister.

Cisoti, Umberto. 1928. *Lezioni di Calcolo Tensoriale*. Milano: Libreria Editrice Politecnica.

Coburn, Nathaniel. 1955. *Vector and Tensor Analysis*. New York: The MacMillan Company.

Craig, Homer Vicent. 1943. *Vector and Tensor Analysis*. New York: McGraw-Hill.

Creanga, Ioan, and Tudora Luchian. 1963. *Introducere in Calculul Tensorial*. Bucuresti: Editura Didactica si Pedagogica.

Danielson, D.A. 1992. *Vector and Tensors in Engineering and Physics*. Boston: Addison-Wesley Publishing Company.

Das, Anadijiban. 2007. *Tensors: The Mathematics of Relativity Theory and Continuum Mechanics*. New York: Springer.

© Springer International Publishing Switzerland 2016

E. de Souza Sánchez Filho, *Tensor Calculus for Engineers and Physicists*,
DOI 10.1007/978-3-319-31520-1

De, U. Chand. 2007. *Differential Geometry of Curves and Surfaces in E3: Tensor Approach*. New Delhi: Anamay Publisher.

De, U.C., Absos Ali Shaikh, and Joydeep Sengupta. 2005. *Tensor Calculus*. Oxford: Alpha Science International Ltd.

Deckert, Adalbert. 1958. *Vectoren und Tensoren, eine Einführung*. Leipzig: C. F. Winter'sche Velargshandlung.

Delachet, A. 1955. *Calcul Vectoriel et Calcul Tensoriel*. Paris: Press Universitaires de France.

Denis-Papin, M., R. Faure, and A. Kaufmann. 1962. *Calcul Matriciel et de Calcul Tensoriel*. Paris: Editions Eyrolles.

Dirac, P.A.M. 1981. *General Theory of Relativity*. New York: Dover.

Dodson, C.T.J., and T. Poston. 1979. *Tensor Geometry: The Geometric Viewpoint and its Uses*. London: Pitman.

Dube, K.K. 2009. *Differential Geometry and Tensors*. Unnao: I.K. International Publishing House Pvt. Ltd.

Duschek, Adalbert, and August Hochrainer. 1946. *Grundzüge der Tensorrechnung in Analytischer Darstellung: Tensoralgebra*. Teil 1. Berlin: Springer.

Edwards Jr., C.H. 1994. *Advanced Calculus of Several Variables*. New York: Dover.

Eisele, John A., and Robert M. Mason. 1970. *Applied Matrix and Tensor Analysis*. New York: Wiley-Interscience.

Eisenhart, Luther Pfahler. 1940. *An Introduction to Differential Geometry with the Use of Tensors Calculus*. Princeton: Princeton University Press.

Eisenhart, Luther Pfahler. 1949. *Riemann Geometry*. Princeton: Princeton University Press.

Eisenhart, Luther Pfahler. 2005a. *Coordinate Geometry*. New York: Dover.

Eisenhart, Luther Pfahler. 2005b. *Non-Riemann Geometry*. New York: Dover.

Finzi, Bruno, and Pastori Maria. 1949. *Calcolo Tensoriale e Applicazioni*. Bologna: Nicola Zanichelli Editores.

Flügge, Wilhelm. 1986. *Tensor Analysis and Continuum Mechanics*. Berlin: Springer.

Frankel, Theodore. 2007. *The Geometry of Physics*. Cambridge: Cambridge University Press.

Gallot, Sylvester, Domique Hulin, and Jacques Lafontaine. 2004. *Riemann Geometry*, 3rd ed. Berlin: Springer.

Gerretsen, Johan C.H. 1962. *Lectures on Tensor Calculus and Differential Geometry*. Groningen: P. Noordhoff N. V.

Gibbs, J. Willard. 1929. *Vector Analysis*. New Haven: Yale University Press.

Golab, Stanislaw. 1974. *Tensor Calculus*. Warszawa: Elsevier PWN-Polish Scientific Publishers.

Grinfeld, Pavel. 2015. *Introduction of Tensor Analysis and the Calculus of Moving Surfaces*. New York: Springer.

Guggenheimer, Heinrich W. 1963. *Differential Geometry*. New York: Dover.

Guillén, Franscisco Javier. 1957. *Cálculo Tensorial*. Mexico: Editora Iglesias.

Hausner, M.A. 1998. *A Vector Space Approach to Geometry*. New York: Dover.

Hawkins, G.A. 1963. *Multilinear Analysis for Students in Engineering and Science*. New York: Wiley.

Hay, G.E. *Vector and Tensor Analysis*. New York: Dover. Reprinting of the 1953 edition.

Hestenes, David. 1996. Grassmann's Vision. In: *Hermann Gunther Grassmann (1804–1877)—Visionary Mathematician, Scientist and Neohumanist Scholar*, 191–201. Dordrecht: Kluwer Academic Publisher.

Hladik, Jean, and Pierre-Emmanuel Hladik. 1999. *Le Calcul Tensoriel en Physique*. Paris: Dunod.

Hodge, W.V.D. 1952. *Invariants of Quadratic Differential Forms*. Cambridge: Cambridge University Press.

Hoffman, Banesh. *About Vectors*. New York: Dover. Reprinting of the 1953 edition.

Horst, Teichmann. 1973. *Physikalische Anwendugen der Vektor-und Tensorrechnung*. Großkrotzenburg: Hain-Druck KG.

Iben, Hans Karl. 1999. *Tensorrechnung*. Leipzig: B.G. Teubner.

Itskov, Mikhail. 2007. *Tensor Algebra and Tensor Analysis for Engineers*. Berlin: Springer.

Jaeger, L.G. 1966. *Cartesian Tensors in Engineering Science*. Oxford: Pergamon Press.
Jeanperrin, Claude. 1999. *Initiation Progressive au Calcul Tensoriel*. Paris: Ellipses.
Jeanperrin, Claude. 2000. *Utilisation du Calcul Tensoriel dans les Géométries Riemmaniennes*. Paris: Ellipses.
Kay, David C. 1988. *Theory and Problems of Tensor Calculus*. Schaum's Outline Series. New York: McGraw-Hill.
Khan, Quaddrus. 2015. *Tensor Analysis and Its Applications*. New Delhi: Partridge.
Kilcevski, N.A. 1956. *Elemente de Calcul Tensorial si Aplicatiile lui in Mecanica*. Bucuresti: Editura Tehnică.
Klingbeil, Eberhard. 1993. *Tensorrechnung für Ingenieure*. Berlin: Wissenschaftsverlag.
Knowless, James K. 1998. *Linear Vectors Spaces and Cartesian Tensors*. Oxford: Oxford University Press.
Korn, Granino A., and Theresa M. Korn. 2000. *Mathematical Handbook for Scientists and Engineers*. New York: Dover.
Kreyszig, Erwin. 1993. *Advanced Engineering Mathematics*. New York: Wiley.
Lagrange, René. 1922. *Sur le Calcul Différentiel Absolut*. Annales de la Faculté des Sciences de Toulouse. 3ª Série, Tome 14, 1–69.
Lang, Serge. 1987. *Calculus of Several Variables*. London: Springer.
Lass, Harry. 1950. *Vector and Tensor Analysis*. New York: McGraw-Hill.
Lawden, D.F. 2002. *Introduction to Tensor Calculus, Relativity and Cosmology*. New York: Dover.
Lebedev, Leonid P., and Michael J. Cloud. 2003. *Tensor Analysis*. Singapore: World Scientific Publishing.
Lebedev, Leonid P., Michael J. Cloud, and Victor A. Eremeyev. 2010. *Tensor Analysis with Applications in Mechanics*. Singapore: World Scientific Publishing.
Lee, John M. 1997. *Riemann Manifolds. An Introduction to Curvature*. New York: Springer.
Levi-Civita, Tullio. *The Absolute Differential Calculus (Calculus of Tensors)*. New York: Dover. Reprinting of the 1926 edition.
Linchnerowicz, André. 1987. *Éléments de Calcul Tensoriel*. Paris: Éditions Jacques.
Loewner, Charles. 2008. *Theory of Continuous Groups*. New York: Dover.
Lovelock, David, and Hanno Rund. 1988. *Tensors, Differential Forms and Variational Principles*. New York: Dover.
Mattews, P.C. 1998. *Vector Calculus*. London: Springer.
McConnell, A.J. 1957. *Applications of Tensor Analysis*. New York: Dover.
Mercier, Jacques L. 1971. *An Introduction to Tensor Calculus*. Groningen: Wolters-Noordhoff Publishing.
Michal, Aristotle D. 1947. *Matrix and Tensor Calculus*. New York: Wiley.
Mital, P.K. 1995. *Tensor Analysis for Scientists*. New Delhi: Har-Anand Publications.
Nayak, Prasun Kumar. 2012. *Textbook of Tensor Calculus and Differential Geometry*. New Delhi: PHI Learning Private Ltd.
Nelson, Edward. 1967. *Tensor Analysis*. Princeton: Princeton University Press.
Neuenschwander, Dwight E. 2015. *Tensor Calculus of Physics: A Concise Guide*. Baltimore: Johns Hopkins University Press.
Oeijord, Nils K. 2003. *The Very Basics of Tensors*. Bloomington: iUniverse.
Ollendorff, F. 1950. *Die Welt der Vektoren*. Österreich: Springer.
Ollof, Rainer. 2004. *Geometrie der Raumzeit*. Wiesbaden: Vieweg. 3 Auflage.
Pathria, R.K. 2003. *Theory of Relativity*. New York: Dover.
Pauli, Wolfang. 1976. *Theory of Relativity*. New York: Dover.
Sánchez, Emil. 2007. *Tensores*. Rio de Janeiro: Editora Interciência (in Portuguese).
Sánchez, Emil. 2011. *Cálculo Tensorial*. Rio de Janeiro: Editora Interciência (in Portuguese).
Santaló, Luis A. 1977. *Vectores y Tensores con sus Aplicaciones*. Buenos Aires: Editorial Universitaria Buenos Aires.
Schade, Heinz. 1997. *Tensoranalysis*. Berlin: Walter Gruyter.

Schey, H.M. 1997. *Div, Grad, Curl and All That: An Informal Text on Vector Calculus*. New York: W. W. Norton & Company.
Schmidt, Harry. 1953. *Einführung in die Vektor-und Tensorrechnung unter besonderer Berücksichtigung ihrer Physikalischer Bedeutung*. Berlin: VEB Verlag Techinik.
Schouten, J.A. 1989. *Tensor Analysis for Physicists*. New York: Dover.
Schouten, J.A. *Ricci-Calculus*. Berlin: Springer. Reprinting of the 1954 edition.
Schroeder, Dieter. 2006. *Vektor-und Tensorpraxis*. Frankfurt am Main: Verlag Harri Deutsch.
Schutz, Bernard. 2009. *A First Course in General Relativity*. Cambridge: Cambridge University Press.
Semay, Claude, and Bernard Silvester-Brac. 2009. *Introduction au Calcul Tensoriel. Applications à la Physique*. Paris: Dunod.
Sharma, J.N., and A.R. Vasishita. 1987. *Vector Calculus*. Meerut: Krishna Prakashan Mandir.
Simonds, G. James. 1994. *A Brief on Tensor Analysis*. New York: Springer.
Singh, Shalini. 2007. *Tensor Calculus*. New Delhi: Sarup & Sons.
Sokolnikoff, Ivan S. 1951. *Tensor Analysis: Theory and Applications*. New York: Wiley.
Solkonikoff, I.S., and R.M. Redheffer. 1966. *Mathematics of Physics and Modern Engineering*, 3rd ed. New York: McGraw-Hill Book Company.
Spain, Barry. 1958. *Tensor Calculus*. Edinburgh: Holland Oliver and Boyd.
Spiegel, Murray R. 1971. *Cálculo Avançado* (in Portuguese). São Paulo: Coleção Schaum, Editora McGraw-Hill do Brasil.
Stephani, Hans. 2004. *Relativity.An Introduction to Special and General Relativity*. Cambridge: Cambridge University Press.
Struik, Dirk J. 1988. *Lectures on Classical Differential Geometry*. New York: Dover.
Synge, J.L., and A. Schild. 1978. *Tensor Calculus*. New York: Dover.
Thomas, Tracy Y. 1931. *The Elementary Theory of Tensors*. New York: McGraw-Hill.
Thomas, Tracy Y. 1965. *Concepts from Tensor Analysis and Differential Geometry*. New York: Academic Press.
Weatherburn, C.E. 2008. *An Introduction to Riemann Geometry and the Tensor Calculus*. Cambridge: Cambridge University Press.
Weinreich, Gabriel. 1998. *Geometrical Vectors*. Chicago: The University of Chicago Press.
Weyl, Hermann. 1950. *Space-Time-Matter*. New York: Dover.
Wrede, R.C. 1963. *Introduction to Vector and Tensor Analysis*. New York: Dover.

Index

© Springer International Publishing Switzerland 2016

E. de Souza Sánchez Filho, *Tensor Calculus for Engineers and Physicists*,
DOI 10.1007/978-3-319-31520-1